UNRELIABLE

UNRELIABLE

Bias, Fraud, and the Reproducibility Crisis in Biomedical Research

CSABA SZABO

Columbia University Press
New York

Columbia University Press
Publishers Since 1893
New York   Chichester, West Sussex

Copyright © 2025 Csaba Szabo
All rights reserved

Library of Congress Cataloging-in-Publication Data
Names: Szabó, Csaba, M.D., author.
Title: Unreliable : bias, fraud, and the reproducibility crisis in biomedical research / Csaba Szabo.
Other titles: Bias, fraud, and the reproducibility crisis in biomedical research
Description: New York : Columbia University Press, [2025] | Includes bibliographical references and index.
Identifiers: LCCN 2024039326 | ISBN 9780231216234 (hardback) | ISBN 9780231216241 (paperback) | ISBN 9780231561181 (ebook)
Subjects: MESH: Reproducibility of Results | Biomedical Research | Bias | Scientific Misconduct
Classification: LCC R852 | NLM W 20.5 | DDC 610.72/4—dc23/eng/20250111

Cover design: Henry Sene Yee

GPSR Authorized Representative: Easy Access System Europe, Mustamäe tee 50, 10621 Tallinn, Estonia, gpsr.requests@easproject.com

To my wife Anita and children Lili and Marcell

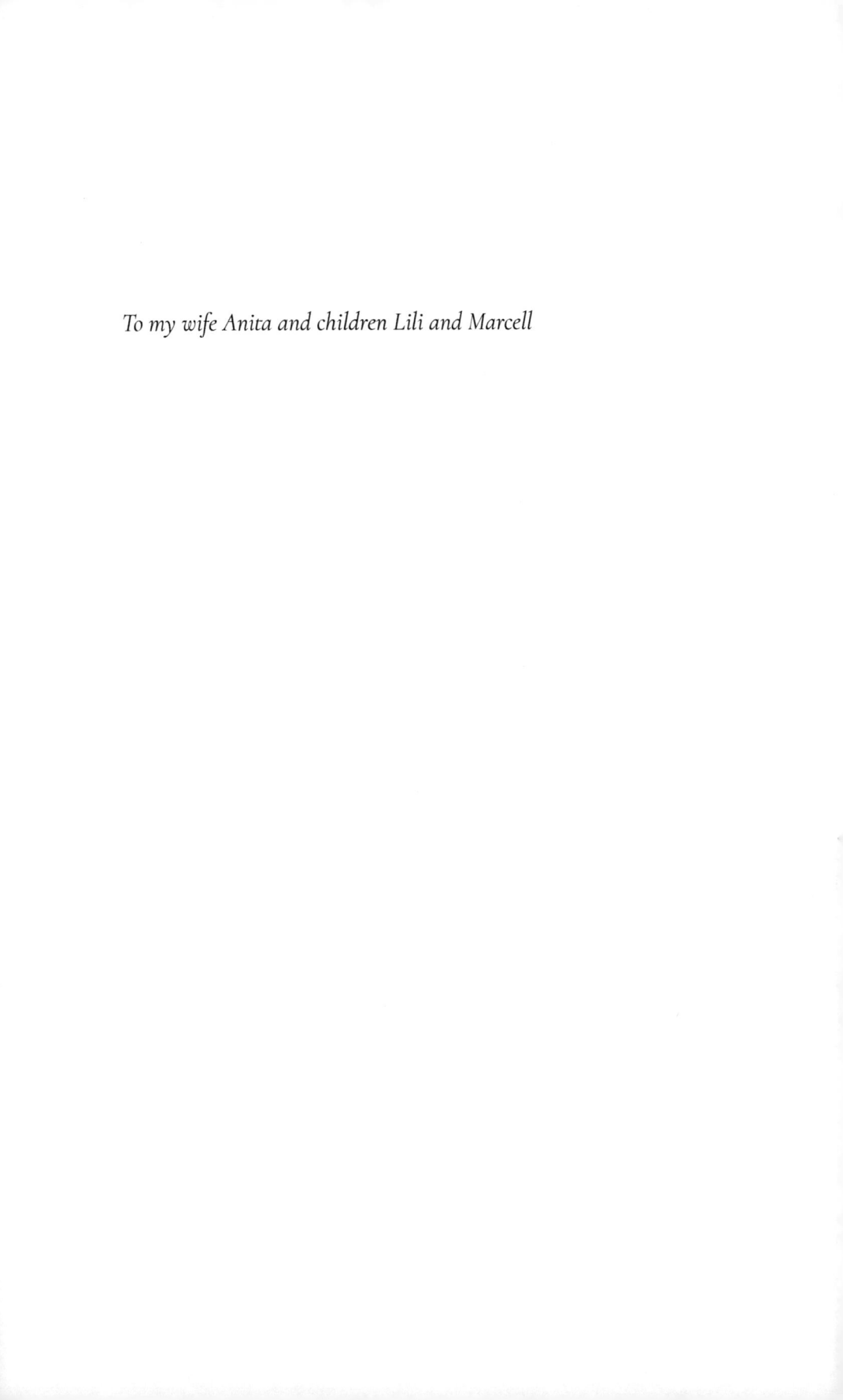

CONTENTS

Preface ix

1 Careers and Career Pressures in Biomedical Science 1
   My Own Path Into Biomedical Science 1
   Reasons to Choose a Career in Science 7
   Academic Pressures 10
   Work/Life Balance 20
   Brain Drain, Working Visas, and the Nationalization of the "Exceptional Scientist" 22
   Alternative Career Paths in Biomedical Science 28

2 Hypercompetition for Research Grants 32
   "No Money, No Honey" 32
   Applying for Grants: The Process and the Odds of Success 34
   Study Sections, Grant Reviews 44

3 "Doing Science": From Hypothesis to Publication 64
   Experimental Models 64
   Hypothesis Generation, Hypothesis Testing, and Interpretation 73
   The Reproducibility Crisis 82
   Variability in Research Materials and Reagents 88
   Variability Due to Substandard Experimentation Practices and Honest Errors 100
   The Abuse of Statistics 104

4 Scientific Fraud—and the Fraudulent Fraudsters 115
   High-Profile Fraud Cases 115
   Easily Detectable Methods of Manipulation 132
   The Extent of Intentional Fraud in the Scientific Literature 143
   Psychological Aspects of Scientific Misconduct 145

5 A Broken Scientific Publishing System 148
   From Manuscript to Publication 148
   "Real Journals" and Their Publishers 156
   Predatory and Fake Journals and Their Publishers 164
   "We Are the Millers": Criminal Publishing Gangs 168
   Plagiarism, Cryptomnesia, Citation Bias 176
   Conferences, Scamferences 181
   Retractions and Their Consequences 185
   How to Read a Scientific Paper? 189

6 The Way Forward 193
   Where Is the Oversight? 193
   Criminalization of Scientific Misconduct? 201
   Traditional Suggestions for Gradual Improvements 204
   Call for Real Reforms 208

   Afterword 225

   Appendix 231
   Notes 243
   Bibliography 283
   Index 301

PREFACE

Almost twenty years ago the American biostatistician John Ioannidis published an article titled "Why Most Published Research Findings Are False."[1] His considered biostatistical opinion is that most of the published scientific literature is incorrect and therefore—by definition—not reproducible. Ioannidis is not a crackpot. He is *the* leading expert in a field that some people call "science about science." He devoted his life to studying publication patterns, biostatistical issues, authorship issues, scientific citations, clinical trial designs, and much more. His research focuses on the hallowed biomedical scientific literature: a massive body of original papers and review articles, which are considered the bread and butter of everyone working in biomedical research and clinical medicine.

This literature is massive: PubMed, the largest database of biomedical literature, contains more than thirty-five million papers; and Europe PubMed Central (Europe PMC) contains more than forty-two million papers, abstracts, and patents. This database grows by about one million new papers per year. Subsets of this immense collection are the starting points for every scientist who wishes to embark on discovering something new; identifying the foundation of new drug discovery and development; or the basis on which a physician can find out if a drug or medical procedure is effective or harmful. If this body of information is faulty or unreliable, how can anyone conduct meaningful research?

Basic research is the foundation of all biomedical science. This type of research is typically conducted in a research laboratory, for example, using cells cultured in a Petri dish or animal models of various diseases. These model systems are our workhorses to study how fundamental biological processes work—for example, how cells become cancerous or how various cells in the brain or the immune system communicate with each other. Some of this information leads to translatable concepts that can be the basis of designing new medicines, which—if things go well—eventually become available for the treatment of patients. But what if the entire body of biomedical literature cannot be trusted? Seems like there is an elephant in the room. The precious few who dare to mention its name meekly call it the *reproducibility crisis*.

The magnitude of the reproducibility crisis had not been tested systematically until recently. When it finally was tested, the results were mind-boggling. In 2011, scientists at the German pharmaceutical giant Bayer made a splash with news that they could not replicate 75 to 80 percent of the sixty-seven preclinical academic publications that they took on.[2] In 2012, the hematology and oncology department of the American biotechnology firm Amgen selected fifty-three cancer research papers that were published in leading scientific journals and tried to reproduce them in their laboratory. In 89 percent of the cases, they were unable to do so.[3] Several systematic reproducibility projects have been launched subsequently. The "Reproducibility Project: Cancer Biology" selected key experiments from fifty-three recent high-impact cancer biology papers. Although the replicators used the same materials and methods as the original paper, 66 percent of the published work could not be replicated.[4] Several other replication projects are currently ongoing, and as you will see later in this book, the results aren't looking too good either.

If you think it's bad enough that scientists can't reproduce other scientists' data, then consider this. It turns out that most scientists can't even reproduce *their own data*. In 2016, the prestigious scientific journal *Nature* published the results of its anonymous survey.[5] More than 1,500 scientists replied, and in the fields of biology and medicine, 65 to 75 percent of the respondents said they were unable to reproduce data published by others and . . . 50 to 60 percent said they had trouble reproducing their own previous research findings.[6]

The implications are scary: it looks like *the published body of biomedical literature is unreliable*. I attempt to get to the bottom of this matter in this book. I cover some of the questionable but rather commonplace practices in research that exist in most published literature (for example, lack of blinding, lack of randomization, and selective publishing). Problems related to the inherent unreliability of some of the methods and reagents used in research are also covered. Questionable—but once again, commonplace—statistical practices (such as outlier exclusion and "p-hacking") are also covered. I look into the dark world of so-called image correction as well as intentional data manipulation—all the way down into the deepest pockets of scientific fraud, to the so-called scientific paper mills where entire scientific papers are conjured up literally from thin air for paying... well, let's call them... customers.

It is important to find out what goes on behind the scenes. What pushes people to resort to dishonest methods and approaches? Most of the public is unaware of the immense pressures and perverse incentives that are commonplace in biomedical research. In this book I examine the responsibility of individuals working at various levels in a research laboratory, but I also expose the significant pressures that the entire biomedical scientific ecosystem—the scientific institutions, the granting agencies, and the publishing industry—places on each individual investigator.

I introduce some of the "science sleuths" (also called data detectives). Official ones, such as the Office of Scientific Integrity (OSI) at the National Institutes of Health (NIH), as well as unofficial ones: the dedicated private investigators who—often working in anonymity and in fear of litigation—devote their time to exposing scientific fraudsters and scammers.

Winston Churchill once said: "Democracy is the worst form of government—except for all the others that have been tried." Similarly, the current system is the worst system to conduct research—except for all the others. The current system has produced major advances and will continue to do so. Some of the brightest minds on the planet work in biomedical science. *It is the only "game in town."*[7] It *is* possible to do good, valid science, and there *are* many investigators to prove it. Even though some observers conclude that the current situation is hopelessly broken, I disagree. Yes, the system is wasteful, redundant, and littered with "scientific garbage." But maybe it is not beyond help. In chapter 6, I propose a few ideas to improve the situation.

Initially, I planned to cover a broader scope of subjects, from basic laboratory studies to clinical trials and drug approvals. I soon realized that basic research alone can easily fill a whole book. This is that book.

Finally, a personal note. I am not a disgruntled scientist who is desperate to take down the institutions of science out of spite. I chose a life in science at a very young age, and I have a fulfilling life in it. I don't have a personal agenda against any of the scientists mentioned in this book. Even with all the dark phenomena detailed in here, I continue to believe in biomedical science.

I just think it can be made better.

UNRELIABLE

CHAPTER 1

CAREERS AND CAREER PRESSURES IN BIOMEDICAL SCIENCE

MY OWN PATH INTO BIOMEDICAL SCIENCE

I attended primary and secondary school in the 1980s in the small town of Győr in the western part of "communist Hungary." The country's political system was officially called "socialism," not communism, but the rest of the world called Hungary—together with Czechoslovakia, Poland, East Germany, Bulgaria, Romania, and Yugoslavia—a "communist" country. It was part of the so-called Soviet Bloc: countries the Russians liberated from Nazi rule during the final stages of World War II but then stayed there for good. Hungarians were instructed to celebrate this "liberation" with street marches and Russian flags every April fourth. The tens of thousands of Russian soldiers in the garrisons were called "temporarily stationed Soviet troops," and they were there to defend the country against "imperialist threats." One of the standing jokes at the time was that the temporarily stationed Russian soldiers had received permanent housing in Hungary. Of course, one had to be careful about making such jokes. If one cracked an antiestablishment joke and somebody reported you to the authorities, the consequences were unpleasant. Such was life behind the Iron Curtain.

But Hungary was viewed as "the happiest barrack" of the Eastern Bloc. The regime, led by János Kádár, instituted a system of "Goulash communism," which allowed the populace certain limited liberties—including limited travel to Western countries—as long as the basic premises of the country's constitution were not questioned and as long as the 1956 revolution was not mentioned. As schoolboys, we all knew that what we were taught about the political system was highly suspect. It doesn't take a genius to realize that elections with a single party on the ticket are farcical. But most of the country assimilated to the prevailing order: some because they were "true believers," some because they were conformists, and some to make a better life for themselves and for their children. Party membership had many benefits. Growing up in the city of Győr we were close to the Austrian border, and I was lucky also because we could get Austrian television channels on our TV. My parents had the good judgment to send me to a German-language kindergarten, so I became a fluent German speaker at a young age. Thus I could always get the Western perspective on things from Austrian television and could compare it to the Hungarian standard political fare that was piped into our brains through state-owned media.[1]

As young people all around the world living under various forms of softer or harder dictatorship, we learned how to enjoy ourselves and how to lead a relatively normal life. Some clubs played Western music, record stores carried records from all over the world, and movie theaters played films from all over the world—including American movies. There was a rich tradition of theater, ballet, and classical music, as well as a very active sports life. Also, a rich selection of books were available in the bookstores. Some books were considered antiestablishment and were on the forbidden list, but there were plenty of excellent books to choose from.

The cultural climate of Hungary was determined by the doctrine of a powerful minister of culture, György Aczél, called "3T," an acronym made up from the Hungarian-language initials of the concept's 3 pillars: support, tolerance, and ban. This meant that some materials were supported, some were tolerated, and some were banned. Luckily for me science and medicine, as well as many other intellectual pursuits from chess to the Rubik's cube (a world sensation at the time[2]), were in the supported category.

Science became my fascination at an early age. A series of books called *Wise Owl* introduced preschool kids to fundamental scientific concepts. My favorite one was called *The Wonders of Blood*, another one had the title *Why Do Trees Have Leaves?* I must have been about ten years old when a book and an associated multipart television program called *Secrets of Inheritance* came out. This was written and moderated by the internationally recognized geneticist-physician Endre Czeizel. Both the series and the moderator himself gained immense popularity in the country. (To be fair, at the time there was just one regular television channel in Hungary; the choice—even for the less science-inclined—was either to watch genetics lectures or not to watch anything at all.[3])

CAREERS AND CAREER PRESSURES IN BIOMEDICAL SCIENCE 3

1.1 On the subject of early-career scientists . . .
Source: Leonid Schneider / ForBetterScience.com, reproduced by permission.

Some American popular science books were translated into Hungarian, including Navin Sullivan's *The Message of the Genes*. I read this book with double fascination because it not only explained how our DNA is structured and how our genes work but also explained *what the specific experiments were* that led to the discovery of various fundamental biological principles. Among other discoveries, it explained the details of the experiments that led to "deciphering," i.e., understanding, that various "triplets" in DNA encode various amino acids and how proteins are assembled based on this code. I became fascinated by this "deciphering process," and already as a schoolkid I decided this was what I wanted to do when I grew up: try to "decipher things" in some area of biology or medicine (figure 1.1). I don't think I knew that there was an official name—"biomedical scientist" or "physician-scientist"—for such people back then.

Scientists were and to this day are revered, lionized individuals in Hungary. Any Hungarian will proudly tell you that Hungary holds the world record for the number of Nobel Prizes per capita.[4] Philipp Lenard, Robert Bárány, Richard Adolf Zsigmondy, Albert Szent-Györgyi, George de Hevesy, Georg von Békésy, Eugene Wigner, Dennis Gabor, John Polanyi, George Olah, John Harsanyi, Imre Kertész, Avram Hershko, Katalin Karikó, and Ferenc Krausz have all received Nobel Prizes. They might also tell you, a bit more quietly, that few of them did their Nobel work in Hungary; most of them worked in the United States. The only exception is Albert Szent-Györgyi, the discoverer of vitamin C, who did his Nobel work in Szeged, Hungary, in the 1930s. He was also a pioneer in many exciting new fields of biology such as muscle function, cellular bioenergetics, and the electron transport function of proteins. He is one of my heroes in science.[5] Selected writings of Szent-Györgyi became available in Hungarian bookstores around the time I went to secondary school, and—once again—I was fascinated by Szent-Györgyi's explanations of *how he arrived at his discoveries.*

As a student, I picked things up quickly, and I read an immense amount of literature: popular science, science fiction, and everything else I could get my hands on. As a young teen, I came across a book on scientific fraud titled *Science–Fake Science*, which was written by the renowned physical chemist Mihály Beck.[6] It featured mind-blowing incidents about a wide range of fake science, from the story of the Piltdown Man to the enigmatic Piccardi tests, as well as various scams and frauds related to magnetism and electricity, old and new versions of alchemy, and, of course, the story of N-rays: enigmatic rays that were completely nonexistent, except in the minds of biased observers. There were some biomedical stories in the book too, including the story of the immunologist William Summerlin, who, in the mid-1970s, used a black pen to paint his mice to create the appearance of a successful skin transplantation. The most lasting contribution of Summerlin's career seems to be that the term "painting mice" is still used to describe fraud and deception in research.[7]

I participated in various science competitions in high school and competed in the national finals in physics and biology. The natural path for me, then, was to go to medical school. I was accepted to the Semmelweis Medical School in Budapest, where I received my medical degree. I already knew that I didn't want to become a practicing physician; I wanted to do biomedical research. In medical schools, an organization called Students' Scientific Circle interfaced with similar organizations in various European countries. Students who were part of this circle could try their hand at actual laboratory research. There were annual competitions, with presentations and prizes, and winners of local competitions could go abroad and present at international students' conferences.

I applied to a basic research institute of Semmelweis University that had the complicated name of Clinical Experimental Research Department and Second Institute of Physiology. The institute was headed by an internationally known physiologist named A. G. B. Kovách. The profile of the institute was the regulation of blood vessel tone and reactivity, with a special focus on blood vessels that supply the brain. Another focus was circulatory shock, a life-threatening condition that can occur, for instance, when somebody gets a severe bacterial infection or loses a lot of blood due to severe trauma. Kovách has serious international connections; he was codirector of a laboratory at the University of Pennsylvania, no small feat for a scientist in "communist Hungary." Many scientists working at the institute would have an opportunity to spend some time at Penn's sister lab.

My direct supervisor at the institute was a maverick professor named Eörs Dóra who had recently returned from a small research sabbatical in Sweden and brought back a series of nifty instruments that enabled us to measure changes in the tone of small blood vessels. At the time, the big question in the field of blood vessel research (vascular physiology) was the nature of a mysterious substance called Endothelium-Derived Relaxing Factor (EDRF). Endothelial cells form a single layer on the inside of all blood vessels. It was already known that these cells can produce interesting local hormones, for example, prostaglandins (a field copioneered by the British scientist Sir John Vane, whom I worked with later). But EDRF was not a prostaglandin. As shown by the American scientist Robert Furchgott in 1980, EDRF was produced by endothelial cells and relaxed the underlying smooth muscle layer. If the endothelial cells were damaged, this relaxation was no longer present. This seemed to be an important cause of blood vessel occlusions, such as the ones responsible for heart attacks or strokes. When a blood vessel becomes atherosclerotic, the endothelial cells become damaged and don't produce EDRF. Instead, they contract, and this can stop the blood supply to the brain or the heart. One of the first English-language scientific papers I read in my life was about this enigma.[8] Nobody had any idea what this mysterious EDRF could be. From the literature it seemed that EDRF had many properties that were similar to those used in a group of heart medications, called nitrovasodilators—the principal member of the class being nitroglycerin, which worked by releasing the gas nitric oxide (NO). But surely NO is not something a biological system could possibly generate—at least this is what everybody thought at the time.

Research reagents from Western countries were difficult for us to acquire in those days. Only a single company in Hungary could order reagents; often they arrived a year later and sometimes never. Our budget was rather limited, and a colleague returning from a research period in Philadelphia was often kind enough to bring some much-needed reagents to us in his pocket. We felt we had no chance to enter the hottest field of vascular biology, which was to figure out what EDRF is. Instead we studied various responses of blood vessels; we explored how various ions, such as calcium and magnesium, affect it, and we tested whether circulatory shock impairs the ability of the endothelial cells to generate EDRF. Even with limited resources, as a third-year medical student I felt I was ready to become part of the scientific deciphering process. We had enough equipment and reagents to try a few things. We could observe certain responses; we could formulate new hypotheses related to how our blood vessels worked and reacted to various conditions; and we had a working method to test whether these hypotheses were correct. It was an intellectual challenge, and it gave us a degree of freedom to think, to hypothesize, to try, and to test various ideas.

I spent countless days, nights, and weekends in the laboratory, often locking up the institute at the end of a long night. Dóra must have seen my enthusiasm, and perhaps my affinity for research, because he supported me strongly. In fact, with his encouragement and with coauthors, I published my first paper in the Swiss journal *Blood Vessels* as a fourth-year medical student.[9] The paper was short, had only a couple of figures, and described an interesting observation: When we placed blood vessels in fluid that had lower-than-normal magnesium levels, the tone of the blood vessels started to oscillate. Low blood magnesium levels were known to create various disturbances in the regulation of the circulatory system; we hypothesized that aberrant motion of the blood vessels of the type described in our paper may potentially contribute to it.

It was the summer of 1987. Most people were on vacation, but I was working in the lab. I still remember the morning Dóra came into the laboratory, his face glowing with excitement. He held in his hand a photocopy of an article, just published in the top scientific journal *Nature*, by a group of British scientists that included Salvador Moncada.[10] The paper claimed that EDRF is, in fact, nitric oxide (NO)—the same substance that is released from the drug nitroglycerin, a molecule everybody was certain human cells could not produce! We instantly knew this was a major discovery; we felt this would be a game changer for vascular biology and possibly even for other medical fields. Sure enough, a few years later three scientists working at the foundation of the NO pathway received the Nobel Prize, including the original EDRF discoverer, Robert Furchgott—but, curiously, not Salvador Moncada.[11]

I remained fascinated with NO and continued working on its various biological roles throughout my career. Later I expanded my investigation to include the roles of other small, diffusible biological regulatory molecules, including various free radicals, oxidants including hydrogen sulfide, and most recently even hydrogen cyanide.[12]

After I graduated from medical school, I obtained two separate PhD degrees. One of these degrees was in physiology and is based on research I did during medical school. At that time this was not called a PhD but rather a Candidate of Science—based on the Russian scientific system's designations. Universities did not issue this degree; instead it had to be submitted to and defended at the Hungarian Academy of Sciences, a large, powerful, Soviet-style scientific institution that held an immense stronghold on everything science-related in Hungary.[13] My second PhD, in pharmacology, is from the University of London, which I defended while I was working at the William Harvey Institute, headed by the Nobel laureate Sir John Vane. I went to the United States and worked in biomedical science for twenty-five years. I became an associate professor, and then a full professor. I also worked in biotechnology as Chief Scientific Officer at several companies involved in the discovery and development of various novel drug candidates. Five years ago I returned to Europe, and now I head the section on pharmacology at the University of Fribourg in Switzerland. I have led many research projects, both in basic and translational science; I have published extensively, received many research grants and various awards; and I have trained many students and postdocs. As I write this book, I have more than thirty years of direct experience in biomedical research.

REASONS TO CHOOSE A CAREER IN SCIENCE

What motivated me to choose biomedical science as my life's pursuit is clear: a fascination with science, the hope to become part of scientific discoveries, and the hope that some of these discoveries would be useful for human medicine. I don't like to use big words, but in this case I will: I cannot think of a more sacrosanct purpose in life. Science is a lifelong intellectual challenge and a life of constant puzzle-solving, and the solutions can advance medicine. Big words, right?

I would like to think that everybody who enters the field of research starts with these motivations. Maybe I am naïve, but when I read biographies of giants in the field from the nineteenth and twentieth centuries, it is clear that James Watson, Linus Pauling, Albert Szent-Györgyi, and many others chose science for the right reasons. Discovery, intellectual pursuit, and some hope that the work will be useful for medicine. Sometimes the usefulness part is de-emphasized; in many fields of basic science, it is difficult to predict when—if anywhere, if ever—the usefulness will come. As Albert Szent-Györgyi said: "When young people come to me saying they want to be researchers because they want to contribute to alleviating human suffering, I prefer to send them to charities. Research requires selfish people who are passionate about 'useless' issues and willing to sacrifice everything, even their own lives, to solve them."[14] Solve it—even if just for the sake of the discovery.

The right reason to enter the world of science is the search for truth, the pursuit of a discovery. For the more tech-minded crowd, it may include a fascination with innovation and the opportunity of working with cool instruments, i.e., cutting-edge technology. There may also be a general draw into the academic environment, a life of continuous learning, a chance for new discoveries, and—for those who also like teaching—the opportunity to teach and mentor future generations of scientists. For a young person with these motivations, there is almost no choice but to enter some field of science regardless of what the odds are for making a truly novel discovery or for reaching the top of the scientific pyramid in terms of an academic position: and for which the odds—as you will see later—are astonishingly low.

Recipients of the prestigious National Science Foundation (NSF) Graduate Research Fellowships from Civil and Environmental Engineering and Computer Science and Engineering were asked about the pros and cons of entering the field of science, technology, engineering, and mathematics (STEM).[15] Generally speaking, the responses are appropriate: the fellows seem to hold fairly realistic views at a rather early stage of their careers (table 1.1). One item on the list looks concerning to me, though. Although the majority of the respondents have indicated that their primary motivation is some form of truth-seeking, only 22 percent of the respondents mentioned "intellectually stimulating work" as a positive feature of the profession and only 12 percent mentioned "intellectual environment or lifelong learning."

TABLE 1.1 Top ten pros and cons of working in STEM academia

Pros of working in a STEM field	Responses (in percent)
Academic freedom	59
Flexible schedule	34
Mentoring students	29
Intellectually stimulating work	22
Teaching	21
Autonomy	14
Intellectual environment or lifelong learning	12
Benefit society or make a difference	11
Job security	8
Create new knowledge	8
Prestige or respect	8

Cons of working in a STEM field	
Constant writing of grants	28
Excessive workload or long hours	28
High stress and pressure	25
Low pay	24
Pressure to publish	22
Poor work/life balance	20
Hypercompetitive	18
Extreme pressure to get tenure	15
Institutional politics	8
Limited job prospects or location inflexibility	7
Excessive bureaucracy	5

Source: Results of a survey of NSF Graduate Research Fellowships holders

Importantly, already at this early stage of their life, more than 60 percent of the fellows interviewed said that "the level of cheating made them rethink their career choice and the people their field was attracting." Hmm.

From time to time I meet early-stage scientists who seem to exhibit many of these motivations. But I also meet many others who do not have a clear answer to why they want to enter a PhD program. And almost none have a clear idea of the demands of the job: intellectual, physical, and ethical. Likewise, most cannot even imagine the degree of hypercompetition they will be facing for the rest of their lives.

ACADEMIC PRESSURES

In the old days a PhD was not an absolute requirement to enter biomedical science. Many people chose biomedical research careers with an MD degree and no PhD. Typically, these individuals had some serious level of laboratory research exposure during their premed days or during medical school or after graduation. For example, Lasker Awardee Solomon Snyder, one of the most respected pharmacologists on the planet, "only" has an MD from Georgetown University. Also, the Nobel laureate Bruce Beutler is a medical doctor and internist yet he conducted laboratory research throughout his career. Kevin Tracey, a pioneer of inflammatory processes and a founder of the field of bioelectronic medicine is a physician and a neurosurgeon—without holding a PhD degree. Anthony Fauci at the NIH has been directing laboratory-based as well as translational research for decades, even though he "only" has a medical degree and no PhD. Yet another MD Nobelist from the NIH is Harvey Alter, the codiscoverer of the hepatitis C virus. Barry Marshall, who has recognized bacterial infection as the true cause of gastric ulcer, also had only an MD degree (more on his work later).

In today's biomedical research environment, "pure MDs" who decide to enter biomedical science are very few. The typical path is to start with a PhD in some field of biology, such as cell biology, molecular biology, genetics, immunology, biomedical engineering, or genetics.[16] In the United States, a PhD takes about five years.[17] There is also a highly competitive MD/PhD combined program at a select group of universities, where both degrees can be completed in a period of seven to eight years. By the end of this period, it is expected that individuals awarded the PhD degree will have become deeply immersed in a subfield of biology or medicine, in which they have already made some novel (and publishable) contributions; plus they will have received training in a broader scope of subjects in biology and medicine, a good selection of experimental methods and techniques, a good dose of statistical and biostatistical approaches, as well as the principles of ethical conduct in science.

Unfortunately, many PhD students are primarily treated as a pair of hands: the focus is on the generation of data rather than on their comprehensive training in science. According to a worldwide survey in 2017, almost 40 percent of PhD students work more than fifty hours a week; about 20 percent work more than sixty hours a week; and more than half of them also work on Saturdays or Sundays (or

1.2 "Apparently he learned his technique from supervising grad students."
Source: Gerard M. Crawley and Eoin O'Sullivan, *The Grant Writer's Handbook: How to Write a Research Proposal and Succeed* (London: Imperial College Press, 2015), reproduced by permission.

both).[18] Even though the declared goal of the PhD program is to teach research skills and scientific thinking, many PhD students are also forced to teach—over and above their laboratory duties (figure 1.2). The level of compensation for PhD students is less than stellar. In the United States, a PhD student earns about $55,000 a year; in some European countries the salaries can be lower (e.g., in Germany or France) or higher (e.g., in Switzerland). All of this is taxable income.

Even though the pay is average and the workload is high, most PhD graduates seem to be fairly satisfied with their choices. According to the 2019 survey conducted by the prestigious journal *Nature*, 37 percent are very satisfied, and 36 percent are somewhat satisfied with the decision to pursue a PhD, and more than 70 percent say that the program meets their expectations.[19] Shockingly, about half of the responders receive less than one hour of one-on-one time with their supervisors each week. This is one hour *per week*, not per day. The top concern expressed by PhD students includes their uncertainty about their career prospects (79 percent). There were also many concerns expressed about the overall low success rates of grant applications and the pressure to publish. About 36 percent had reported that they have sought help for anxiety or depression related to their PhD. In most institutions, the acceptance of at least one peer-reviewed publication, preferably with the candidate being the first author, in a reasonably well-respected scientific journal is the expectation.

After the PhD is defended, the next step in the process is postdoctoral training, or postdoc. Although PhD students appear to be anxious about their job prospects, in my experience it is rather easy for a well-trained PhD candidate to find a good postdoc position. In fact, today's challenge appears to be for a faculty member, even at a brand-name university, who has a well-funded, ongoing project to find a good postdoc. *Science* magazine recently posted an article about the problem of postdoc shortages.[20]

Let's assume that the academic laboratory hires a qualified postdoc. Postdocs are expected to have a degree of independence, but at the same time they are supervised and mentored by a senior faculty member in the group. The expectation is that they conduct experiments and write papers, or at least contribute with experimental data that they generate, analyze, and plot for publications. Ideally, they also draft or write various sections of the paper. Postdocs are given a substantial degree of independence, which enables them to grow, develop, and shine; but perhaps more often they have this independence because their supervisors are busy with teaching, grant writing, and administrative chores. In the ideal case, postdocs can devote most of their workday to actual research as opposed to, say, teaching medical students or writing grant applications. In a well-funded laboratory, at least in theory, postdocs spend their time advancing science and their own careers as well.

These postdoctoral positions last for a predefined period, usually three to five years. One of the principal measures of a postdoc's success is the quality and quantity of the publications produced. (As discussed later, scientific journals have a well-calibrated hierarchy; scientists are often measured more by the perceived quality of the journals they publish in than by the actual quality of the study itself.) The name recognition of the institute and the scientific prominence of the supervisor are important factors for the future of postdocs, but publications seem to matter the most.

A postdoc's next position quite often is yet another postdoc. Many young scientists end up in a seemingly never-ending cycle of postdocs: a second one, and often even a third one. These sideways moves happen because the number of available faculty positions is only a fraction of the number of new PhD graduates year after year. In 2011, approximately 36,000 science and engineering PhDs were awarded in the United States in a single year, but only about 3,000 new academic positions were created in the same year.[21] This is a ratio of less than 1:10. A Royal Society of Chemistry study found that only 3.5 percent of PhD graduates will get *any type* of permanent position, and only about 0.5 percent of all PhD graduates will become full professors.[22] These are odds ratios of 1:30 or 1:200. Other estimates arrive at odds of 1:5 to 1:10 for a PhD to land in *any* academic position.[23] The odds are clearly not in favor of the young scientists.

For the sake of argument, let's say that the postdoc makes it out of the postdoc phase relatively unscathed and is offered a junior faculty position. The steps are typically instructor, assistant professor, associate professor, and full professor (figure 1.3). The names of these positions vary from country to country, but the

1.3 On the subject of academic career progression and publications . . .
Source: Leonid Schneider / ForBetterScience.com, reproduced by permission.

principle is the same: one has to satisfy various expectations—as measured in dollars and cents (i.e., the amount of extramural grant support)—and progress through the promotions committees to reach the next level.[24]

Most academic systems distinguish between tenure track and nontenure track positions. The former is a relatively straightforward path to the top of the pyramid, at least in principle. The latter is a sort of academic Land-of-Nowhere, in which the person is covered by internal funds or on another investigator's grant funding until something comes along. To advance in science, one thing is essential: extramural money. The more, the better. Important scientific discoveries and publications in good journals are nice things to have, and the department head will probably say thank-you for them. If the candidate is excellent at teaching, this is okay too—but it is not viewed as anything special. The key is money, that is, grants.[25] The type of grants preferred also bring significant additional funds (called overhead) to the institution.[26] Suffice to say that (a) grant writing is hard work; (b) the skill sets required for writing grants (to "sell" the science) are not the same skill sets required to do good science; and (c) the odds of getting grants are not at all in the applicant's favor, to put it mildly. Without grants academics languish in poorly funded laboratories because most institutions provide little or no money for faculty members to do research—not even institutions with enormous endowments and large charitable contributions.

Extramural grants are expected to cover not only the cost of research supplies and reagents but also the salaries of all laboratory personnel: postdocs, research assistants, younger faculty members who don't have their own grants, and in most institutions a significant part of the investigator's own salary as well (figure 1.4).

Grants are essential for doing science and for advancing up the ladder, but they are also needed to maintain one's current academic position. One trick for squeezing some research money out of an academic institution is to move to a new place. If a new institution recruits a new faculty member, the new position is usually one step higher on the ladder than the prior one. The newly recruited investigator may also receive a one-time rainfall called start-up funding. It may be used to buy a new instrument or to pay a technician or postdoc for a year or two. After that new investigators are on their own: extramural funds must be brought in. Scientists quickly learn that they are most marketable at the beginning of a new grant cycle. If a positive grant score comes in, ambitious faculty members immediately start to look around for the next position at the next institution.[27]

1.4 "Spare a dollar for some lab consumables, buddy?"
Source: Crawley and O'Sullivan, *The Grant Writer's Handbook*, reproduced by permission.

When a scientist manages to succeed in the world of grants and advance through the entire ladder—welcome to the top of the academic world. The successful scientist can now be called the principal investigator or PI (figure 1.5).[28]

At some point the scientist may even rise to the level of tenured full professor, and possibly also to Chair of a department. The job of a PI is to be an excellent scientist, to discover the mechanisms of how cells and molecules work and interact in our body, how diseases develop, new ways to predict, and how to diagnose and treat them. In addition, the PI is expected to manage a group of individuals, mentor them, find a way to advance them to the stage where they can start their own fields, catalyze their careers, and of course bring in more grant money (figure 1.6). Furthermore, the PI is also expected to do classroom and graduate-level teaching and to be a useful member of the so-called academic community, which means administration, budget meetings, promotion committees, PhD thesis committees, space committees, extramural grant meetings, and countless other types of meetings. Each scientist is also supposed to

1.5 On the subject of laboratory hierarchy . . .

Source: Leonid Schneider / ForBetterScience.com, reproduced by permission.

1.6 On the subject of academic recruitment . . .

Source: Leonid Schneider / ForBetterScience.com, reproduced by permission.

spend significant amounts of time acting as a referee and evaluating manuscript submissions that colleagues send in for publication in scientific journals. PIs are also expected to attend scientific meetings to present scientific progress and to network with colleagues. It does not hurt for a PI to maintain good relationships with popular science writers who spread (although in many instances *hype* is a better word) the new science to so-called lay audiences.[29] While all of this is going on, PIs are expected to be up-to-date with every new paper in their field; they are expected to come up with novel, exciting, and experimentally testable hypotheses on a regular basis; and they are expected to review, analyze, and troubleshoot every experiment that the group conducts. Of course, PIs are also expected to write (or at least edit and correct) all of the scientific papers that the group produces: after all, the PI is the main (senior) author on the paper, the most distinguished one among all authors. Naturally, the PI is expected to maintain the full scientific rigor and integrity of all of the above processes. Every time a PI puts a signature on a submitted paper it certifies that everybody listed on the paper, at all times, maintained the highest standards of scientific integrity. In other words, with each submission the PI declares that everything in the submitted paper is 100 percent "kosher."

Bench scientists working in academic institutions are expected to publish at a regular rate. Everybody has heard the adage "Publish or perish"—preferably in the so-called top journals. The more, the better (figure 1.7). For a practicing physician, perhaps it is enough to practice the profession well; but when it comes to rising on the career ladder in an academic institution, "Research Is King." Institutions clamor to hire scientists who have multiple active grants; they are less than excited about hiring ones who have limited current funding—even though they have had an excellent track record of research and funding. Past performance, in this case, does not guarantee future results.

The expected publications do not have to be about high-level laboratory research. They could be about "only" clinical research, some compilation of clinical case reports, or a retrospective analysis of some data buried in patient files. Even better is participation in a company-sponsored drug trial, which will ultimately yield only a lowly coauthorship in a journal—perhaps one author out of fifty or one hundred—but will bring some money into the institution. Any kind of publication describing any type of research is okay—and the more the better. This is expected of the so-called academic clinician even if that person does not have much training, affinity, or interest in research. Almost everybody

1.7 On the subject of pressure to publish . . .
Source: Nick Kim / scienceandink.com, reproduced by permission.

is expected to "do science" these days. Amazingly, the institution expects publications even from residents, fellows, or staff members in a small county hospital. Something—almost anything.[30]

This trend in which almost everybody is expected to be a scientist is a worldwide phenomenon. In China, not only the physician-scientists of the leading academic hospitals but also doctors in a county hospital are expected to "produce" scientific publications. It may not be demanded, but it certainly gives a major boost to a person's career (figure 1.8). In many Chinese institutions, a successful publication in a major journal is rewarded with a significant cash bonus. By 2016, the incentive Chinese universities offered first authors had increased to $150,000 for a paper in *Science* or *Nature* and proportionally less for lesser publications.[31] In 2020 an official ban was placed on this cash reward system.[32]

In addition to the work outlined in the previous sections, the PI is expected to bring in extramural grant money with the same regularity. Even at the top of the ladder—*especially* at the top of the ladder—a Chairperson may receive some

1.8 On the subject of incentives in life science research...
Source: Leonid Schneider / ForBetterScience.com, reproduced by permission.

discretionary funds from the institution but usually not enough to support any actual research. (Did I mention that in most fields of biology every item, every reagent—even water—is super expensive?) The money to buy research supplies and to pay the salaries of all research personnel, including some of the investigators' own salary, must continue to flow at a steady (or preferably increasing) rate—*all from the outside*. Even in tenured positions, which are supposedly secure and lifelong, most universities attach a stipulation demanding that some percent of their salary must be covered from the outside. If this does not happen, year by year, little by little, the university decreases the person's salary. They can also take away parts or all of the laboratory of the "nonproductive investigator" and transfer it to a better-funded colleague. There are ways to compensate for this salary deficit, such as by taking on additional teaching or administrative duties. But once someone goes down this path, not only the available physical resources and personnel but also the time available to do research and to generate preliminary data for a new grant application will diminish. If someone is considered a "failed scientist" (which does not mean the person is a bad scientist or does not have a good track record; it simply means a person who is not bringing in grant money *today*), eventually the university may try to force this individual out by offering some sort of one-time settlement and an early retirement package. (In Hollywood, directors are considered only as good as their last movie. Scientists, in the eyes of their administrator overlords, are only as good as their last grant.)

Does it sound like there is some degree of pressure (figure 1.8) at all stages of the academic ladder that is essentially baked into the system? If so, is it possible that all of this pressure contributes to the shenaniganry and fraud detailed in later chapters?

WORK/LIFE BALANCE

In a 2016 opinion article, Dahui You, a biomedical scientist with a PhD from Louisiana State University, does not mince words. She began the article with this statement:

> Let me start off by saying—there is no such thing as "work–life balance" in a biomedical research career. I have not yet encountered a successful scientist with a laid-back lifestyle throughout my research career from 2004 (when I became a graduate student) to present day. In fact, in my experience, most successful scientists struggle with busy schedules and lack of work–life balance.... When I was a student or postdoctoral trainee, I dedicated my life to research. Some of my experiments were inherently time-consuming, so I had to stay up till 03:00 a.m. the next day although I started at 07:00 a.m. At times that I returned home at regular hours, I read papers, books or tutorials related to my research. I did not have a life outside of my research, and at that time I did not feel that I needed one. Research was fun and the excitement from a successful experiment, publication, or presentation for a national meeting was satisfying ... until my son Arthur was born 3 years ago. Naturally, as a mother I want to spend time with him, and do not want to miss a single step that he makes as he grows up. This change in my life inevitably affected my dedication to my career. All the extra hours that I used to commit to research are gone; I suddenly need a life away from the bench. The only option I have is to achieve a work–life balance, even if it seems a "mission impossible."[33]

She then gives some advice on how to achieve some sort of work/life balance by prioritizing, exercising, and other tricks. But it seems that she was right about it being mission impossible. When I checked her LinkedIn page in 2018, two years after publication of her article—with more than ten years of experience in biomedical research, with $7.2 million in federal funding, and thirty-three publications—she chose the industry track. She went to work for Bausch, then Takeda; now she works at Bristol Myers Squibb.

This is also what happened to the last five postdocs who graduated from my laboratory—all extremely bright, hard-working individuals, with excellent training in science. One of them is working in the publishing industry and manages an open-access neuroscience journal. Another is working in a grant office for another Swiss university. The third is working as a sales representative for a biomedical supplies company. None of these three do any laboratory-based research work anymore, which is a shame. In their current lines of work, they continue to utilize some of the scientific knowledge they have acquired—but only a fraction of it. The other two postdocs are doing yet another postdoc: one in Switzerland and one in the United States. N=5 is a small sample size, but it may be indicative of something. After the postdoc is completed, many young scientists exit the game or choose a science-adjacent career that may offer a comparable—or better—salary, a higher level of job security, and a more predictable daily schedule. In short, a better work/life balance.

In the 2019 *Nature* survey mentioned previously, a common concern of PhD students—a close second after their career prospect uncertainty (79 percent)—was their difficulty in maintaining a work/life balance (78 percent).[34] In a recent survey of 215 postdoctoral fellows, 62 percent experienced work/life conflicts; these conflicts, unsurprisingly, correlated well with the level of anxiety the respondents experienced.[35] Almost 90 percent of the respondents believe that careers outside of academia offer a better work/life balance. One of the respondents stated:

> I actually had to detach myself from that PI because I was, like, "you're pressuring me too much on working these 12 h days." I'm getting sick. I don't want this. This is not for me.

According to another respondent:

> This was the kind of lab where you were just expected to produce, and if you weren't producing, you were going to get in trouble. Any time the PI had a whim, you were going to do it and there was no regard for your interests or burnout or anything, you know?

A third postdoc said:

> There was not even one weekend that I stayed home and didn't come to work, not even one. I was working on President's Day, all the vacation days I was here in order to be able to manage to do all these things.

Some disagree, though. For example, the Nobelist Randy Schekman declared at a Lindau Nobel Laureate Meeting that he does not subscribe to the commonly mentioned "long-hours culture":

> If you organize your day in advance—and do not spend time posting pictures of pets on the web—you can get things done. I see no barrier to having a successful career and a family life.[36]

So there you have it. Just cut out the pet photos and your life will return to full harmony.[37] I think part of the solution to the work/life balance "problem" is not viewing the work part as work at all but rather as a "mission in life"—as part of your actual life.[38] But this requires a different kind of life philosophy. It is definitely not for everyone.

BRAIN DRAIN, WORKING VISAS, AND THE NATIONALIZATION OF THE "EXCEPTIONAL SCIENTIST"

It is well-known that the United States and many countries in Western Europe are welcoming young scientists from all over the world. This influx—universally termed "brain drain"—is a principal life force that maintains the scientific and technological superiority of the brain-drainer countries. This is nothing new; a prime example is the Manhattan Project, in which top U.S. military brass supervised a group of physicists and mathematicians from Germany, Hungary, Austria, Czechoslovakia, Poland, Russia, Italy, Canada, Switzerland, Denmark, and Great Britain. It is estimated that about half of them were foreign-born.[39]

According to the American Immigration Council, foreign-born workers in the United States make up almost one-fourth of all STEM workers in the country. This number represents more than ten million people, i.e., about 7 percent of the total U.S. workforce. In life sciences, the percentage of foreign-born workers is estimated at 20 percent. The adage goes like this: The official language of biomedical science is broken English.[40]

Most foreign PhD and postdoc scientists arrive on some type of training visas, such as the J-1, a nonimmigrant visa that supports various visitor exchange programs.[41] In 2022, the U.S. State Department issued more than 280,000 J-1 visas, with an approval rate of approximately 90 percent. If a university (which has special offices called Office of International Scholar Services or something similar) initiates a J-1 visa, it is almost guaranteed that the visa will be issued and the individual can enter the country to participate in a program, but only at the institution where the visa process was initiated. The declared purpose of these visas is to support a kind of catch and release: to provide scientific training opportunities for foreign scientists, after which they are supposed to be released back to their natural environment, i.e., return to their home countries. An important stipulation attached to the J-1 visa is something called a two year home residency requirement. This, of course, is not the true goal of most participants in the process. Most J-1 recipients go to the United States to make a career in science: returning to their home countries for two years would put them at a major disadvantage. Also, many PIs at the recipient institutions would prefer that their trainees—at least the best ones—stay and continue their research. J-1 visas max out after five years. By this time most students have become adjusted to their new country and envision their future there. Ladies and gentlemen, let the Visa Games begin.

Many lawyers in the United States specialize in "helping" scientists with J-1 visa issues. The most common way to deal with the two year home country requirement is to have the student's home government consent to it through a no-objection letter. There are also other ways, for example, if returning to the home country would present "exceptional hardship to the fellow's U.S. citizen or permanent resident spouse or child."

But it is not enough to obtain a waiver. One also needs to obtain a next-level position, which is typically a faculty-level appointment, such as a position of assistant professor. As discussed previously, positions of this type are difficult to find and applicants are expected to have already secured some sort of grant funding for their research. The institution that offers the position must be enthusiastic enough about the applicant to sponsor a next-level visa application, an H-1. Technically, these are still called nonimmigrant visas, and their purpose is to fill specialized technical positions that cannot be readily filled by American citizen applicants. ("The purpose of the H-1 is to employ foreign workers in specialty occupations which requires the application of specialized knowledge and a bachelor's degree or the equivalent of work experience.") Just like the J-1, they also specify a maximal duration, which is six years. There is a limit (cap) on the number of H-1 visas the United States is allowed to issue each year, which is around 85,000 visas. Every year, at the beginning of the year, a new allocation is released, and the U.S. government begins to receive applications on January 1. Often, by the middle of the year, all of the allocations are gone; in 2022 all H-1s were gone by August. In situations like this, prospective applicants must wait until the beginning of the following year. In recent years, some changes were made in the online application process, which created a monstrous process called the H-1-lottery. (The system randomly selects approvals and rejections from the pool of applicants.) About half of the applicants were abusing the system by submitting multiple H-1 applications through several different prospective employers. It was reported that for the 85,000 visa slots over 750,000 applications were submitted in the first three months of 2023—at which point the system was shut down for the rest of the year to stop the flood of further applications. An official fraud investigation is ongoing to identify the bad actors and to prevent this type of problem from happening again.[42] A lot of this abuse was committed by tech (i.e., software) companies and satellite companies (H-1 visa mills) that specialize in trying to obtain visas. While this is going on, other prospective applicants (e.g., universities trying to recruit a professor of biology or medicine) suffer too because their chances of obtaining a visa are decreasing.

After (or shall I say if or when) the H-1 visa is obtained, scientists have several years to immerse in their research, and in an ideal situation they may be able to rise to the level of assistant professor, associate professor, or even full professor. But as the end of the visa period approaches, they face the next challenge: How to stay and continue the work? Once the H-1 expires, the visa holder is supposed to leave the country within thirty days. After a one-year foreign country stay requirement, the applicant can reapply for a new H-1. It is not difficult to imagine the kind of disruption this would be for the scientist and the scientist's home institution. For example, if someone started on a J-1 and then moved on to an H-1, the person would have had ten years of a continuous career in science. And then the scientist is suddenly told that "you have a few days to pack up and leave."

The quest continues, now for the green card, which is permanent residency status and enables the person to take up regular employment in the United States. For a foreign-born scientist, this is the first opportunity to choose employment without being intricately connected to the single sponsor institution (in reality, in many cases, to a single principal investigator and laboratory). If the employer likes the investigator and wishes for that person to stay, the employer will sponsor the application and start the process, which can take anywhere from six months to two years or more. In theory, an excellent scientist has a good chance of processing reasonably quickly and efficiently because the "first and second preference" is to support "employees who demonstrate extraordinary ability" and to support people with "advanced degrees or exceptionally skilled workers." A slick lawyer can prepare a strong application even for below-average scientists. A couple of recommendation letters from well-established professors and department heads is usually a big help. Officially, these letters are supposed to prove the "Alien's Original Significant, Scholarly Contributions of Major Significance in the Field."[43] In most cases, the conversion from H-1 to green card is relatively straightforward, although we regularly hear about instances when things did not go as planned.

As a green card holder, the scientist is allowed to stay and work in the United States for ten years. After ten years, the card can be renewed. Most of the green card holder scientists I know apply for U.S. citizenship at some point, and sooner or later they will become naturalized U.S. citizens.

Without a doubt, this process is lengthy and complicated. It takes away precious time from actual science. It certainly does not decrease the stress level, nor does it improve the scientist's work/life balance. Along every step of the way the scientist is measured, evaluated, and compared with peers in similar positions. As previously explained, publications are important, but even more important is the individual's success in bringing in what else?—extramural grant funding.

It is plain to see how this process can put a young immigrant scientist in a precarious situation. First, with the J-1 the opportunities to switch positions or move to other universities are limited. The scientist is tied to the institution that supported the visa. In many cases, the research fellow arrives at a university without having had a chance to properly interview or visit the future place of work; many interviews happen over the phone or online. In an ideal situation, the person finds a welcoming environment and a supportive boss. In less than ideal situations, research fellows may be stuck in an environment in which the reality is much different from their hopes and original expectations. But they only have two choices: suck it up and stick it out or return to their home country. After the J-1 phase, the pressure intensifies: now the young scientist faces the dual challenge of finding a faculty position and securing an H-1 visa. Once again, almost everything depends on "productivity," and in later stages of the career also on "extramural funding."

I don't want to jump ahead of myself in the book, but the best way to illustrate what happens if the precarious employment situation of scientists is exploited is to include the witness testimony of a postdoctoral fellow from a laboratory that was closed down due to massive scientific fraud. (Head of the lab: Piero Anversa, Harvard University, now retired. His story is covered in much detail in a later chapter of this book.) This testimony was submitted to Ivan Oransky, who runs a website called Retractionwatch.com (more about this later). The title of the section of the testimony I cite here is "Information Segregation + Machiavellian Principles = Successful Lab."

> The day-to-day operation of the lab was conducted under a severe information embargo. The lab had Piero Anversa at the head with group leaders Annarosa Leri, Jan Kajstura and Marcello Rota immediately supervising experimentation. Below that was a group of around 25 instructors, research fellows, graduate students, and technicians. Information flowed one way, which was up, and conversation between working groups was generally discouraged and often forbidden.
> Raw data left one's hands, went to the immediate superior (one of the three named above) and the next time it was seen would be in a manuscript or grant. What happened to that data in the intervening period is unclear.
> A side effect of this information embargo was the limitation of the average worker to determine what was really going on in a research project. It would also effectively limit the ability of an average worker to make allegations regarding specific data/experiments, a requirement for a formal investigation.
> The general game plan of the lab was to use two methods to control the workforce: Reward those who would play along and create a general environment of fear for everyone else. The incentive was upward mobility within the lab should you stick to message. As ridiculous as it sounds to the average academic scientist, I was personally promised money and fame should I continue to perform the type of work they desired there. There was also the draw of financial security/job stability that comes with working in a very well-funded lab.
> On the other hand, I am not overstating when I say that there was a pervasive feeling of fear in the laboratory. Although individually-tailored stated and unstated threats were present for lab members, the plight of many of us who were international fellows was especially harrowing. Many were technically and educationally underqualified compared to what might be considered average research fellows in the United States. Many also originated

in Italy where Dr. Anversa continues to wield considerable influence over biomedical research.

This combination of being undesirable to many other labs should they leave their position due to lack of experience/training, dependent upon employment for U.S. visa status, and under constant threat of career suicide in your home country should you leave, was enough to make many people play along.

Even so, I witnessed several people question the findings during their time in the lab. These people and working groups were subsequently fired or resigned. I would like to note that this lab is not unique in this type of exploitative practice, but that does not make it ethically sound and certainly does not create an environment for creative, collaborative, or honest science.[44]

This type of pressure may not be a rare exception (figure 1.9). Indeed, when 434 cancer researchers at the MD Anderson Cancer Center were surveyed, about

1.9 On the subject of supervisory pressure...
Source: Leonid Schneider / ForBetterScience.com, reproduced by permission.

30 percent of them stated that they have "noted pressure from a mentor to prove his/her hypothesis correct, even though the data may not support the hypothesis," and about 20 percent of them felt "pressure to publish findings" about which they have had doubts.[45]

ALTERNATIVE CAREER PATHS IN BIOMEDICAL SCIENCE

A few alternatives to the grants-based, academic rat race are open to people who wish to conduct scientific research (figure 1.10). One of them is joining a national research institution. Institutions of this type are supported by government or federal funds, and investigators who work there do not need to bring in extramural grant support. In the field of biology and medicine, the main one in the United States is the National Institutes of Health (NIH). Most of its institutes are located in Bethesda, Maryland, not far from Washington, DC. The NIH employs about 20,000 people, about one-third of whom are at the PhD level or higher and one-third at the postdoc/trainee level. The NIH is

1.10 On the subject of potential alternative career paths for life scientists . . .
Source: Leonid Schneider / ForBetterScience.com, reproduced by permission.

considered to be a major center of scientific excellence with significant physical and intellectual resources; intramural NIH scientists are involved in virtually all aspects of biomedical research; more than 1,200 principal investigators run their independent research groups; the NIH intramural scientists have made significant discoveries over many decades; they publish extensively and in premier journals; and many Nobel laureates did all or part of their work as part of the NIH intramural research program.[46] There also seems to be no strict age limit at the NIH; investigators who are capable and willing are allowed to continue their research as long as they remain productive. Too good to be true? Well, there are a few catches.

First of all, NIH researchers are federal employees, so there is a significant administrative burden compared to what's commonplace at a research university. Obtaining approvals for travel, completing all the various levels of training and continuing education, navigating administrative procedures, and dealing with red tape are the main sources of frustration to my colleagues working at the NIH. Even though scientists working at the NIH do not need to apply for grants from the outside, they still need to pass through regular Progress Reviews, which are conducted by extramural, independent scientist colleagues. (My colleagues at the NIH regularly complain about these reviews, although I cannot imagine that this process is comparable to actually obtaining extramural grants in an academic laboratory and maintaining constant funding for decades.)

There may also be some limitations at the NIH on choosing projects, and there are some limitations on the resources an NIH investigator can utilize at any given time. However, my understanding is that most NIH institutions are very well funded, and a lot of work can also be done in intramural collaborations, i.e., without the need for outside funds to pay for it. Most laboratories are allowed to grow to a certain size, but mega-groups are rare. The titles of NIH investigators (e.g., Principal Investigator or Section Head) are less glamorous than the often lofty-sounding academic titles.[47] Also, some investigators may not like the salary caps the NIH maintains even though these caps are rather generous for scientists at the top of the ladder. Others may not like the fact that NIH investigators face various restrictions if they want to start a spinoff company and usually do not benefit much commercially from their inventions. Finally, NIH investigators must train junior scientists but typically do not teach regular university courses—some may view this as a problem, others may see it as a blessing. Overall, in my opinion, the good aspects outweigh the bad ones, and I am surprised that the NIH is not attracting even more exceptional talent, especially in today's hypercompetitive research environment.

There are national institutions in many other countries where research can be conducted using intramural funds. In addition, there are research institutes founded by billionaires—typically focused on a relatively narrow scope of research—where the funding is relatively secure, and "all" the scientists working there need to do is cutting-edge science. For example, the Chan-Zuckerberg Biohub in San Francisco focuses on various areas of genomics and infectious diseases. In Seattle, the Allen Institute for Brain Science, founded by the late Microsoft cofounder Paul G. Allen, works to understand how the human brain functions through large-scale, open science initiatives. The Parker Institute for Cancer Immunotherapy in San Francisco, California, focuses on cancer immunotherapy research. The Howard Hughes Medical Institute maintains a state-of-the-art research center called Janelia Farm in rural Virginia that focuses on neuroscience. Another player in the field of neuroscience is Elon Musk's Neuralink project, which seeks to develop human brain/computer interfaces.

Some of these institutions operate, principally, as scientific foundations to support open-ended exploratory/basic research, and others resemble biotech companies that focus on more concrete—commercial—goals. One step further in the commercial direction are the various scientific opportunities within the biotech and pharmaceutical industry. These research facilities—as well as the support systems to help with work/life balance—tend to be pretty nice as well, and the salaries and the benefits are generous. Working in these companies is considered relatively secure, although the smaller companies—many of them funded with venture capital—can run out of money and shut down on short notice.

Larger companies can quickly change their direction, and entire research campuses or research groups could be dissolved on short notice. Of course, in a company the vast majority of researchers don't have much say in the direction of the research or the therapeutic area concerned. They work on a given project toward a given goal. They have some freedom in choosing certain approaches, and sometimes they run into unexpected or unintended discoveries, but these are the exceptions and not the rule. The company's direction can turn on a dime, a person working on a receptor blocker for stroke might find out one morning that they are now working on a different disease or a different approach, such as gene therapy for a rare liver disease.

The major complaints of people working in biotech and pharma are that (a) "nobody knows" their contribution to a certain project on the outside and that (b) the company does not allow publication of their findings in a scientific journal. The data may end up in a company patent, and the scientist who did the work may end up as a coinventor on the patent. (The commercial rights to the patent, of course, are owned by the parent company.) This is the price the industry scientist "pays" for the higher salary, relative job security, and the nice work environment. The no publication policy, however, is not absolute. Experiments related to projects that do not work out and do not progress further into the development funnel can still be published—although often only with significant delays. An adage in the pharma industry states that a surefire way to know a company gave up on a drug target or development candidate is when they publish their findings in the general scientific literature. Some of the larger companies also allow limited exploratory academic time for some of their employees, and sometimes biotech and pharma companies publish truly innovative findings that advance science. But certainly there is no explicit pressure to publish in biotech or pharma.

There are, however, other types of pressures, related to time constraints to complete projects, or pressure—stated or unstated—for the project to "work," and at later stages of programs, to *prove* that they are "safe and well tolerated" before a drug candidate progresses into the clinic and then to *prove* it is "safe and effective" in the clinical trials. An egregious example of biotech pressure is, by now, very well publicized. The now defunct medical diagnostics company Theranos—which at one point was valued at over $1 billion—did not have a working machine to diagnose diseases from a single drop of blood. But under the immense pressure exerted by the company's founders and executives, the company's employees perpetrated fraud and deception on an astonishing scale. Theranos bigwigs Elizabeth Holmes and Sunny Balwani are currently serving their ten-year+ prison sentences. The whole sordid affair is extensively covered in the best-selling book *Bad Blood* and in the mini-TV-series *The Dropout*.[48]

CHAPTER 2

HYPERCOMPETITION FOR RESEARCH GRANTS

"NO MONEY, NO HONEY"

It was the summer of 1992. I was barely out of med school, but I already had three or four years of hands-on research experience working in the Students' Scientific Circle. By that time, I had also spent a few months doing research abroad, one time in Sweden at the University of Lund, and another time at the University of Pennsylvania in the Cerebrovascular Research Center. A few of our studies had been published, some in international journals such as *Blood Vessels*, *Stroke*, and the *Journal of Physiology*. My work focused on the endothelium-dependent responses and nitric oxide, which were hot scientific topics at the time (and still are). We found some interesting things, including how the endothelial responses are impaired in shock, and how nitric oxide maintains blood flow to various parts of the brain. I had submitted my PhD-equivalent dissertation at the same time that I received my MD degree in 1991.

So I set out some big plans for myself. I wanted to join one of the best labs on the planet in my field, which was the famed William Harvey Research Institute in London, led by the Nobel laureate Sir John Vane. Sir John received his Nobel for demonstrating how aspirin and related drugs exert their effect by inhibiting the biosynthesis of a labile group of mediators called prostaglandins. The focus of the institute was on endothelial-cell derived biological factors such as prostaglandins, endothelin, and, of course, my favorite molecule, nitric oxide. His group published exciting, novel findings, usually in top journals. They had an amazing array of methods and techniques. This was the ideal place for me to learn, grow, and develop as a scientist. As it happens, Sir John gave a big plenary lecture at a meeting on prostaglandins in Vienna, Austria, that summer. I took a train to Vienna, somehow managed to get inside the lecture hall for the afternoon of his lecture, and after the session I summoned up all of my courage and walked up to him with my printed curriculum vitae (CV) in hand. I explained who I was, what my research field was, and how I had dreamed of joining his lab as a postdoc. He seemed surprised, maybe annoyed, and did not say much in terms of a reply, but he took my CV. After returning to Budapest, to my surprise, a letter came saying that he would be happy to have me in his lab—*if I can bring a grant that covers my salary.*

A few months later an announcement was posted on my research department's notice board notifying prospective applicants that postdoctoral research fellowships were available to work in the field of biomedical science in England. The grant was a joint project between the Hungarian Academy of Sciences and the Royal Society in England. It all sounded ideal to me. But there was a slight—and distinctly Hungarian—twist in the story. The grant was announced on the notice board *on the exact same day as the submission deadline.*[1] I did what I could, first contacting the William Harvey Research Institute in London and having the required "acceptance letter" faxed to me (all of this was before the internet, you see), then I wrote a grant application overnight, and I hand-delivered it with my application package to the ornate building of the Hungarian Academy of Sciences as soon as it was humanly possible—one day *after* the official deadline. Somehow I managed to convince the administrator to add my application to the rest of the pile of submissions that were already on her table.

Time passed and I only found out later what had happened, directly from Sir John Vane. Apparently, the applications were evaluated, and mine was one that was deemed fundable, according to both the Hungarian and British reviewers. However, my application was disqualified by the Hungarian side because it did not satisfy the submission criteria—it was received one day too late. Luckily for me, on the British side Sir John had access to both versions of the rankings: the original one (with my project listed as fundable) and the final one (from which my project was deleted).[2] This whole affair infuriated him so much that he decided to hire me anyway. I arrived in London a few months later eager to start a new chapter in my career. Then, when already in London, I submitted another fellowship application, this time to a British grant-giving body, the Lloyd's of London Tercentenary Foundation, and I was funded. This is how I managed to get my first research grant.

The whole affair taught me a few things about grants early in my career. First, it showed me that grant funding is essential, and even the most prestigious laboratories and institutions depend on it, or at least prefer to use them rather than spending their internal funds. Second, it showed me that this whole grant business is a highly competitive activity, and that unfair competition may not be uncommon.

APPLYING FOR GRANTS: THE PROCESS AND THE ODDS OF SUCCESS

Over the next thirty years of my career, I wrote countless grant applications.[3] Most of them were rejected, but some were funded. Between 1994 and 2018 I was part of the United States scientific and granting system—first as an NIH grant applicant, and later as a regular NIH grant reviewer (a so-called Study Section member) as well. I had the opportunity to observe how the system has undergone changes and how the funding percentages declined over the years, and I have seen countless colleagues become frustrated and desperate in the face of repeated grant rejections. I have also seen laboratories that received significant funding on a regular basis: many have utilized the funds well and advanced biomedical science, but others were engaged in questionable research practices and had to correct or retract some of their papers. I have also participated in many other grant-giving bodies, as an applicant and sometimes a recipient or as a grant referee. One of these was the Marie Curie grant mechanism, part of the European Union's Horizon Program, which awarded exchange grants to early-stage scientists. For the last five years, my research has been supported primarily by the Swiss grant system, where the primary source for life sciences funding is the Swiss National Science Foundation (SNSF).

On the surface, the grant process is fairly simple. The investigator is expected to identify an area of science where some thing or some things are not known. Imagine a blank area on a geography map. Then the investigator is expected to come up with a working hypothesis about what might be going on in these unknown areas. For example, how a certain reaction pathway plays a role in a certain biological process or disease process. To support the hypothesis, most grant-giving bodies expect (or mandate) some experimental results as part of the application. Typically, these so-called preliminary data are expected to be "novel"—i.e., not yet known to anybody else on the planet and not yet published in a scientific journal. The reviewers of the grant application must sign a piece of paper that they will keep the application, including all of its ideas and data, strictly confidential and that they are not going to use them to foster their own scientific thinking or in their own future research. (*Word of Honor!*) Then applicants are supposed to convince the reviewers that they are eminently qualified to perform the proposed research tasks and have all the necessary know-how and equipment at their disposal (most grants don't provide enough money to buy any significant new equipment, so all large equipment must already be in place when the application is submitted).[4]

For NIH grants, it is usually also essential for the applicant to convince the grant reviewers that the proposed project is significant in a practical sense and can have some medical implications in the future. However, the NIH also supports applications in which the goal is to discover fundamental biological principles for which the medical applications are not yet clear. Other grant-giving bodies, such as the U.S. National Science Foundation (NSF), also have mechanisms to fund more fundamental (basic research) activities.[5]

It helps if the grant application is neatly organized, convincingly written, and is organized by a prescribed set of sections, such as these: Specific Aim, Introduction, Experimental Plan, Preliminary Data, and Discussion. The applicant's CV must also be presented in a highly prescribed and organized format, with strict length limitations. Every detail counts, even the font and the line spacing chosen. If you use a smaller font or more lines per vertical inch than specified, the application won't even be entered in the race. The investigator must then submit a new, properly formatted one at the next available deadline. Although your university is not supposed to be an evaluation factor, it does not hurt for the application to be submitted from a brand-name university such as Harvard or Stanford.

The applicant must then wait while the grant goes through the evaluation process. In large granting institutions such as the NIH there are multiple institutes, and each of them sets up multiple groups of evaluators, called Study Sections, based on the topic of the application. In these Study Sections, expert reviewers (in effect, the applicant's peers, usually working in the same field—meaning, often the applicant's potential or actual direct competitors) will decide whether they like it or not, and if they like it, how much they like it. Bizarrely, the NIH tells applicants fairly early on who the Study Section members are, but they also instruct applicants never to discuss the application with them. The applicants are not prohibited, however, from talking to the reviewers about other subjects, such as soccer, or even becoming friendly with them in general. A common game played in academia is to invite Study Section members to give plenary lectures at the applicant's institution: wine and dine them—you get the picture.

Only a few of the twenty or more Study Section members actually read each grant thoroughly, and the applicant does not know which three from the group will do this—although one can guess the assignments based on the research field of the reviewers and the topic of the proposal. When the results of the evaluation are received, verbal designations as well as numerical scores are made regarding the quality of the project. The difference between a grant deemed "outstanding" and one only rated "excellent" may decide whether the project will be supported or rejected. The verbal comments are organized as "strengths" and "weaknesses" of the proposal, and various scores are assigned, the most important of which is the "percentile number." This number tells the applicant what percent of the grant applications, in that particular cycle and in that particular evaluation group, were ranked higher (i.e., better) and what percent were ranked lower. For instance, if a grant percentile is sixteen, this means that if the grant was evaluated in a batch of fifty applications, then seven grants were found to be stronger and forty-two were found to be weaker than the submitted application. (In which case, very sorry, but the project will *probably* not be funded. If the field is cancer research, then with this score the application will *definitely* not be funded. More on that later.)

There is also the so-called triage process, which is supposed to ease the burden of the reviewers. If the referees feel that the grant belongs to the bottom 50 percent, it won't even be discussed at the Study Section and is swiftly rejected. Some colleagues seem happy to get any percentile score the first time they submit a grant. If they revise their form, they argue that the project may reach a fundable score. (Successful funded revisions of originally triaged applications are almost unheard of.)

There are several grant cycles every year, in the NIH cycle there are three of them. Sometimes the NIH issues special topics on a focused area and invites applications related to certain mechanisms of diseases that are considered a national priority. These Requests for Application (RFAs) sometimes have a better chance of being funded than regular grants. In the old days, only early-stage applicants were eligible to apply for some grants, which were called

R29 grants. The funding chances were slightly better in this system than in the regular grant system.

Common NIH grants are the regular investigator-initiated grants, and the research topic can be freely chosen. The prototypical one is the R01, typically a four or five year project with a total "direct budget" of approximately one million dollars.[6] "Direct" money means money that the investigator can use for salaries, research supplies, and maybe a small piece of equipment. (There are also "indirect" costs, which are discussed later.) One million dollars sounds like a lot of money, but if the institution expects 50 percent of the investigator's own salary to be covered from this extramural funding, there may only be enough money left for a postdoc's salary and some research supplies. The $200,000 to $250,000 per year does not reach much further than this.[7] Therefore, having only one R01 is not enough to support a research group.

In 1995, when I submitted my first NIH grant from the University of Cincinnati, the R29 system was still in place for early-career investigators. The page limit was twenty-five pages, and if not funded, one could respond to the reviewers' comments, revise the application, resubmit it, and hope for the best. A second revision was also permitted. I was funded after my second revision/resubmission. A few years later I received my first R01—I think after the first revision. At that time, the R29 funding rate was somewhere around 25 to 30 percent in the NIH institutes where my research belonged (NIGMS and NHLBI). These rates were much lower than the rates when the NIH grant system began (in the 1960s award rates were 50 percent or higher), but they were still reasonable. Later the application process was simplified to reduce the burden on the applicants and the reviewers. Applications was cut to twelve pages, and only one revision was allowed.

In the mid-1990s, the Clinton administration and Congress decided to *double the NIH budget over five years*. It seemed that a new Golden Age of biomedical research would begin. But, paradoxically, grants and funding chances started to go downhill. Many universities built massive new biomedical research campuses, hoping they would be filled with excellent—and well-funded!—investigators. A hiring boost started. The number of people working in the American university research infrastructure skyrocketed.

The administrators' enthusiasm for creating more research facilities was not exactly selfless. It is agreed that research will benefit all mankind. But, as it turns out, research is also a very lucrative business for universities. This is due to a gravy train called "indirect cost" (or "overhead"). The official term is Facilities &

Administrative costs (F&A). Every institution that receives NIH grants negotiates and establishes an NIH overhead rate, which is a percent of funds the university administration receives over and above the direct money the scientist's lab receives. The indirect money is awarded to pay for facilities costs, electricity, heating, building maintenance, and (importantly) much of the university administrators' salary. The F&A income received by research institutions can be enormous. In the 1990s, most universities negotiated federal overhead rates of 40 to 50 percent. At one point in 2012, the Boston Biomedical Research Institute had an overhead rate of 102 percent: for each $100 the investigator received to do research, the administrators received an additional $102 to do . . . well, administration. Brand-name places such as Harvard and Stanford had overhead rates of 70 percent. After a few juicy stories of universities using indirect cost funds to pay for yachts and decorations in the president's house, indirect cost rates decreased slightly for a few years, but then they started rising again. Today a 60 percent overhead rate is considered normal for an average American research university.[8]

If the NIH pays about $30 billion per year for extramural grant awards, we can guestimate that at least $10 billion per year is, in fact, paid for university overheads. According to some estimates, in addition to the overhead money that goes into the administrative budget of research universities, U.S. taxpayers also subsidize the research interests of billionaires (so-called philanthropists) to the tune of $7 to $10 billion each year.[9]

In 2018, President Trump floated the idea that the government should direct more funds to the scientists who do the actual research and reduce overhead rates to a generic rate of 10 percent. An unprecedented rebellion ensued—orchestrated by university administrators and probably executed by their lobbyists—and the proposal was quickly buried, *never to be mentioned again.*

After the doubling that occurred in the Clinton years (from about $12 to $24 billion per year), university administrators expected that the NIH budget would just keep rising and rising, perhaps forever. Instead, the NIH budget plateaued around 2001. In some years it even decreased—both in terms of actual numbers and especially in terms of purchasing power.

Meanwhile, the newly recruited scientists were—as before—fully expected to bring in the grants. Consequently, the number of NIH grant applications increased to unprecedented levels (from about eighteen thousand per year in 1995 to about seventy thousand per year in 2022), and, naturally, the funding success

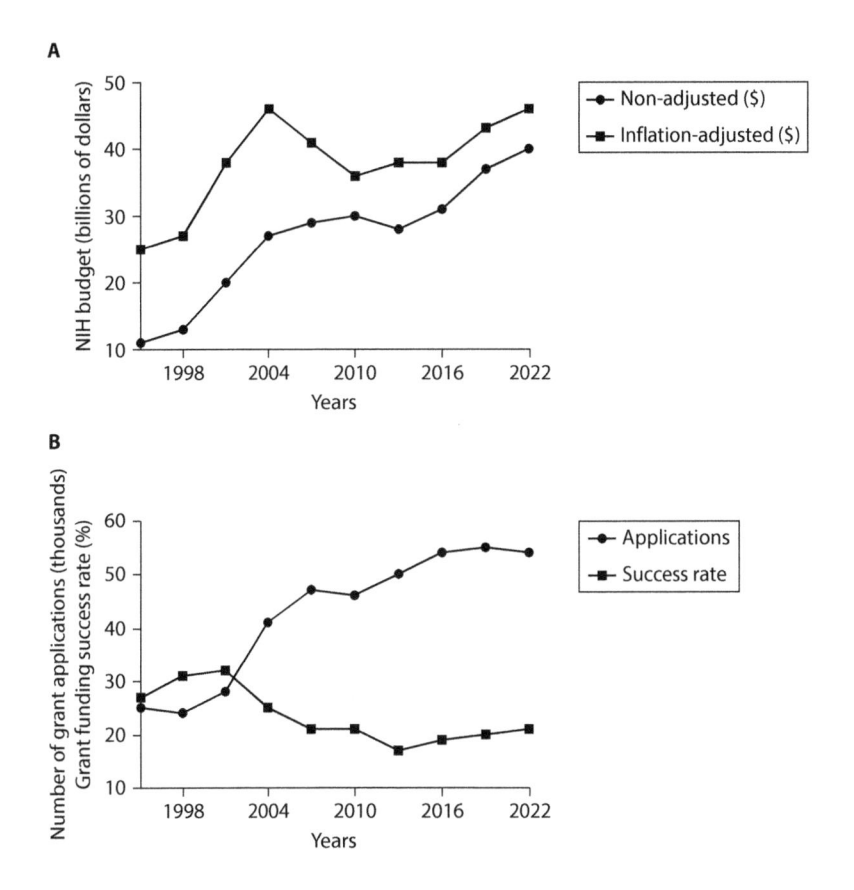

2.1 (A) NIH annual budget (non-adjusted and inflation-adjusted, in billions of dollars) and (B) the number of R01 grant applications the NIH receives (in thousands) and the rate of success over time (%).

Source: "The NIH Data Book," National Institutes of Health, Washington, D.C., https://report.nih.gov/nihdatabook/.

rates decreased: from more than 30 percent in 1997 to about 18 percent in 2016.[10] Today it hovers around 20 percent when regular R01 applications from all fields are considered (figure 2.1).[11]

But these are only the *average* numbers. Depending on a scientist's research field, one could do somewhat better, or one could be much worse off. One would think that cancer research is somewhat important because it is one of the main causes of death today, but the National Cancer Institute's (NCI) R01 funding rate hovers between 8 and 11 percent. The NCI issues more than $2.5 billion of grant money on cancer research each year—no small potatoes by any means. But if someone submits an R01 project on cancer research, the statistical chance of being rejected is about 90 percent. The applicant then has one chance to revise the application, and the chances of being rejected are, once again, about 90 percent.

If the 90 percent rejection rate has caught your eye, let's follow it up with a small statistical exercise. Let's assume that a particular investigator's expertise is in cancer research, and the university expects the investigator to maintain two R01 grants in parallel, to support a small research group of four or five people. As previously explained, each time the investigator submits an R01 there is a 90 percent statistical chance of being rejected. Let's say this particular investigator happens to be an excellent scientist, much better than other scientists. Let's say that this investigator can constantly come up with exciting, groundbreaking ideas. Let's say this investigator has learned the grantsmanship lessons and knows how to write convincing proposals. For the sake of this exercise, let's say that this particular investigator is three times better than the average applicant. Let's increase the expected chance of grant funding from 10 percent to 30 percent with each submission. With these types of odds, this applicant will still need to submit six grant applications each year to achieve a 90 percent chance of having at least one grant funded.[12] There are three submission cycles each year, so two parallel grant applications must be submitted for each cycle. It takes three to four weeks to write a solid grant application, so 50 percent of this investigator's waking time will be spent grant writing. The next year the same process must be repeated. Once again, six grant applications must be submitted, and once again, there is a 90 percent chance of at least one of them being funded. The chance of the investigators' efforts being equally successful in both years and the investigator securing the expected two parallel-running R01 grants is about 77 percent.

In other words, almost all the applicant did—who, in this example, was three times better than the other scientists and had a three times higher chance of grant success—was grant writing, but there still is a 1:4 chance that the university administrators' expectations won't be met. (This means no promotion, maybe a salary cut, maybe personnel or lab space will be taken away, probably the investigator will be assigned more teaching duties, and so forth.) And even if the investigator succeeds and has two parallel-running R01s, in a few years' time, the process must start again: with the same statistical odds.[13] Chances are that—sooner or later—even the most successful investigators will experience a period in which they may fall in between grant funding.

Unfortunately, many people lack the skills necessary to make probability calculations, especially when they apply to real-life situations.[14] The field of game theory is a relatively recent area, which was created by John von Neumann in the late 1920s.[15] I attended many grant-writing seminars and met many university employees in various grant offices. Many of these grant seminars go into infinitesimal detail: things like which fonts not to use, right-justify the pages or not, and which words to avoid. However, at no point have I seen any funding chance calculations being presented. Some people may simply look at the situation as follows: "The chances of funding are maybe 10 to 20 percent. I am much better, so probably my chances are 33 percent. If I write three strong applications (3 × 33 percent is 99 percent), I will be okay." And others, who may be better at math, are better off staying quiet: What purpose could it serve to demoralize the prospective applicants? At the University of Texas, I showed my probability calculations to my department Chair, and I asked how he feels about them. "I am not the one who made the rules" was the answer.

Here is one more thing to consider. Even if one could physically perform the above outlined, heroic, constant grant writing, this activity would not only distract the scientist from doing actual research but would also not be considered an acceptable activity for an NIH-grant-funded investigator. In other words, NIH grants must not be used to pay people to write more NIH grants (figure 2.2). A few years ago the famed Scripps Institute in San Diego found this out the hard way. According to a report published in the *Times of San Diego*, Thomas Burris, a tenured professor of the Departments of Molecular Therapeutics and Metabolism and Aging at Scripps, together with at least forty other faculty members, was "caught up in a high-pressure system that led to the alleged fraud—forcing staffers to secure 100 percent of their salary via grants."[16] Burris estimates that, over a five-year time period, he spent between 20 and 50 percent of his working time on grant proposal activity—including compiling data for and writing and submitting eighteen applications. Other faculty and research staff spent additional, significant amounts of their time assisting him in the same activity. Burris blew the whistle. In the subsequent lawsuit, the Department of Justice determined that Scripps was misallocating NIH funds. The institution paid $10 million to settle the claims. As is customary, "the claims resolved by the settlements are

2.2 On the subject of grant applications . . .
Source: Ralph Hagen / CartoonStock.com, reproduced by permission.

allegations only" and "there has been no determination of liability." Case closed, end of story. Although I seriously doubt that Scripps Institute is the only place on Planet Earth where scientists are writing new applications while being fully or partially paid on NIH (or other) grants. In addition, I know for a fact that Thomas Burris is not the only scientist on the planet who spends comparable time on grant writing and on performing actual research.[17]

How can anyone maintain continuous funding under these circumstances? Everybody is trying their very best. Hard work, stamina, and good intellect (and excellent typing skills!) certainly help. To have a productive and successful track record also helps, and—even if, officially, this is not supposed to be factored into the evaluation process—being at a famous, "brand-name" institution seems to help as well. Some fields of life sciences and medicine are better funded than others: many investigators try changing fields and sometimes find "greener pastures" in new areas.[18]

Although the R01 is the gold standard and is almost the only grant that matters for promotion, there are other grant mechanisms, and some have better success rates. Sometimes several investigators from various departments or institutes join forces and try for larger joint grants called a "Program Project." Scientists who start spin-off companies are eligible to apply for technology transfer and development grants called Small Business Innovation Research (SBIR) and Small Business Technology Transfer (STTR). Yet others try to get by on shorter-duration pilot grants called R21s, which are sometimes easier to get. Dual-use (military and civilian) projects may be funded by other branches of the U.S. government, such as the Office of Naval Research (ONR) or the Defense Advanced Research Project Agency (DARPA). And the government is not the only source of funding. Many foundations focus on specific diseases and offer grants—typically smaller than the R01. Among them are the American Heart Association, American Lung Association, Michael J. Fox Foundation, and the March of Dimes Foundation. Most nongovernmental grant sources have low overhead rates or none at all. It should not come as a surprise to you by now that they are frowned upon by the university administrator types and do not count as full achievements in the eyes of academic promotional committees. In certain fields, it is also possible to find some income through contract research or industry-sponsored or collaborative projects.

Many investigators have joint affiliations with other systems, where grants may be available. For instance, the Veteran's Administration (VA) or the Shriners Foundation support specific areas of research (e.g., burn injury research and orthopedics), with somewhat better odds of funding than the typical NIH grant.[19] Also, as previously mentioned, many scientists play the "moving game," which is almost the only way known to mankind to extract internal research funds from one's own research institution in the form of one-time start-up funding.

Some institutes have internal pilot grant programs (seed money) to help the investigator gather preliminary data that can later be used to apply for a larger, extramural grant. All in all, many scientists manage to scrape it together, but the percentage of investigators who are in between R01 grants is steadily increasing. And many investigators leave the field, or take early retirement, or move to other countries in the hope things may be better there.[20]

STUDY SECTIONS, GRANT REVIEWS

Over my career, I have not only been a grant applicant (and, on occasion, recipient) but also a grant reviewer. If one receives regular R01 funding from the NIH, sooner or later the scientist will be recruited into the "referee pool." The process goes something like this. First, the investigator is invited as an "ad hoc" Study Section member. Everybody understands that this is a "trial run." After a while, if the new member seems to perform well, a "regular" Section Membership may be offered.[21]

Study Section membership is considered a very big deal. Promotional committees appreciate it; and university administrators love it too. One benefit of Study Section membership is that grant deadlines no longer apply to them: they can now submit their own grants any time it fits their schedule. This process is called continuous submission. In addition, Study Section members' own grants are reviewed at separate Special Study Sections, where the chance of funding tends to be higher.[22] Definitely, there are some incentives. But for the most part, it is grueling work, a huge responsibility, and an activity that rarely brings new friends but can definitely produce new enemies.

It is important to emphasize that Study Section members do not directly evaluate all grant applications that are assigned to their section. Only a fraction of the submitted applications are assigned to each member, although all submissions are made available to every member and all members are "encouraged" to read all applications—if they can. (Most likely, they can't. The workload is massive, even to do a proper job for the assigned ones.) Primary, secondary, and tertiary reviewers are designated. The "job" is to read the application, to look up the background and the context of the project in the literature, and to determine its merits (strengths) and demerits (weaknesses). All of this happens remotely—weeks or months before the actual meeting (figure 2.3).

When the Study Section finally meets in person, the primary reviewer introduces the application to the whole room (most of whom have not read a single word of it before), the secondary and tertiary reviewers discuss the application with the primary reviewer, and then everybody else can chip in. In the end, the primary reviewer recommends a priority score (or a range of scores) to the whole Section. Verbal designations such as "Exceptional" or "Very good" or "in the Excellent to Outstanding" range are also specified.[23] This designation is based

Reviewer #1	Reviewer #2	Reviewer #3
Groundbreaking... Extremely novel... **Definitely fund.**	Good idea... Clearly presented... Strong resume... **Fund.**	Unoriginal... Flawed... Weak track record... Poorly written... **Decline.**
OUTSTANDING (9/10)	EXCELLENT (8/10)	POOR (2/10)

2.3 On the subject of grant reviewers . . .

Source: Gerard M. Crawley and Eoin O'Sullivan, *The Grant Writer's Handbook: How to Write a Research Proposal and Succeed* (London: Imperial College Press, 2015), reproduced by permission.

on a table provided in advance to the reviewers (table 2.1). Everybody knows that only "High Impact" ones stand a chance of funding. Everybody knows that "Very Good" means "really bad" and "Satisfactory" means "unsatisfactory."

Until recently the scoring criteria listed in the table had to be applied to five categories: (1) Significance, (2) Innovation, (3) Approach, (4) Investigator, and (5) Environment. In 2025, the NIH will introduce a different system, in which only two categories will be scored: Significance and Innovation will be pulled together into a category called "Importance of the Research," and Approach will be renamed "Rigor and Feasibility." The last two categories (Investigator and Environment) will be merged into a single category called "Expertise and Resources." This category will not be scored: it will either receive a yes (meaning that the environment is sufficient for the proposed research) or a no (in which case the reviewers must justify why it is not acceptable). The latter change in the evaluation system serves to minimize something called *reputational bias*.[24]

TABLE 2.1 The NIH R01 grant impact scoring system

Degree of Impact	Impact Score	Descriptor	Additional Guidance on Strengths/Weaknesses
High	1	Exceptional	Exceptionally strong with essentially no weaknesses
High	2	Outstanding	Extremely strong with negligible weaknesses
High	3	Excellent	Very strong with only some minor weaknesses
Moderate	4	Very Good	Strong but with numerous minor weaknesses
Moderate	5	Good	Strong but with at least one moderate weakness
Moderate	6	Satisfactory	Some strengths but also some moderate weaknesses
Low	7	Fair	Some strengths but with at least one major weakness
Low	8	Marginal	A few strengths and a few major weaknesses
Low	9	Poor	Very few strengths and numerous major weaknesses

For NIH grant applications, the guiding principle for referees is—or rather is supposed to be—the project's "impact" rather than its "excellence." There may be an excellent application on the table, but if the referees feel that it won't make an "impact," the application may never make it into the fundable range.[25]

Even a slightly lower opinion of only one of the three reviewers can be enough to push an application down into the not fundable zone. This will be the case even if the person with the low opinion has no significant expertise in the area in question, or even if this minority lower opinion comes from the tertiary reviewer (who is not required to write full comments about the application, and, as a consequence, often reads the application only superficially). After a consensus score recommendation is reached, every section member votes separately and in secret. Each member is supposed to raise a hand and notify the group in the rare instance that the member's score falls outside the generally agreed range. A Study Section member is allowed to vote outside the recommended range, but if a member does it too often, it is considered impolite. In the end the scores are averaged, and—based on the average of these scores, with no outliers excluded—all the grants evaluated in the study section are ranked from top to bottom. Although the room is full of people, and everybody in the room had access to the full application from the beginning, the fate of the grant is largely determined based on the opinion of the three original reviewers. Rarely a nonassigned person in the room may bring up a valid point that can influence the recommended score—typically downward. Also, each Study Section has separate ranking, and grants in different sections will not be compared later on. There may be weaker batches and stronger ones, depending on the topic and the Study Section.

The applicants have *one single chance* to revise and resubmit their applications. They usually end up in the same Study Section, but quite often they are assigned to different reviewers. Some reviewers were only ad-hoc the first time, and others may have rotated off from the roster in the meantime. Even if the same reviewer gets the revised application, there is absolutely no guarantee that the revised version—however "responsive it is to the original reviewer comments"—will be scored better than the original grant. If the revised grant goes to different reviewers, the chances of finding new faults or even disagreeing with the prior reviewers' comments and disliking the changes made for the revised applications are real possibilities. Here is what an early-career investigator posted to the NIH nexus website:

> I am a New Investigator struggling with NIH. I had an A0 application that received a 29 percentile ranking, with high scores for innovation. I resubmitted the A1 at the next New Investigator deadline. This A1, which was better than the first one and included a new last author publication in a very good journal, dropped its percentile by 10. All the reviewers of the A1 application were clearly new, and didn't care about comments of the A0 reviewers.[26]

If a revised application is rejected, further revisions are not permitted. A new, different grant application may be submitted later; by the time this happens, the applicant investigator may have completely run out of funding and may have to let their team go.[27] From that stage, it will be even harder to restart the lab, even if—by some stroke of luck—eventually the new funding comes in.

There is a mythical process called "appeal." (In the EU grant system, it is called "redress.") During the appeal, the applicant whose grant was rejected can submit an official complaint and ask for a reassessment. The appeal must be based on specific criteria: (a) evidence of bias on the part of one or more peer reviewers, (b) conflict of interest on the part of a reviewer, (c) lack of appropriate expertise within the review panel, and (d) factual error(s) made by reviewers that could have substantially altered the outcome of the review. Although the process sounds straightforward and reasonable, in my thirty years of working in science, I never heard of a single successful NIH R01 appeal, nor have I ever met any colleague who has heard of one. Let's leave it at that.

In practice, the process goes like this. Members of the Study Section are summoned to some hotel, usually in Washington, D.C., usually for two days. By now everybody was supposed to have completed their homework, evaluated their assigned grants, and entered their comments into the online system, including their preliminary priority scores, which will be discussed later and may be altered during the actual meeting. Section members arrive the night before. Next morning, they meet in a crowded, over-air-conditioned windowless conference room, and sit around a table, with laptops in front and piles and piles of grants and papers everywhere (figure 2.4). Some reviewers can't make it to the meeting in

2.4 Deidentified image from a "model" Study Section meeting. The picture does look like the real thing. Imagine an average temperature of, say, 65 degrees Fahrenheit, continuous buzzing from some electrical system, the noise of incessant typing on multiple keyboards at all times, the smell of stale tea and coffee . . . and you get the full picture.

Source: This image is courtesy of the research team at the Center for Women's Health Research at the University of Wisconsin-Madison, led by Molly Carnes, in collaboration with Cecilia Ford, Anna Kaatz, Josh Raclaw, and Elizabeth Pier.

person; they call in over the speakerphone, don't hear the room very well, and it is rather hard to make out what they are trying to say. The evaluation goes on for two long days, after which everybody is free to catch their plane home, only to get together again a few months later for the next section, and the next batch of grants. All very proper, all very tightly regulated. If a grant of a colleague who works at the same institution or a grant of a collaborator is discussed, the member is asked to leave the room and will be excluded from the discussion. If a member feels that any conflict of interest exists with any of the applications, it can be declared and the member is excused from the discussion of the particular application.[28]

As I mentioned previously, the bottom 50 percent get thrown out, right off the bat; this process is called triage. This is a significant decider of the fate of the proposal. If an NIH grant application is triaged, statistics show that there is roughly a 50 percent chance that the revised application will suffer the same fate. In contrast, applications that are discussed but not funded have an almost 90 percent chance that the revised application will at least be discussed.[29] Of course, resubmission and rediscussion does not guarantee funding, but the chances are generally somewhat higher the second time around. All applications that survive the triage process are discussed and scrutinized down to the tiniest minutia.

The grant discussion uses a specialized, somewhat arcane language, that I call *grantese*. It is a universal grant reviewer language, and its roots are in English. In grantese, scientists are called "investigators," and the words "he" and "she" are replaced with "the investigator."[30] In grantese, the word "ambitious" does not mean anything good: its true meaning is "unrealistic." "Outstanding" means something drastically different from "excellent." The precious few top-tier projects, which the reviewers truly like and try to support, must be called "exceptional" in this language.

Although reviewers are supposed to be realistic and are encouraged to equally consider the strengths and weaknesses of each application, the exercise often degrades into a hunt for the tiniest of mistakes or problems. It really feels like a chopping block, in which solid, well-designed projects can be dismissed as "incremental" and exciting ideas can be dismissed as "pie-in-the sky" projects (figure 2.5). A feeble parody of the malicious grant reviewer comments I have encountered over the years can be found in appendix 1.

Albert Einstein once famously asked: "If we knew what it was we were doing, it would not be called research, would it?" With this attitude, Einstein would have had no chance for an NIH grant. Applicants must pretend that

2.5 "Is it just me, or are these review panels getting a lot tougher?"
Source: Crawley and O'Sullivan, *The Grant Writer's Handbook*, reproduced by permission.

they know *exactly* what they will be doing four or five years from today. The so-called expected findings must be specified in detail, and various "alternative approaches" must be presented as well. Budgets must be planned four or five years in advance. Sometimes the scientists are even asked to predict/ estimate how many papers will be published from this work, and in which journals (!!!). Everybody knows that this is not how research works, and yet everybody goes along with this fantasy. One time I tried to be honest in my R01 application. I outlined, to the best of my abilities, the four years of the proposed project. I then explained that—depending on the results obtained in the first four years—I would spend the fifth year further exploring the most interesting findings, and I presented a few possible scenarios. My application was flatly rejected, and I was particularly ridiculed for what I wrote about the fifth year. (Lesson learned.)

When funded investigators later submit their annual Progress Reports— which are not reviewed externally—they can inform the NIH about so-called deviations from the original plan. This is not a problem as long as the investigator stays close to the overall field and direction of his originally proposed research and remains productive in terms of publications. Similarly, after completion of the entire grant period, when requesting the next tranche of funding—these are called Competitive Renewals—the reviewers consider the quality of the research completed, and very rarely do they dig into the original application. If the applicant, in general, did "good work" in the field, usually there won't be any penalty for deviating from the original plan and actually discovering something new and exciting. In other words, *everybody knows how research really works*. It's just that everybody suspends this knowledge —for those couple of days in the Study Section rooms.

It is amazing that attempts for true discovery are frowned upon. Suppose somebody proposes a screening project in which the goal is to try to find a genuinely new mechanism or target, or a new inhibitor of an important enzyme. These types of studies begin with the validation of a model system, and then use libraries (for example, diverse collections of small molecules, or silencing RNAs that can individually inhibit the expression of single genes) and test them to find out which ones have a certain effect. With sufficiently large libraries, the chance of finding something interesting is almost guaranteed.[31] But time and again I have seen efforts of this type declared as "not hypothesis-driven" or ridiculed as a "fishing expedition." A few times I tried to argue, unsuccessfully, with the Study Section, suggesting that there is nothing wrong with a fishing expedition if the captain and the crew are qualified and the expedition brings back the fish.[32]

Proposals related to repurposing are no Study Section favorites either, even though drug repurposing—i.e., finding new actions and medical applications of existing drugs—has been proven, time and again, to revolutionize medicine.[33] Methotrexate was originally introduced as an anticancer drug; later it found a new life as an effective therapy for rheumatoid arthritis. Thalidomide's original, horrific history as a sedative is well-known: almost everybody has heard about the embryonic developmental defects it caused when taken during pregnancy. Yet this scary drug found several new lives: first to treat leprosy, and later to treat multiple myeloma. The list of successful repurposing efforts is long. One would think repurposing should be a particularly attractive area for academic scientists because it is a simpler, easier, and cheaper process than coming up with a brand new drug candidate and progressing it into the clinic. (The latter is a long and costly process, way beyond the scope and budget of R01 grants, and it is beyond the expertise and means of academic research centers.[34])

However, repurposing efforts are viewed as lacking novelty and sometimes even viewed with suspicion. (Why does this applicant have this weird attraction to this particular drug? Maybe the applicant wants to make money off this project?)

One of my own research areas is in the biological roles of an enzyme called poly (ADP-ribose) polymerase (PARP). After decades of basic and translational research, several PARP inhibitors have been brought to the clinic for the treatment of various forms of cancer. Multiple lines of research suggest that the same class of inhibitors may have utility in a variety of other diseases, from sepsis to acute lung injury and from stroke to neurodegeneration.[35] At one point, I cowrote a major review—perhaps more of a position paper—that outlined the scientific justification for this type of repurposing and even laid out a path forward. Many significant investigators who work in the field of PARP coauthored it: from the Lasker laureate Sol Snyder from Johns Hopkins University to the original pioneer of the field, Nathan Berger, from Case Western University.[36] I think we made our case fairly convincingly. And, of course, I have tried, repeatedly, to secure an R01 grant to support this repurposing effort. No such luck. My applications were rejected on the basis of low priority. Eventually I was able to get some grant funding for it—but not in the United States—in my current place of work in Switzerland.

Let's get back to these cold and crowded Study Section rooms in the United States. Almost all of the emphasis during evaluation is placed on two attributes: *novelty* and *innovation*. But, as previously shown, only the type of novelty and innovation that the applicant can *completely guarantee* in advance is considered.

Relevant to the subsequent chapters in this book is the fact that, in my entire career as a grant reviewer, I have never seen an application focusing on the independent confirmation of a prior finding. Not one. Not in the United States, not in Europe. In theory, replication efforts should be supportable through the R01 grant mechanism because the NIH emphasizes the critical importance of *impact*. Isn't it impactful when one confirms and thus solidifies the findings published in an exciting new paper by showing that it is real and reproducible? In fact, one could even argue that it would be equally impactful if someone demonstrates that some published novel findings—let's say an effect reported in a flashy new paper—are, in fact, not reproducible. Such a negative discovery could stop everybody else from chasing a mirage and wasting time and money. And yet grant applications focused on replication are virtually nonexistent. Is this because everybody in the granting universe takes it for granted that the published body of literature is reliable and no confirmation or replication is necessary?[37] Or is it so because most scientists are driven by the goal of new discoveries and are not interested in confirmatory work? After all, no confirmatory study was ever rewarded with a Nobel Prize. However, a Nobel Prize has been awarded for scientific concepts that later turned out to be completely incorrect or for medical interventions that later turned out to be harmful. But these incorrect or harmful concepts had to be *novelly* incorrect or harmful concepts at the time.[38]

It is also important to mention that confirmatory studies are difficult to publish in good journals. Or could it be that most scientists already know that grant proposals related to replication would be swiftly dismissed for being derivative and not hypothesis-driven? The NIH RePORTER site lists more than 870,000 projects that include the word "study," but only about 1,300 projects (0.15 percent) even mention the word "replicate." The word "confirm" or "confirmatory" is uttered in less than 2,000 instances, and the word "reproducible" is used even less—and almost never as the main focus of a project.[39]

Let's say the applicant knows the system and does not propose anything radical that may annoy the reviewers (such as trying to discover something truly novel) or something unusual that the reviewers would dismiss right away (such as trying to repurpose something or confirm something).[40] With current funding rates, perfectly reasonable grant applications are rejected on a regular basis. Let's say the application comes from a well-respected university, a respectable research laboratory, from an investigator who publishes regularly in good journals, has good expertise in the field, and just wants to continue this research. Let's say the applicant presents plenty of convincing preliminary data, and pretends that a five-year time line and workplan can be predicted. Even when the applicant plays it completely safe, a super-strong and totally rule-compliant proposal may not make it to funding in the current environment. A skill called grantsmanship—i.e., an applicant's ability to sell this science—may be the difference between a funded and a rejected application. This rather questionable skill is, to some extent, teachable.[41] Even though grantsmanship skills have little correlation with actual productivity, innovation, or scientific thought, many institutions have set up programs to educate their faculty members in this "art form."[42]

When Study Section members leave the room, they don't actually know which grant ultimately will be funded and which will be rejected. That is determined by NIH administrators through a separate, subsequent process, which, of course,

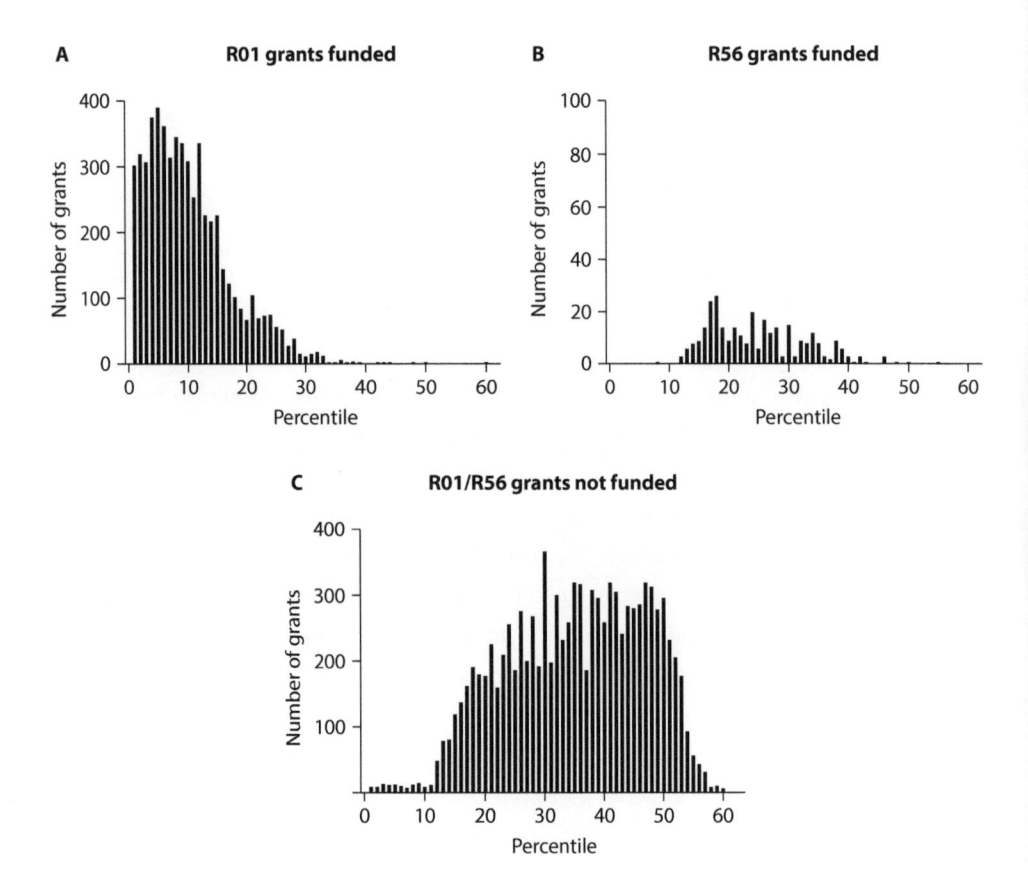

2.6 R01 and R56 funding as a function of the percentile score. Note that some R01s are funded even in the mid-thirties percentile range (A), and some grants are not funded even in the single-digit percentile range (C).

Source: These figures are based on year 2022 NIH data. "The NIH Data Book," National Institutes of Health, Washington, D.C., https://report.nih.gov/nihdatabook/.

considers the score and ranking of the project but also looks at other factors, such as programmatic priorities and the particular individual aspects of the applicant and is probably subject to some degree of subjectivity.[43] In 2022, if an investigator's project was ranked in the first and the 10th percentile, funding was almost guaranteed (although, amazingly, a few percent of applications remained unfunded). If the percentile was sixteen, the chance of getting funded was about 50 percent (figure 2.6). At higher percentile scores, the funding chances rapidly diminish. Yet some—obviously very special—grants were funded with percentile scores in the thirties and forties.[44]

This figure also includes some information on the funding rate of Short-Term Project Awards (or R56s), which I call "pity grants." Investigators may not apply for R56 grants; they are awarded at the administrators' discretion for shorter periods of one to two years for new or Competitive Renewal R01 applications with scores that fall just outside the funding limit. With the introduction of the R56, the NIH attempted to decrease the monotonous rise in the number of investigators who were between grant funding.

NIH administrators have a certain degree of discretion in deciding which grant is ultimately selected for funding, so it does not hurt to be respected by these administrators. In the 1990s and early 2000s, one of my core areas of research was in critical illness (circulatory shock). This topic belongs to an institute called the National Institute of General Medical Sciences (NIGMS). The main grants administrator of the institute, a rather level-headed and reasonable fellow named Scott Somers (now retired), took a decidedly hands-on approach. He attended every annual meeting of the Shock Society, where investigators who worked in the field of shock gathered each year and, of course, regularly applied for NIH grants to support their research. From what I remember, Somers never missed a lecture; he even attended the poster sections. It was clear to everyone at the meeting that Somers is the main gatekeeper of the money; so you can imagine the kind of reverence and ingratiation that surrounded him.[45] He seemed visibly and completely uncomfortable with all of this adulation, and without any sarcasm I must say I am pretty sure that this did not influence his granting decisions. Let's hope, for everyone's sake, that Somers's integrity is shared by every grants administrator on the planet.

As mentioned previously, each grant application is supposed to outline an interesting new research direction, in which most of the proposed work has not yet been started and definitely not yet published. The *feasibility* of the project should be supported by preliminary data, but the project should outline *future* activities. In reality, very often this is not what happens, and literally everybody who is part of the process—the reviewers, the grant administrators at the granting agency, and of course the applicant and the applicant's institution—knows this. This resulted in various whimsical cartoons circulating among scientists that explain the difference between theory and common practice (figure 2.7).

In essence, this cartoon illustrates that investigators tend to use existing grants to develop future projects, and they often submit grant applications to seek funding for projects that are all but complete. (At least this way the investigator will have an easy time predicting the results, right?) By now you can guess some of

It is supposed to work like this:

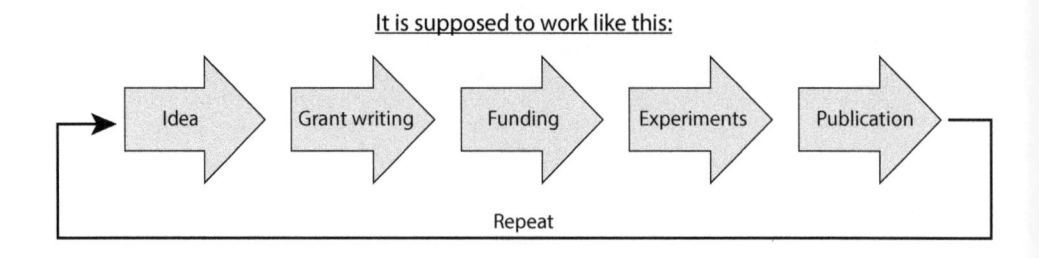

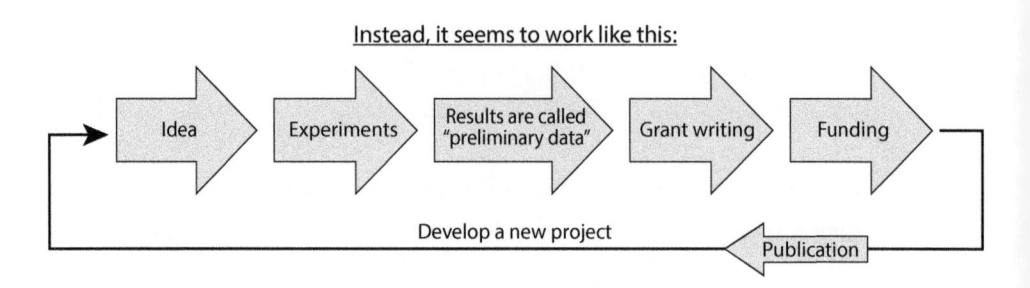

Idea → Grant writing → Funding → Experiments → Publication → Repeat

Instead, it seems to work like this:

Idea → Experiments → Results are called "preliminary data" → Grant writing → Funding → Publication → Develop a new project

2.7 The grant cycle in theory and practice.
Source: Based on various parody cartoons that circulate in scientific circles.

the reasons many scientists "creatively modify" the grant cycle in such a manner. First, in most universities it is difficult to find money to test a new idea that is outside the scope of one's funded research. There are some seed intramural grant mechanisms in some institutions, but they are oversubscribed and the amount of funding is limited. And—as you may have guessed by now—good luck trying to get the administrators to give up some of the overhead they are pocketing. Second—as you will see later—most scientists get burned at least a few times during their career when they disclose their latest findings at meetings or grant applications, only to see them crop up under other people's names. Many scientists become paranoid—or at least extremely cautious—about disclosing any unpublished data to potential competitors.[46] Many investigators feel that the best strategy is to time submission of a grant application to when the corresponding paper is under review by a journal, some of the journal reviewers' comments are already available, and the scientist is reasonably certain that the paper will be accepted.

And so it goes. Little games are being played: everybody acts politely, sits around the table, keeps their mouths shut, and acts like the whole process is functioning perfectly well. From time to time strange things can happen. Sometimes I have seen Goorgrowdian-American reviewer colleagues acting as outliers and enthusing about mediocre applications where the applicant also happened to have Goorgrowdian roots.[47] Sometimes two referees recommend triaging an application, and a third one has a better opinion of the project and insists on the full discussion process. In this case, the third reviewer is supposed to go along with the majority, but sometimes that person does not. It would make sense if the dissenting reviewer would justify why a marginal project should be discussed in the section. The official rule, however, is that the referee does not need to state a specific reason. I have seen reviewers staunchly insisting on discussion of mediocre applications. Thus the grant *will* be discussed; it probably will have a very low score, but at least it will have *some* score. This is better than the triage alternative, and in the revision it might slip into the fundable range.

By the way, even though the whole exercise is about funding or not funding applications, the words "funded," "not funded," "fundable," and "nonfundable" are distinctly no-no at these meetings, and reviewers who misspeak are often scolded for this. The Study Section "recommends," but the NIH has the final word on funding.[48]

For me, the most demoralizing part of the process has been the futility of it all. I did not mind the hard work, I did not care that the coffee was undrinkable, and I believed that grant reviewers provide an important public service.[49] The depressing part was knowing that the vast majority of applications discussed had no chance of funding given the realities of the paylines. I saw the immense amount of work, effort, and time poured into these applications, only to see 80 to 90 percent of them rejected—grant cycle after grant cycle. I sat through seemingly endless discussions about the strengths and weaknesses of projects we all knew would *never* make it into the fundable range. Occasionally, my mind wandered. Sometimes I thought of the history of science: in the old days science was not a way of making a living, doing science did not "pay." Rather science was conducted as a pursuit that *cost money*; it was supported or sponsored through other means. Supported from inheritances for giants such a James Maxwell, Charles Darwin, John William Strutt, Tomas Jefferson, Lord Kelvin, or others. Or supported by donations by friends, admirers, or various charitable sources, such as in the case of Alfred Russel Wallace or Peter Mitchell.[50] In the old days some scientists did their science as a hobby, such as Gregor Mendel, the father of modern genetics, who lived in a monastery and whose day job was as a monk.

As it turns out, Mendel most probably had also "beautified" his data: the statistical chance of plant hybridization experiments turning out *exactly* the way he put it down in his notebook is infinitesimal.[51]

Oh yes, the beautification of data—this immediately brings me back to the reality of these Study Sections. I have never seen so many "perfect" graphs, convincing differences, tiny standard errors, and ultra-tight correlations as were found in some of these applications. It was all rather amazing, especially considering that these preliminary data are supposed to represent unpublished, often still in progress experiments. Of course, even mentioning the subject of beautification is a big no-no at these meetings. We are told at the beginning, in no uncertain terms, that any concerns or issues related to potential scientific irregularities or misconduct must be discussed separately, *and only with the administrator*. This rarely happens. We are not detectives after all. We are colleagues, investigators, who are supposed to assess each other's work in good faith even if we work in a certain field and know that a particular method has a much larger standard error than is shown on these figures. Or even if a revised application comes in that shows the applicants have performed the additional supporting experiments the reviewers suggested but it is obvious that this was physically impossible to do in the few months' time that passed between the original submission and the revised submission. So, as directed, the applications will be reviewed *in good faith*.

Later, sometimes *much later*, we find out that applicants had submitted "doctored" data in their grant applications. Such findings are regularly released by the Office of Research Integrity (ORI), the NIH's watchdog office. The ORI website contains many findings (Case Summaries) of this type.[52] For a long time, anybody who was interested could learn about prior misconduct cases. However, the ORI website has recently been altered, and now the only cases listed are those in which the investigators have been found to have committed misconduct and "CURRENTLY have an imposed administrative action against them." Today only about two dozen such cases are listed. For example, in 2019 Erin Potts-Kant at Duke University was found to have used fabricated data in nearly 120 figures linked to sixty grant applications. The latest ORI announcements also reveal that Carlo Spirli, from Yale University, engaged in research misconduct by "knowingly, intentionally, or recklessly falsifying and/or fabricating data" in three NIH grant applications (and several papers); and the neurotrauma researcher William Armstead of the University of Pennsylvania was found to have committed research misconduct by "knowingly and intentionally falsifying and/or fabricating 51 figures, methods, data, results, and conclusions in 3 grant applications" (and in many papers).

When read one after another, these cases look pretty depressing. But all things considered, the total number of investigators caught cheating on NIH grant applications seems fairly low, especially when compared to the slew of fraud and deception prevalent in the scientific literature presented in later chapters of this book. ORI has only completed and disclosed about fifty misconduct investigations in the last twenty years, with about ten new cases added each year. From 2008 to 2023, the NIH received more than one million R01 grant applications (and about the same number of other types of applications). It would be really nice to believe that only one in fifty thousand NIH grant applications contains problematic preliminary data or plagiarism. Call me a pessimist, but I think this assumption is highly unlikely. It is more likely that fudging is more difficult to detect in grant applications than in published papers, and there are probably several reasons for this. First, much of the data presented in grant applications are called preliminary data, and it is understood that they can be based on a small sample size and can be, well, preliminary. If subsequent, larger studies don't confirm the preliminary findings, it's okay; it happens. Second, the data detectives who are the first to reveal scientific fraud have full access to published papers, but they have no access to grant applications.[53] And third, the ORI—with its two dozen staff members—must be severely understaffed and overworked. It cannot and will not scrutinize the vast majority of the seventy thousand annually submitted applications. It appears that ORI investigations usually begin only when a whistleblower alerts them to a problem.

It should also be mentioned that the penalties for the investigators caught cheating on a grant application seem fairly mild. Eventually something called a Voluntary Settlement Agreement is reached, in which the investigator "voluntarily agrees" to some combination of the following: (a) The so-called volunteer will be excluded from applying for federal grants for a period of three to five years. (b) In addition, or alternatively, the "volunteer" will undergo close supervision—typically conducted by a committee of several senior faculty members—according to a detailed supervision plan, for a few years. (c) During the above periods, the "volunteer" will be excluded from serving in any advisory or consultant capacity to the NIH, for instance, on a grant review committee.[54] (d) Sometimes the "volunteer" will also "agree" to retract the NIH-funded fraudulent publications from the literature. That's all. No financial penalties, let alone any criminal consequences. None of these penalties sound overtly scary to a nonscientist. But in reality, the loss of grant funding and the dim perspective of receiving future grant funding represent a severe blow to the career of the "volunteer."

But here comes the real kicker: with all this complicated process, with all these arcane rules and regulations, and with all the time and money spent on grant review, an astonishing degree of randomness is baked into the process. The NIH grant process, after all, was developed in the 1970s, when the overall number of grant applications was a fraction of what it is today and the funding success rates were as high as 50 percent. The same process that may work well with a high granting success rate can fail miserably at the current 80 to 90 percent rejection rates. (And most of us sitting around the table know about it and never mention it—we are in polite circles, as I mentioned.)

This randomness has been studied using proper statistical methods. It has been proven, time and again—not only for the NIH grant system but also for a multitude of grant systems in North America and Europe—that the grant reviewing process has an unacceptably high level of randomness and unpredictability.[55] Famously, in a study published in the prestigious journal *PNAS* led by Elizabeth Pier at the University of Wisconsin, after replicating all relevant aspects of the NIH peer review process (including the crowded rooms and probably the undrinkable coffee as well), the "inter-rater reliability for grant application reviews" was determined by two different statistical methods to be . . . wait for it . . . zero.[56] Two randomly selected ratings for the same application were, on average, just as similar to each other as were two randomly selected ratings for different applications. Different reviewers of the same application, in effect, were not in agreement on anything: not on the strengths, not on the weaknesses. Other studies found strikingly low nonzero numbers ranging from 0.15 to 0.3. Any agreement number below 0.6 can be considered terrible; the desired number would be above 0.75.[57]

Let me rephrase the conclusion: different referees will have wildly different views of any proposal. Statistically speaking, the applicant will be funded if luck strikes and all three randomly selected reviewers happen to like a particular proposal. I have asked Pier if her study has received any official follow-up or discussion from the NIH or any other agency. The answer was "no, not really." The investigators who conducted the landmark *PNAS* paper on the randomness of the NIH grant review process have drawn their own conclusions. Pier now works at a nonprofit institution that supports U.S. public schools. The center where the research originated is no longer in existence, and the researchers involved have moved on to other institutions, retired, or, sadly, passed away.

In my own experience, there may be a somewhat better level of agreement at the top. Occasionally, one sees a top project, by a top investigator, looking at an exciting question. Some grants seem to jump off the page. In cases like this, reviewers tend to agree more.[58]

To summarize, the way scientific grants are being selected for funding, in the United States and probably all around the world as well, is *unreliable*. This leads to an extremely scary conclusion: the only way to increase the chances of grant funding is to submit multiple applications; in effect, the applicants must buy as many "lottery tickets" as possible.[59] It may sound wild, but there are, in fact, serious calls to convert the entire granting process into a *real* lottery.[60]

Given the high degree of randomness, it should not come as a surprise to anyone that today's granting mechanisms (in the NIH system but also worldwide) are poor at predicting the future output and eventual real-world impact of projects. This is a contentious topic, which was famously raised about ten years ago by the biometrics experts Joshua Nicholson and John Ioannidis, in an article titled "Conform and Be Funded."[61] It appears that the big idea projects that could potentially make a significant impact are the ones that don't make it through the funding institutions—no matter how many times the word "impact" is featured on the reviewers' instruction sheets. This revelation has annoyed the administrators at various granting organizations, and many pro and con articles have been written on this subject. My own meta-analysis of the relevant literature concludes that Ioannidis was correct in 2012: he is still correct today, and everybody else is wrong. The grant evaluation system currently used all over the world does not predict the eventual societal benefit of the projects funded, and for the most part it does not even predict the future productivity of the supported investigators.

Part of the root cause of the problem is an extremely high degree of randomness, which is shown by real-world statistical data and modeling studies. But another root cause may be the ultraconservative attitude of the grant-giving bodies and the reviewers they use. I have already discussed the attitude of a typical grant reviewer when it comes to what they like to call "fishing expeditions." As the Nobelist Roger Kornberg states:

> In the present climate especially, the funding decisions are ultraconservative. If the work that you propose to do isn't virtually certain of success, then it won't be funded. And of course, the kind of work that we would most like to see take place, which is groundbreaking and innovative, lies at the other extreme.[62]

Even though the granting process is clearly outdated and random, and even though it does not correlate well with scientific quality or research productivity, it is one thing for sure: *quantifiable*. University administrators love to add up the total number of grants received, and they revel in the rankings their institution enjoys in terms of NIH grant money. These grant numbers are important factors in determining the overall ranking of an American university, and, of course, they are proudly displayed on university websites. Anyone can easily find out that the Johns Hopkins University was number 1 in 2022. It received more than $839 million in grant awards. A close second was the University of California, San Francisco with $823 million, followed by the University of Pittsburgh with $675 million, and then Duke University with $672 million, and so forth.[63] These figures and rankings seem to receive much more emphasis on these websites than the actual discoveries and medical advances produced by the recipients.

In light of these enormous figures, the penalties paid by some of the top-ranked universities for occasional instances of research fraud seem minuscule. For instance, the rather sizable $119 million settlement Duke University had to pay in 2019 to settle a case related to the research fraudster Erin Potts-Kant—who, over a period of seven years, had fabricated data linked to as much as $200 million in federal research grants—amounts to less than 20 percent of the grants-based annual income of Duke University.[64] In relative terms, it's pocket change.

So far my discussion has focused on the American granting system and NIH grants in particular, in part because these systems are the most transparent: a lot of public information and analysis is available. The inner workings of granting systems in many other countries are more closely guarded. The process of evaluation for other United States grants tends to be similar: the funding rates are also comparable, and in some cases are slightly better. I don't have the space or expertise to describe all grant mechanisms in all countries in the world. For the

European Union, the European Commission, headquartered in Brussels, runs a major grants organization: the current ongoing life science/biomedical program, a principal source of funds for EU countries, is called Horizon Europe. As opposed to the NIH system, which is largely based around a single individual or a small team, much of the European grant money is allocated to support multicenter collaborative grants and the "mobility" of research personnel between institutions. The submission forms are different and the evaluation process is slightly different, but the funding rates—hovering between 12 and 16 percent for the main program—are comparable to the NIH grant system. Although less information is publicized about the shortcomings of the EU grant system, I suspect that people are people everywhere, and the evaluation processes are similar, so the problems must be similar too. Accusations of improper EU granting processes tend to crop up from time to time; one of the latest is evidence that grant reviewers appear to be suspiciously partial when it comes to funding applications submitted from their own institutions.[65]

All European countries have their own national granting agencies as well. Non-EU countries sometimes participate in the EU system, either with full or partial access.[66] Non-EU countries have full national grants programs that support biomedical science. Some of these countries spend significant parts of their national budget on them. Switzerland appears to be one of these big spenders, in relative terms; in recent years the main granting agency, the SNSF, spent more than one billion Swiss francs per year on grants. It funds about 30 percent of the regular grant proposals received. So there are some regional differences. Overall, however, the odds of funding in most countries haven't seemed to improve over time.

CHAPTER 3

"DOING SCIENCE"

From Hypothesis to Publication

EXPERIMENTAL MODELS

Although for an uninitiated reader the previous chapters may suggest otherwise, but the true purpose of doing biomedical science is *not* to get grants, publish papers, or advance on the academic ladder. The purpose of science—any science—is, of course, to extend our understanding of the world in which we live. For physics, it's about understanding the nature of the physical universe. For life sciences, it's about trying to understand how living things got to be the way they are and how they function. For biomedical sciences—which makes up about 20 percent of all published life science literature—it is mainly about how the human organism works on the level of molecules, cells, tissues, and the whole organism; how diseases develop; and how could we prevent diseases or improve their diagnosis and treatment. For example, my own research is mainly in physiology (focusing on figuring out how a healthy biological system works), pathophysiology (focusing on how diseases develop on a cellular and molecular level), and pharmacology (the study of how drugs affect biological systems and how to discover and characterize new drugs).[1]

Depending on the country and the medical school, these "basic" topics are thought of as part of the medical curriculum, either by separate departments over the first few years of school or as part of the so-called integrated medical education, which focuses on organ systems and integrates these topics with diagnosis and therapy of diseases. The first approach is used in most countries in Europe as well as in many other parts of the world, and the second one is an American invention that seems to have spread into many countries over the years.

Fundamentally, research in life sciences, especially in biomedical research, starts with identifying a question or a problem that is not understood or is controversial and is worth investigating. For physician-scientists, these topics usually come from studying or practicing their own discipline. For example, a neurologist may be interested in finding molecules (called biomarkers) to predict or diagnose Alzheimer's disease early on. Or a critical care physician may be interested in the mechanisms that lead to the failure of various cells or organs in patients with circulatory shock whom he is trying to treat every day. For PhD biomedical scientists without a medical background, the topics they study may come from a review of the literature, and quite often they are continuing research on some aspect of the topic they became familiar with during their PhD research.

For people who do laboratory-based research (also known as bench research), the next phase of the process is to find a so-called model system to try to find an answer to their question. At which point these scientists label themselves "investigators"—no Columbo-style trench coat needed, though. These model systems are reductionist, which means they are simplified versions of the real situation. I describe a few such systems in detail in the following sections. All of the details I describe will be important when it comes to the central topic of this book—reproducibility—so please bear with me.[2]

When I started my own research in medical school, we wanted to study how blood vessels work. The reductionist system we used was a so-called blood vessel chamber, in which a small part of the blood vessel was hung between two steel wires, stretched to a certain tension, surrounded by a liquid that had ions and sugar resembling that of blood, at 37°C (to approximate the temperature of the body), and was bubbled through with a mixture of oxygen and carbon dioxide, similar to how fish get their oxygen in an aquarium. Then we added various substances (for example, vasoactive hormones) and measured how the tension of the blood vessel changed in response.

In this model system, everything *resembled* the real thing, but nothing was *exactly* like the real thing. In real life, blood vessels are not swimming alone in a colorless liquid. They are surrounded by nerve endings and fat cells that, as it turns out, regulate their function, and which were carefully removed prior to the experiment as part of the preparation process. In real life, the blood vessel does not have the exact same fluid on the inside and the outside: in the inside it has blood, and on the outside it has another type of liquid called extracellular fluid. In real life, the blood vessel is pressurized evenly, from the inside. In the model system, two wires, placed into the inside, were pulling it in two opposite directions, and the shape of the vessel was no longer exactly ring-like. (The whole preparation is referred to as a vascular ring system. Even the name of the system is an approximation, of course.) The temperature of the system is not exactly the normal temperature: all our experiments had a slight fever. The gas mixture used was not the same as what we have in our blood, which is just as well because this system did not have any red blood cells to deliver oxygen to them. All of this modeling or approximation happened before we even started the actual experiment. In the experiment, various substances were added to the system at concentrations that often exceeded what these blood vessels would normally experience in the body.

After all, this was a model system. And yet this system was way more advanced for the study of blood vessels than many previous versions, for example, a system called vascular strips. With model systems this simple, Nobel prize winning observations were made in the 1970s, 1980s, and 1990s—sometimes, in part, by accident. For example, Robert Furchgott's famous observation (which earned him a Nobel) was that if one is careful with the preparation of the vascular system, then after addition of the hormone acetylcholine the blood vessel will relax. But if the preparer stretches and mangles the blood vessel too much during the preparation process, the blood vessel will behave differently: it will contract in response to acetylcholine. Furchgott discovered that the mangling disrupts the internal layer of the blood vessel and damages the innermost layer of cells, called endothelial cells, and hypothesized that the endothelial cells must produce a substance that is responsible for this effect, which he called "EDRF."[3] (According to another Nobelist, Albert Szent-Györgyi, "Discovery consists in seeing what everyone else has seen and thinking what no one else has thought."[4] Similar model systems were also used by other investigators, including the Nobelist Louis Ignarro, another pioneer of the nitric oxide field. An earlier version of the same system (cascade superfusion bioassay) was used by Sir John Vane to study prostaglandins—and was heavily featured in his Nobel lecture.[5]

Another commonly used model system utilizes cultured cells. "Cultured" in this case does not mean that the cells have a particularly refined taste or manners (figure 3.1). It means that they are stuck to the bottom of a plastic dish, surrounded by a specialized fluid, and are bombarded with various insults and substances.

Let's say an investigator wants to study a life-threatening condition called septic shock in a laboratory setting. Septic shock is an extremely complex disease,

CULTURED CELLS   UNCULTURED CELLS

3.1 On the subject of cultured cells . . .
Source: Benita Epstein / CartoonStock.com, reproduced by permission.

which happens when bacteria overwhelm a patient's immune system to fight them. It is associated with fever, systemic inflammation, and multiple organ failure, which means that various organ systems—such as the heart, the liver, kidney, even the blood vessels—give up and stop working. In most hospitals, septic shock is associated with mortality rates of 30 to 40 percent, and to this day no specific therapy that would significantly reduce mortality rates has been identified. A highly reductionist model of septic shock would mean that one doesn't study the whole body but only one selected cell type, for example, an immune cell called macrophage. The investigator isolates this cell and places it in a dish (culture), maintains it in a special liquid, and uses it for various experiments.

Isolated human cells are difficult to get, and many of them don't like to divide and multiply in a dish. So more often than not the investigator switches over to mouse cells. One can buy mice from a commercial supplier, and thereby have a reliable source of cells to study—no need to wait for human donors or patients. Also, the investigator does not have to spend a long time writing and revising human ethics protocols to receive authorization to work with volunteers or patients.

If the investigator does not want to get into the repeat isolation of cells from a mouse, it is possible to switch over to what's called an immortalized cell line. These cell lines are available from various repositories such as the American Type Culture Collection (ATCC). Cell lines, like a J774.2 mouse macrophage line, are easy to maintain in a dish. They grow fast, they divide, and in a day or two your cell culture dish will be full of them. They also stick to the bottom of the plate, which makes the investigator's life easier (cells that float in the liquid phase are hard to work with). After they cover most of the plate, they will be split, meaning the investigator scrapes them, or uses an enzyme to get them off the bottom of the dish, and divides them into several aliquots and places them in new dishes, where they continue their growth and division. It is important to know that the responses to various experimental interventions can vary, depending on how densely the cells are packed next to each other; this variable is called confluency.

Of course, the very reason these immortalized cells keep growing and dividing forever, which allows the investigator to study them easily, is due to the fact that *they are not completely normal cells.* They are, in essence, cancer-like cells. Normal cells don't do you the favor of dividing freely, so the investigator is between a rock and a hard place. One choice is to use cell lines, knowing that they are not exactly normal. The alternative is to use primary cells, knowing that they are harder to work with and can differ from isolation to isolation, depending on the donor and the particulars of the isolation process.

When cells divide, as they progress from an earlier "passage" to a later "passage," they switch various genes on and off. A P10 cell (after the tenth passage) can behave markedly different from fresh, young P2 cells. After many passages, cultured cells reach retirement age in a process called senescence. Investigators can use these cells as a model of aging. For investigators who are not interested in aging, the best option is to go back and restart with an earlier passage cell and carry on the work that way.

Cells in a dish are surrounded by a liquid called "medium."[6] It has various substances, such as sugar, amino acids, and ions. These media can be purchased from major research suppliers such as Sigma/Aldrich or Fisher. Different cells like different types of media, and different companies sometimes sell many different media. Standard media that have a certain name (such as DMEM) are supposed to be identical regardless of where they were bought. Media can be purchased in a liquid form, or it can be bought as a powder and made into media in the lab by adding purified water. Most cells also need a magical substance called FBS (fetal bovine serum), which has growth factors that the cells like to have so they can "behave normally."[7] FBS is exactly what its name says: a yellowish, viscous liquid purified from the blood of baby cows. It can be purchased from various suppliers, and it is added into the medium at a certain ratio, for example, 10 percent. FBS can also come from various suppliers, and all FBS, from all over the world, is *supposed to be* equivalent. A baby cow is a baby cow, right?[8]

Going back to the topic of septic shock. Now a cellular model system has been established. Obviously, this is only one type of cell, not a whole organism. But it represents an important cell for the study topic because we know that immune cells are the ones making various proinflammatory hormones, called "mediators," that flood the circulation during septic shock and are considered root causes in the ensuing organ failure. How, then, can an investigator model your septic shock in the dish? One cannot add bacteria directly, so the investigator must choose one single component of the bacterial wall, which can be an important stimulator of the macrophage. Many such components exist, but a commonly used one is called bacterial lipopolysaccharide (or LPS), a bacterial component of the cell wall. This fluffy white substance can be purchased from various suppliers, who purify it from killed bacteria. A common one is called *Escherichia coli*. If a commercial product is *Escherichia coli* LPS, then—regardless of where you buy it—it should always be exactly the same material, right?[9]

As it turns out, many laboratories already have LPS, even if they have never spent a single dime to purchase it. All investigators are supposed to keep their instruments clean and their cell cultures sterile, but bacteria can be everywhere, and LPS is a common contaminant of everything in a laboratory—including laboratory glassware. It is difficult to get rid of it: the standard autoclaving procedure won't remove it. LPS is a potent molecule, and it can have dramatic effects at minuscule concentrations or doses. It is also difficult to measure its levels with precision; a highly sensitive LPS measurement method may detect minuscule levels of LPS in a laboratory—almost everywhere—but we hope not in the tissue culture room, which is supposed to be completely sterile.

Now let's say that the reductionist model is ready to be used. The investigator can add some LPS to the cultured macrophages and watch what happens. Changes in gene expression can be measured, the release of various factors can be detected, and various drugs can be added to see if they modulate the response. The investigator can look at the cells under the microscope, DNA or RNA or proteins can be extracted, and countless numbers of variables can be measured, for example, via an approach called "omics."[10] The possibilities are endless.

But be aware that the investigator is working with another reductionist model system of septic shock. Instead of a whole organism, only one cell type is used, and instead of a human cell, a mouse cell is used. Instead of a normal cell, a cell line is studied that is suspiciously cancer-like. Instead of blood, a liquid is used with various components in which different animal species are mixed together: mouse cells are surrounded by blood components from baby cows. Then the investigator adds one single membrane component of one single bacterium and studies a certain effect or response *that has been picked in advance.* All of this may sound artificial or even primitive, yet cell cultures of this type are common workhorses in biomedical science all over the world. And—just like in the case of vascular systems—they have led to major discoveries. Several investigators in the late 1960s and mid-1970s used a system like this to discover a very important factor of inflammation called tumor-necrosis factor alfa (TNFα). Bruce Beutler and Anthony Cerami later discovered that a previously known hormone known to induce severe weight loss (i.e., cachexia) in various organisms that carried cancer cells—called *cachectin*—is, in fact, TNFα. They discovered that TNFα activates macrophages and other cell types by binding to a protein on their membrane called TLR4. Bruce Beutler's Nobel lecture contains a nice picture in which LPS-activated macrophages (in fact, only the macrophage-like cells called RAW 264.7) secrete cachectin/TNF.[11] Kevin Tracey and Cerami later demonstrated the role of TNFα in septic shock.[12] In later projects—once again, using model systems of the type I outlined previously—Tracey and his colleague Haichao Wang identified several other factors that activated macrophages release, one of them being HMGB1 and a more recent one being procathepsin-L.[13] Initially, it seemed that HMGB1 acted on its own, but later it became clear that it binds LPS and "delivers it" to cells. (Remember what I said earlier about this pesky LPS? It could be everywhere in the lab at small concentrations and is very hard to get rid of. In other words, in the case of the HGMB1 mechanism, the effect observed in the experiment was *jointly* caused by the protein under investigation and a laboratory contaminant.)

Even with its obvious limitations, my group has been using cell-based models of this type for the last thirty years, in literally hundreds of papers, often in combination with other types of studies. We have used cells to define the promoters of various genes, including one form of nitric oxide synthase.[14] We have used cells to model the effects of various oxidants and free radicals in tiny, reactive cell-damaging species that are important components of shock, inflammation, and many other diseases.[15] We used endothelial cells placed in high sugar concentrations to mimic the blood vessel complications of diabetes and—even though this model system may sound decidedly primitive—we were able to find some new underlying mechanisms of the associated cell dysfunction.[16]

We have used various cancer cells to study cancer cell growth, division, and signaling and to discover new pathways of cancer and formulate anticancer concepts.[17] We have even used cells and tested "libraries" of various clinically used drugs to discover new effects and mechanisms. In recent years, we have used cells from Down syndrome individuals to characterize the bioenergetic defects they experience.[18]

Cell culture studies are staples of biomedical research, and everybody understands their limitations—as well as their many advantages. There is hardly any condition or disease that has not been modeled using cell-based systems. Another "model system" that is gaining popularity is a worm called *Caenorhabditis elegans*. Instead of cells or cell lines, this tiny soil worm is used to model different conditions. There is hardly any condition or disease that has not been modeled in "The Worm" as it is affectionately called.

Countless other model systems are used in biomedical science, but I don't have the space to introduce more examples. Take it from me: many of them feature similar reductionist approaches to the two examples I have discussed. Many of them are *in vitro*, meaning "in a dish," and many of them are *in vivo*, meaning they use an animal model of disease.[19] If you are someone who is strongly against animal experimentation of all types, then I am sorry but I am afraid you haven't done your homework properly. Despite all the efforts to reduce the number of animals used in research and drug development, in many cases, various animal models are the only known way to make scientific progress. Animal models typically use purpose-bred animals (mostly mice) that come from major suppliers such as the Charles River Laboratories. Specialized animals (for example, ones that lack certain genes) are available from specialized companies such as Jackson Laboratories. One can buy and breed their own animal colonies, and sometimes breeding is unavoidable due to the nature of the experiment. But often animals are simply purchased and used in an experiment.

The animal models used are, of course, once again, reductionist. For example, heart attacks in humans happen when a blood vessel first becomes atherosclerotic, and then an atherosclerotic part of the vessel develops a clot (thrombus) inside the blood vessel, and the blood vessel closes up, and blood cannot go into a certain part of the heart, and the cells in that region die. Patients are rushed into the hospital, and the cardiologist opens up the blood vessel, either by infusing a drug that dissolves the clot or by inserting a catheter into the occluded blood vessel and reopening it. (This is a vital process, but the reopening of the vessel can also cause some damage on its own, and this is called a reperfusion injury.) All of this is modeled in the laboratory by opening up the chest of an animal, placing a piece of string around one of the blood vessels in the heart, and closing the blood vessel. Then, after some time, the vessel is reopened by removing the string. One can, then, study the heart's function and structure and test various interventions aimed at mitigating the damage. Or one can leave the ligature in place, close the animal, and wait for weeks or months while studying the problems (chronic heart failure) that develop gradually. As with vascular rings or with cells in a dish, we have a reductionist system. It is an *animal model of a disease*, and—once again—it does not mimic all of the relevant human parameters. For instance, most investigators who study heart attacks are not using diseased (e.g., atherosclerotic) animals or animals of old age; they are inducing an insult to an *otherwise healthy, young animal* that roughly approximates the disease to be studied. For heart attacks, the investigator occludes a blood vessel in the heart. For stroke, one of the blood vessels in the brain is occluded. For diabetes, the animal receives a compound that destroys the insulin-producing cells in the pancreas. For alcoholic liver disease (well, that one might be close enough to the real thing), the investigator forces the mice to drink alcohol for prolonged periods of time.

In addition to the fact that the experimental model is only an approximation, subject variability is another important matter. Every person is different in real life based on their background, genetics, age, sex, and countless other factors. In the lab, however, investigators, for the most part, are working with inbred mice of the same strain. This does not mean working with a group of animals who are brothers or sisters. Due to the amount of inbreeding that laboratory animals have, from a genetic standpoint, in essence, the investigator is *working with the exact same animal*. A four-month-old male Black 6 Mouse (official name: C56BL/6) from Harlan in one mouse cage is supposed to be the exact same as another four-month-old male Black 6 Mouse that arrived at the laboratory door on the same delivery truck. This means that the reason for the variability in the experimental findings—the difference between the reaction of one animal versus another to some experimental intervention—by definition *cannot* come from any inherent variability between the genome of these mice.[20] They all are supposed to be the exact same animals, almost like clones—the same way the famous cloned sheep Dolly is *the exact same animal* as the sheep from which the cells were collected to produce it.[21]

HYPOTHESIS GENERATION, HYPOTHESIS TESTING, AND INTERPRETATION

Let's assume that the model system is ready to go. The experimentation can begin. But first the investigator needs to generate a *working hypothesis*. In an abstract sense, a hypothesis typically comes from an observation, and an observation typically comes from an experiment—but scientists are not supposed to start any experimentation without some sort of hypothesis.[22] How to break this Catch 22? There are several options. Some hypotheses may come from simply reading the literature and thinking about the possible next step in a certain field. This approach seems like a nice, clean start. There are several problems with this approach, however. First, the literature is often conflicting or contradictory on a certain topic. Second—as you will see later in this book—a lot of what is published turns out to not be reproducible. An investigator may set up a model, try to repeat what is published, and the results may not show what was supposed to happen. (In cases like this, the investigator might try to publish the negative findings, but it will be difficult to obtain a research grant based on a finding that shows that a certain biological factor—in spite of what's published in the literature—*does not* affect a certain biological process.) Third, when an investigator enters an active scientific field with a question, the people already working in the field may have considered the same. They may have even done the corresponding experiments. Perhaps their paper, outlining their results, is already under review. By the time the newly arrived investigator gets to an answer, it is possible that the experts who were active in the area will have produced the answer and published their paper.

For physician-scientists, another way to break the cycle is to start from a clinical observation. Many significant discoveries have been made in this manner. The one I usually tell my medical students about is the story of Barry Marshall, who, as a pathologist—without a PhD degree!—working in a county hospital in Australia, observed a certain strain of bacteria (called *Helicobacter pylori*) in the stomach of patients with gastric ulcer and hypothesized that these bacteria are a possible cause of the disease. As it turns out, he was not the first person to see these bacteria; they had been featured in pathology books decades earlier. It's just that nobody before Marshall thought they could be alive, because everyone "knew" that the inside of the stomach is so acidic that no bacteria can survive there. So nobody before Marshall considered the idea that these bacteria might have an *active role* in the development of ulcer disease. Although it took some time and convincing—including Marshall drinking some *Helicobacter pylori* juice to prove that these bacteria can cause gastric inflammation—the medical community eventually embraced the concept.[23] Now we have a fast and permanent cure for most ulcers: it is the eradication of the bacteria using antibiotics.[24] Not every physician-scientist is Barry Marshall, and not every clinical observation will produce a Nobel prize. But even today simple observations in the clinic, or a systematic review of historical patient charts to find some patterns, could be a possible starting point for a new research project.

In practice, most investigators who do biomedical research have some background interest in a certain field, often from their work during their PhD years. Many of them will generate a hypothesis based on this background knowledge. Also, the reality is that the average first R01 recipient is more than forty years old! Scientists start their career in somebody else's laboratory, and the first hypotheses often come from supervisors. A good supervisor will start their scientists in a direction that is worthwhile and in which the findings are expected to be novel and interesting.

In theory, it should be easy to identify a subject in biology or medicine that desperately needs additional research efforts—or else it will remain unexplored, possibly forever. The human genome project has identified about twenty-five thousand genes, each encoding a corresponding protein. It is estimated that the function of at least six thousand of these genes/proteins is basically unknown.[25] The function of the rest of the genes and proteins is only partially explored. Plus there are myriad physiological processes, a massive number of diseases, a multitude of biological factors (hormones, mediators), and drugs with known or partially known modes of action that haven't been explored.

One would think it should be easier to find an unexplored question than an explored one. And yet a lot of research seems to cluster around certain topics and areas. Some of this is dictated by fashion: there are always hot topics that attract the imagination of scientists—and, quite often, they are also generously supported by funding agencies as fashionable "priorities." In some cases, these trends are linked to emerging (or reemerging) hopes to revolutionize the therapy of some disease, with cancer immunotherapy being a recent example. The idea that one could fight cancer by activating the cancer patients' immune system was once considered "revolutionary" (late 1800s), then from "off the beaten track" to "have you lost your mind??" (most of the 1950s and the 1970s), then "maybe interesting but probably of limited value" (1990s), and then again "revolutionary and Nobel worthy" (in the last decade).[26]

Because a lot of what is done (what *can be* done) in biomedical research is dependent on various research methods or tools, some of these research waves are dictated by technological advancements such as the emergence of a new method or technique, which includes the Western blot, flow cytometry, mass spectrometry, the polymerase chain reaction (PCR), transgenic or knock-out animals, animals with cell-specific gene deletion, whole-genome sequencing, stem cell research, cloning, single cell analysis, CRISPR-Cas9 genome editing, optogenetics, and next generation sequencing—just to name a few. Many biomedical scientists are technology buffs and love to work with the niftiest (usually rather expensive) new machine they can get their hands on. Many of these same scientists are reviewers or editors of journals, so papers that utilize new tools tend to have an easier path for publication in prestigious journals.[27]

The waves of exploration are usually self-amplifying. It is difficult to study a completely unknown gene and its protein because a lot of the tools are not readily available. More likely than not, the investigator who ventures into a new field will not be able to find a vendor who sells an antibody against an obscure protein, so the investigator cannot do Western blots, and knockout animals probably are not available either. The investigator would have to create some new research tools—a time-consuming and expensive endeavor. Also, without "preliminary data" and proper "justification," it is an uphill struggle to find grant funding for an "exploratory" project of this type. In the end, the adventurous investigator might find that the gene or protein chosen is not all that important after all. However, if a previously unbeaten path begins to look promising, various tools will appear as more and more investigators enter the field, which facilitates follow-up research in an emerging research area.

Due to a combination of these factors, the progress in biomedical research is uneven. At any given time, some areas are crowded and new information is generated at an astonishing rate, and other areas may interest almost nobody. Some diseases are intensively studied, and others are largely ignored—both by academic scientists and by the pharma and biotech industry. There is *some* correlation between the socioeconomic importance of a disease and the intensity with which it is investigated. But it is not a perfect correlation; diseases in which the discrepancy is very large—i.e., insufficient research is being devoted—include heart disease, stroke, and suicide. From time to time granting agencies such as the NIH may allocate resources in certain areas to try to balance things somewhat, but these efforts lag behind the changes in the practice of medicine and the changing needs of society.[28]

If an investigator has identified a "problem" that seems worthy of exploration, it goes without saying that the investigator will consider this "problem" interesting and the experiments to explore it "critically important." (I have yet to see a scientific paper or grant proposal in which the authors begin by saying that their topic is uninteresting or unimportant.[29])

The experimental phase of the research usually begins with model validation. The investigator must determine how reproducible the model itself is: Does the system *always* respond to certain situations in the same way? and How much variability is there in the day-to-day or experimental run to the experimental run or animal-to-animal variability? Is the variability different when the lead investigator or a postdoc or a technician with thirty or more years of laboratory experience performs the experiment? In theory, the smaller the variability, the better the experimental model. If the variation is too large, the investigator won't see an effect, even if there is a true effect in the experiment. However, if the variability is very small, the investigator may end up reporting an effect that is *statistically significant* but *biologically* (or practically) unsignificant. At this point, all those lectures the investigator sat (slept?) through in biostatistics may become super handy.

The investigator may begin with what's called a *pilot experiment*—a trial run (similar to the pilot episode of a TV series). At this stage, it is a good idea to include *positive controls*, which are supposed to produce a certain effect, and maybe even *negative controls*, which are not supposed to affect the system. Based on the results of the pilot study, the investigator may decide to proceed with the study or may abandon the idea and try some other idea instead. If the pilot experiments don't look "promising," it is usually the hypothesis that is revised rather than the experimental system. This sort of early stage selection of what is expected to be interesting and positive—coupled with the fact that once a negative series of studies is completed it will be difficult to find a publisher—contributes to the fact that statistically significant and positive results are preponderant in the published literature, whereas studies with statistically nonsignificant and negative findings are relatively rare. This phenomenon has a name, it is called *dissemination bias*.[30]

Eventually the real experiments can begin. In theory, the investigator is supposed to know the variability of the system by now. From this information, the number of replicates should be predetermined by a statistical method called *power calculation*. (The number of replicates include the number of animals used in each group or the number of "experimental days" required, i.e., how many times you need to repeat the study.[31]) The investigator must decide what parameter (typically, only one at a time) will be varied in the experiment and what other parameter will be measured.[32] If at all possible, the experiment should also be *blinded* so the person doing the experiment does not know which group is the control and which is treated with a drug or intervention. Also, the person who does the analysis should not be the person who did the earlier stages of the experimentation.

The analysis of the results, in principle, is relatively straightforward. The collected data are entered into a database and analyzed using appropriate statistical methods. In the end, a determination is made. It often takes the form of a statement like this: "hmm . . . if I'm not mistaken, it looks like XX (hormone, drug, etc.) increases (or decreases) YY parameter." This sentence should continue with "in our model of ZZ." This last part, unfortunately, is often omitted.

Next the investigator designs follow-up experiments, for example, to explain the mechanism of the effect just discovered. With further experimentation, the contours of a *working hypothesis* begin to emerge that explains how the studied biological process may work. Most investigators start to visualize it as a cartoon in which names of proteins or pathways are placed in boxes, and one thing points to another, this second thing points to a third thing, and so on. It may begin as a simple scheme, but as the work continues, the cartoon becomes more complex. Almost every paper and grant application contain such schemes. At the same time, everybody involved in the process knows that these cartoons are only approximations: they are probably only applicable under specific conditions, and they are subject to future revisions or modifications.

This limitation tends to be ignored, however, when the investigator draws a conclusion by going back to the original topic (e.g., biological mechanism or disease of interest) and making a statement on how the study applies to it: i.e., the actual biological or medical relevance (or importance) of the findings. By the time the investigator writes a paper or grant application, diagnostic, therapeutic, or public health policy implications may also come into the discussion.

But there is no need to get super excited. Even though the investigator did everything correctly, performed the experiments carefully, and analyzed everything by the book—based on first principles—some scientific "fudging" has already happened. Remember, in the preface of this book, I started out with the disturbing statement that most published literature is "wrong." Some of the reasons are already baked into the process itself and can be summarized as cherry-picking, followed by reduction, followed by introduction of error, followed by subjective analysis, followed by simplification, followed by generalization, and often culminating in hype. All of this happens in real time in the investigator's head as the experiments unfold—before writing a scientific paper or a grant application. Let's reexamine the whole process from this standpoint.

The investigator selected a topic, believing the subject matter is interesting (or else why study it?). In addition, the investigator is probably convinced (and later will try to convince others) that the topic is medically important, perhaps even unique.[33] (Maybe this is true, maybe not.) In any case, the investigator did some cherry-picking right off the bat. Then the investigator reduces the question to the level of a simplified experimental model, and the question is studied under decidedly artificial conditions. Selection of the model system and the measured parameters carries some level of subjectivity. The investigator chooses certain things to measure, which means the investigator also excludes myriad possible variables that could have been measured.[34] During the experiment, the investigator often changes one single variable at a time and then measures various parameters in response to the change. This approach—although it enables systematic experimentation—is quite different from how things work in real life, of course, where multiple parameters tend to change simultaneously. In addition, due to the nature of biological systems, all experiments contain some level of noise and error that the investigator has to live with. Maybe the system is simply not sensitive enough, or the noise in the system is too much, which may result in an effect not being detected even if the overall hypothesis is correct.

Typically, an investigation is conducted through dozens, or perhaps hundreds, of experiments, which are done in parallel or in sequence. Randomization is important. In essence, it is a process that ensures that each subject in a study has an equal chance of being assigned to any group (for example, control or treated). The goal of randomization is to ensure that the various groups in a study contain comparable groups of subjects. It is preferred that the control and experimental groups are investigated concomitantly rather than, for example, completing all control experiments one week and all experimental subjects several weeks later. (In reality, perfect randomization is not always possible.) Also, in theory, it is

3.2 "Double Blind Study"
Source: Lamberto Tomassini / CartoonStock.com, reproduced by permission.

best to conduct everything in a blinded manner, such that the examiner does not know which experimental subject belongs to which group, for example, control or treated with an experimental agent (figure 3.2). Blinding is especially important when there is some element of subjectivity in the measurement, for example, a visual scoring or grading of various tissue specimens. This is often easier said than done because many experimental teams are small. Often a single student and a postdoc are responsible for all aspects of the work, from design to experimentation and analysis.

In each experiment, there is some chance that what looks like a statistically significant difference between two groups is, in fact, a difference caused by pure chance.[35] Most investigators use a significance level of $p<0.05$, which in simple terms means that there is less than a 5 percent chance that the perceived difference between two groups only occurred by random chance. (There are complicated mathematical/statistical considerations in this regard, which I will not get into in this book.) Suffice to say here that as the investigator performs more and more experiments the chance that some of the effects seen are simply due to chance increases. If an investigator reports twenty different effects, each apparently statistically significant at the $p<0.05$ level, it is probable that one of those effects is not real but happened purely by chance. This, and many other factors, will introduce some "error" into the experimental phase, as an integral part of the whole process.

In the next the phase, the investigator will have to determine what the data are (or, if the investigator is a philosophical type, "what nature is") "trying to say."[36] The investigator is supposed to remain objective. At the same time, investigators are incentivized to find some sort of answer. With all the work and effort invested, it feels natural that the investigator would like to collect a reward for these efforts in the form of an answer. I believe this need, principally, comes from human nature.[37] (Yes, it can be amplified by various external factors or even pressures, such as the expectations of the boss or the wish to publish or get a grant.)

When the hypothesis was created, some expectations came with it.[38] But the results may or may not turn out that way. Peter Medawar expresses this elegantly in his book *Advice to a Young Scientist*:

> I cannot give any scientist of any age better advice than this: the intensity of the conviction that a hypothesis is true has no bearing on whether it is true or not. The importance of the strength of our conviction is only to provide a proportionately strong incentive to find out if the hypothesis will stand up to critical evaluation.[39]

Even if the investigator maintains the proper distance from their data, the conclusion will be subjective to some extent.[40] The selection of the topic and the methods and the experimental design have already introduced subjective elements. Another level of subjectivity comes during the investigation. Just as Detective Columbo had some hunches and interrogated certain suspects more than others, a biomedical investigator also explores certain directions more intensively than others when designing (picking?) follow-up studies. When the results are in line with the investigator's expectations (biases?), it is less likely that further experiments will be conducted to challenge the findings. In contrast, when the results are contrary to the investigator's hypothesis, it is human nature to see if somebody in the lab may have "messed up something" during the experimentation. Throughout this entire process, investigators are limited by the constraints of reality, such as the time and money available—neither of which is ever enough to explore every possible question in detail. Even if the money were sufficient, the matter of competition is still there: the project must be completed as soon as possible because the team does not want to get scooped by a faster competitor—perhaps one that has cut some corners to get ahead.

When it comes to interpretation of data, yet another level of subjectivity creeps in. An investigator might find that some sets of results or effects are "more relevant" than others. It may turn out that nineteen perfectly well-designed sets of experiments point to a certain conclusion, but the twentieth experiment may be neutral or may even point to another conclusion. Investigators will, to the best of their abilities, come up with a conceptual framework in which the data—or, at least, the vast majority of them—"make sense."[41] It is easier to discount an outlier than to explore it, even though it may lead to something exciting, possibly to a discovery that could be more important than the original hypothesis.[42] An investigator may use the principle of Occam's razor and try to find conclusions that "require the fewest assumptions." Most people believe that this approach is usually correct, but it is not always correct.

Let's not forget one more important thing. During all of these processes, the investigator is using a biological instrument called the human brain, which has perceptual limits. It evolved on the African savannah and is designed to make quick decisions and is prone to crude generalizations. Its primary function is not to deduct complex truths but to react quickly to sustain survival. Our brain did not evolve to analyze terabytes of data.[43] The human brain is also prone to playing tricks on us, for example, in the form of cryptomnesia that can contribute to inadvertent plagiarism, which I discuss later.

It may sound surprising, but the goal of an investigation is *not* to determine what is happening in the studied biological system—at least not with any sense of finality. The goal is merely to come up with a model that describes the system as well as possible based on the actual findings, in the context of the current level of our understanding. The goal is to achieve the so-called state of the art.[44] In biological systems, where myriad things regulate each other, any conclusion made is probably, still, only a *crude simplification*. It is naïve to believe that the neat schemes (with boxes and arrows pointing this way or that way) actually represent the "full story."

After the investigator is done with the simplification, it's common practice to turn around and *generalize*. Typically, a scientist doesn't say that the experiment demonstrated some type of effect *in their particular model system*. It is rarely stated that "hormone X exerted a particular effect on parameter Z in, say, J774.2 macrophage line, under my particular experimental conditions." Instead, the scientist proudly announces that "X regulates Z in the immune system" ("regulates"—in the present tense; and "in the immune system"—in general. Not in one particular type of mouse immune cell line). The investigator thus concludes that a biological principle has been identified. And not just any principle but one that is *broadly relevant* to the original question or problem that the experimenter proposed to examine.

In this way, the investigator—advertently or inadvertently—may start an avalanche of *hype*: after all, what was found has global relevance for some important human condition. If the findings are published in a top journal, the hype avalanche will be started and propagated by science journalists and bloggers. Eventually regular journalists will join in. By this point phrases like "may have" will become "has"; and an effect of some compound in a cellular model of senescence (cellular aging) will be portrayed as follows: "newly discovered drug reverses aging." A laboratory compound that reduces cancer cell proliferation in cultured cells or delays tumor growth in mice will be touted as "a new drug was discovered to treat cancer." An article may have a title such as "Powerful Antibiotics Discovered Using AI."[45] But if you get down to the nitty-gritty of it, you may find that the antibiotic is not all that powerful and the contribution of AI to the process was much less than advertised.[46]

The topics discussed here do not belong to the category of scientific misconduct. They are part of normal scientific research as it is being conducted today—with the various infrastructural, financial, and human limitations that are facts of life. And yet, right off the bat, the process entails a baseline level of subjectivity and error that contributes to the unreliability of the literature.

THE REPRODUCIBILITY CRISIS

As explained in the preface, a significant part of the published literature is irreproducible. In other words, *unreliable*. Unreliability, however, must not be confused with obsolescence. In all fields of science, as new information comes to light, it can supersede the information contained in earlier papers. One may discover new mechanisms or new biological phenomena that provide a better explanation than earlier observations, or even disprove prior theories. This is a normal process and something that every investigator knows. Every scientific paper is the product of its time, and the biomedical literature does have a certain half-life, which is typically seven to nine years. Older papers are cited less and less frequently in recent papers; if the older set of information is referenced, it takes the form of review articles that summarize the status of a given field.

This phenomenon is not unreliability, it is just normal science. Unreliability or irreproducibility means something entirely different. It means the following: if an investigator reads a paper in the literature and goes to the laboratory to repeat the experiment, the replication attempt will be unsuccessful. The attempt may find a smaller effect, no effect at all, or sometimes even an opposite effect. As you will see later, *irreproducibility is not the exception* but, shockingly, *the rule* in biomedical science.

Reproducibility—even in those cases where it can be found—is not everything. It does not mean biological validity or biological or scientific importance let alone universal applicability or translatability into medical therapy or practice. Even if a finding described in a publication turns out to be technically reproducible, the reason for the reproducibility may be because both groups—the original and the reproducer—make the same mistakes or use the same faulty or nonspecific reagent or use a machine that consistently produces some sort of error. Even if the finding is biologically valid, i.e., it describes a true mechanism or action *in the particular experimental system*—let's say, a particular line of cultured cells—this does not necessarily mean that the same phenomenon will also be true for a different cell line or for a different species.

An experimental finding may be biologically valid but limited in its applicability. A cell is not an organ, an organ is not an organism, and a mouse is not a human. And—if I want to depress you even more—I could add that just because a biological finding may be reproducible and applicable to multiple biological systems, including humans, it does not mean that it can be turned into a diagnostic or therapeutic application in medicine. For example, there are biological mechanisms (drug targets) that cannot be used to build a drug development program around them for various reasons. In some cases, the target mechanism may play important biological supporting roles as well as a role in a disease. In this case, it may be difficult to build a translational program in which the effect and the side effect can be separated. In other cases, certain proteins or other targets remain stubbornly resistant to attempts to make a drug to target them. Reproducibility is only the first step toward something ultimately useful in medicine. It is a necessary but not a sufficient step. If this first step is missing or if this first step is unreliable, then what? Everything else that follows is doomed to failure.

There are several different forms of replication attempts. The most common form—and this is how the term is generally used in this book—is direct replication (or exact replication), in which the replicator intends to do every experimental step exactly the same way as stated in the original publication. There is also the concept of systematic or conceptual replication, in which the essence of the original finding is replicated but perhaps in a slightly different experimental system. These efforts relate to the concepts of validity and general applicability introduced in earlier paragraphs.

It is difficult to estimate what percent of the biomedical scientific literature is irreproducible. A widely accepted estimation is 50 percent. This comes from a 2015 analysis by Freedman and colleagues in which the authors compared the percentage of irreproducibility figures from various small-scale studies (including the Amgen and Bayer reports mentioned in the introduction) and also included some *estimations* (i.e., not true physical reproducibility studies).[47] Table 3.1 summarizes the results of the true (*physical*) replication attempts only; it is clear that the success rate is much less than 50 percent. (It is ironic that in 2024 one of these direct reproduction studies, the one in psychology, had to be retracted, for . . . wait for it . . . "lack of transparency and misstatement of hypotheses and predictions." Apparently, this is where we are now: even the reproducibility studies are irreproducible.)

The last line in the table regarding the study of the SPAN Network, published in 2023 in the journal *Science Translational Medicine*, requires some qualification.[48]

TABLE 3.1 Success rate of the replication studies available in 2024

Replicator PI	Replicator institution	Year	Field of research	Replication attempts	Successful replication (%)
Prinz	Bayer	2011	Various	67	22
Begley	Amgen	2012	Oncology	53	11
Nosek	Center for Open Science	2015	Psychology	100	39
Errington	Center for Open Science	2021	Oncology	158	13
Lyden	SPAN Network	2023	Neurology	6	17

This study started out with the aim of confirming the effect of six therapies that—according to multiple publications, from multiple laboratory studies—were supposed to improve the outcome in various animal models of stroke. This project is an example of conceptual replication—in contrast to exact replication, in which the replicator intends to repeat the exact conditions used by the original group—which attempts to check the overall effect and potential clinical translatability of the concept across several different, well-validated model systems. Patrick Lyden and colleagues applied a stepwise approach, whereby the first stage of the replication set a decidedly low bar of success, a 6 percent improvement in a parameter called "corner turning," which detects the animal's tendency to turn in a particular direction as a result of its brain injury. This initial testing was conducted in young healthy mice. All six treatments passed this low threshold test—so on one hand, the rate of replication may be called 100 percent. At the same time, none of the six treatments showed a reduction of brain infarct size—a gold standard parameter used in stroke models, including most of the publications that the SPAN Network intended to replicate. So on the other hand, the replication rate could also be called 0. As the experiment progressed and more and more stringent criteria and additional models were applied—e.g., animal models with comorbidities typically found in stroke patients such as atherosclerosis or high blood pressure—more and more treatments were found to be ineffective, until only one of the six approaches—an antioxidant named uric acid—was found to show a functional benefit throughout the different conditions and two different rodent species (mice and rat). This is how we end up with the 17 percent replication figure in table 3.1. This study highlights the obvious fact that the success of the replication—among other factors—is highly dependent on the experimental model selected by the replicator.

As mentioned in the preface, John Ioannidis has elegantly demonstrated that some of the irreproducibility is purely based on statistical considerations. It is also likely that the literature is not homogeneous. Significant differences may be based on field, methodology used, and perhaps country and institutions. Ioannidis also proposes two interesting additional corollaries. In one, he estimates that "the greater the financial and other interests and prejudices in a scientific field, the less likely the research findings are to be true," and in another he surmises that "the hotter a scientific field (with more scientific teams involved), the less likely the research findings are to be true."[49] Both of these corollaries sound likely but remain to be tested in the future.

Even if we use the ultraconservative estimation of 50 percent, the loss of money is enormous. In the United States alone, about $60 billion is spent on preclinical (i.e., laboratory-based) research annually; half of this money—likely more—is supporting unreliable research. If we also include clinical research and consider research expenditures worldwide, the amount of lost money multiplies.[50] We are then talking about $30 to $60 billion wasted each year. Half of the countries on the planet have GDPs less than this.

The consequences of the replication crisis are not only financial. They also include the loss of trust—of the public but also of policymakers—in the biomedical research enterprise. In addition, the biotech and pharmaceutical industry, where much of the translation from basic research discoveries into new diagnostics or therapeutics happens, will also become (or, I should say, has already become) highly suspicious of the published body of literature. There are many instances in which a drug company acquires a technology from a university and later on the clinical development program that is building on the basic research will have to be stopped or terminated. There are many reasons a program may be stopped in a pharma company; sometimes it is related to relative priorities within the company or to some unexpected problem with a drug candidate at a later stage of development. However, in some cases, the reason can be an inherent unreliability of the original sets of data. One high-profile case involves a series of compounds related to a natural compound called resveratrol, which is present in red wine and has been implicated in the widely popularized health benefits of the daily consumption of low amounts of red wine. In 2008, the Harvard spin-off company Sirtris Pharmaceuticals was bought by Amgen for a whopping $720 million. The compounds were aimed at activating an enzyme called SIRT1, which—according to the published papers—has beneficial effects, including protection of cells from various insults, has beneficial effects in diabetes, and potentially even promotes cellular rejuvenation. The resulting clinical program had to be stopped: it turned out that some of the underlying fundamental science was unreliable. At least some of this was due to the reliance on a fluorescent probe, which was later found to be unspecific and gives false positive results. In other words, the compounds were not really activating the intended target, SIRT1. It also turned out that previously published effects of a "SIRT1-activator" compound in a model of diabetes could not be reproduced in a pharmaceutical industry laboratory.[51] Because of repeated fiascos of this type, biotech and pharma has become more careful in in-licensing.[52] They have placed more emphasis on early replication attempts, prior to in-licensing a patent from a university.[53] Ten years ago Bruce Booth from Atlas Venture talked about an unspoken rule in the investment community stating that about half of the published studies, even if published in top journals, "can't be repeated with the same conclusions by an industrial lab."[54]

There are some positive signs in the field of reproducibility and replication. First, the topic is gaining more attention in the scientific literature and in the news. Second, the number of direct studies focusing on replication are increasing. In 2015, the NIH published a novel set of reproducibility guidelines that immediately received significant push-back from various institutions, including, sadly, the Federation of American Societies for Experimental Biology.[55] Brian Nosek at the Center for Open Science, Charlottesville, Virginia, has been running replication studies in recent years—at least in part with support from the Arnold Foundation. Although the percent of successful replications is not looking much better in his replication attempts than what had been noted previously in the literature, a lot can be learned from his work. For example, he has highlighted the importance of commitment for the replication from the original authors, and he proposes a system in which the replication attempts are preceded by a so-called *pre-commitment* phase from the original group. In essence, he proposes a joint effort, conducted almost as a collaborative effort, along the lines of mutually agreed replication terms that may include providing more detailed methodological advice to the replicators.[56] The "Reproducibility Project: Cancer Biology" was one of the highest profile projects in recent years. In this project, fifty effects from twenty-three papers were selected for replication. Already at the start of the attempted replication, it turned out that none of the published papers contained sufficient information to allow replication; in every case the replicators had to contact the original authors to request clarifications and further information. Even with this extra help, the previously reported positive effects were found to be clearly reproducible in only 13 percent of the cases.[57] From the twenty-three papers examined, confirmation was fully successful based on all previously set criteria in only two papers. Even when a previously published effect was confirmed, more than 90 percent of the replication effect sizes were smaller than the original.[58]

The Brazilian Reproducibility Initiative is severely behind schedule; it has been hampered from the beginning by the unwillingness of many authors to share information with the replicators.

As the Nosek group's replication work progresses, a few other important things are becoming abundantly clear: replication is not only thankless (as evidenced by the antagonistic attitude of many original authors; and about one-third of the original authors were characterized as "not at all helpful") but is also time-consuming. In Nosek's cancer study replications, the replicator team needed 197 weeks (!!!) to replicate a study, and it was expensive: more than $50,000 per experiment, about twice what was originally planned.[59]

A key problem of replication, in general, is highlighted by Nosek:

> Researchers who have published high-profile papers have little to gain from participating in confirmatory analyses, and much to lose. Replication attempts are often seen as threats rather than as compliments or opportunities for progress.[60]

In the next sections, some of the underlying causes are examined, strictly focusing on experimental biology or preclinical studies, which are typically cell-based and animal-based studies.

VARIABILITY IN RESEARCH MATERIALS AND REAGENTS

A lot has been written about the fact that the variability of reagents and research materials is an important contributor to the irreproducibility of the published body of scientific literature. In previous sections, I have introduced three methods, the vascular ring system, the cell culture system, and in a broader sense, animal studies. Let's look at some of the variability issues in each of these systems.

The vascular ring system is a fairly simple method, but it uses many components. For example, the pieces of blood vessels studied are immersed in a specialized liquid called Krebs-Henseleit solution. This solution is made up of a standard mix of various ions and sugar. There may be differences in the purity of some of the reagents used. Although all the reagents come from commercial laboratory companies, no reagent is 100 percent pure. In fact, in many instances, the companies are selling the same reagent or material at various levels of purity and at various prices: the purer, the more expensive. There may be instances when the components have impurities that cause unintended effects in a biological system. Even the purity of the water can be different in different laboratories; there are water machines that create laboratory grade water using distillation, filtration, deionization, adsorption, or ultraviolet light exposure. These machines have different waters as a starting point (typically tap water, depending on your location); as a consequence, their end product, laboratory water, can have different ions and various contaminants in them. Sodium, calcium, magnesium, iron, bicarbonate, chloride, and sulfate are all present in tap water. Volatile organics such as lower hydrocarbon trace pollutants, and contaminants introduced during water treatment processes (trihalomethanes, chlorination by-products) could also be there. A lot of tap water contains synthetic estrogens from contraceptives because a lot of water is recycled from wastewater, and estrogens do not get fully removed. Estrogens, in fact, are extremely potent substances in many biological systems; if some of the estrogen from the tap water ends up in laboratory water, it could conceivably create biological effects.

Although bacteria and viruses are typically killed by water treatment processes, active by-products and fragments can remain. I previously mentioned the problems associated with endotoxin or LPS (or, in broader sense, *bacterial pyrogens*).[61] If your starting water has micro-organisms in them, simply killing them (e.g., with ultraviolet light) gets rid of their ability to replicate, but this may break them up into smaller pieces, potentially with potent biological effects. In fact, a lot of the changes that happened after long-term incubation of blood vessels in ring preparations were created by experimental artifacts caused by minuscule amounts of LPS present in those systems. This lurking LPS triggers the expression of a new form of the protein in the blood vessel, which makes large amounts of nitric oxide.[62]

Another common water contaminant is iron. Some tap water is visibly rusty, even to the naked eye. Lower levels of iron are not visible but are still present. Laboratory water machines may remove most of it, but perhaps not all of it. Iron is a biological catalyst. One famous reaction is the so-called *iron-catalyzed Fenton reaction*, which can create reactive species called free radicals from oxygen. In the vascular ring system, where a high concentration of nearly pure oxygen is bubbled into the system, you might end up with free radical associated processes.[63]

To give you an idea of the extent of this topic, the NIH has published a twenty-two-page (!!!) treatise just on the quality and contaminants related to laboratory water.[64] Many large laboratories use a central water source, and specialized personnel are assigned to it to maintain high quality control standards. Smaller labs rely on their own water machine. Theoretically, one can also buy ultrapure water at a high cost from a commercial supplier, but most laboratories cannot afford it.

The contamination of LPS is, in fact, a problem that is not discussed enough. Let's say an investigator wants to study the effect of a certain hormone in a vascular system (or perhaps in a cellular system). This hormone is usually not purified from an endocrine gland but is produced in a bacterial system, where the gene for the hormone is artificially inserted. (This process is now used to produce various human medicines, including insulin.) However, at the end of the process, the hormone, produced by the bacteria, must be removed and purified. And bacterial components—including this sticky, pesky LPS, which is a very potent biological substance—can remain in the final product as contaminants and may contribute to the results seen in the experiments.

When using a cell culture model system, the considerations discussed previously apply even more. The purity of water, and the contaminants in the water, cell culture medium, and the various reagents and research materials used become even more important. In addition, the cell culture medium has FBS—often a heat-inactivated form of it. The source of the serum, the variability in the mother cows (depending on where they are from, what they eat, perhaps their age, perhaps the time of the year the collection was made) could affect the biological response. Different suppliers sell different FBS preparations, and it is common laboratory knowledge that cells can have variable growth curves and responses depending on the FBS.[65]

The various substances applied to a cell culture—hormones, drugs, inhibitors of various pathways, silencing RNAs, etc.—may have variabilities, depending on their sources, their purity (declared and actual), and their storage (see below). When one adds reagents to a sterile cell system, the typical step is to pass the solution through a filter to remove contaminants. Typically, the 0.2 μm filter is used, which has holes that are 1/5000th of a millimeter. It sounds like a very fine filter, but it will not filter out the pesky LPS. It won't filter it out even if LPS is the only contaminant in the solution, and it especially won't filter it out from a protein solution because this sticky stuff likes to stubbornly attach itself to proteins.[66]

Almost twenty years ago, while at the University of Cincinnati, Andy Salzman, the head of my department at the time, came up with an interesting project for a PhD student named Tonyia Eaves-Pyles. The goal was to try to see if *E coli* bacteria release any new proteins that may activate epithelial cells (the cells that line the inside of your gastrointestinal tract). The medium of bacteria was collected, fractionated, and added to epithelial cells, and the activation of the cells was measured—assessed as the production of—what else? Nitric oxide. Various fractions were tested, and the active ones fractionated and refractionated. Eventually, Eaves-Pyles discovered that this protein is nothing but flagellin, which had been known for decades. It is a monomeric component of the flagella of the bacteria (the tail-like part that rotates around and allows them to move around). Flagellin looks like a corncob, and each individual flagellin protein could be imagined as a single corn grain. These findings were really exciting. But how could we be sure that the effect is not due to LPS contamination in the system? We started the purification from bacterial supernatant soup after all. Our group had completed many experiments, but the one that was most convincing to us was that flagellin remained effective in a strain of LPS-resistant mice called "C3H/HeJ Mice." Due to a rare random mutation event, these animals had lost one of their membrane proteins (TLR4) that the cells use to recognize LPS. The investigator could give them massive doses of LPS, and these animals won't bat an eye (in normal control mice, LPS causes fever, inflammation, and low blood pressure). When the C3H/HeJ mice responded to flagellin the same way as the normal control mice—with a comparable degree of systemic inflammation—we knew that the findings were not related to LPS contamination. We published the paper in 2001.[67] At the time, we still could not figure out what protein receptor the flagellin was using.

LPS aside, the purity of laboratory reagents—whether they are proteins, small molecules, or anything else—is also variable. There are different suppliers, with different batches, different grades of reagents with different price levels, and different impurity profiles. They are all supposed to contain, mainly, whatever the main ingredient is, which is written on the bottle. But if the bottle says "Purity: >99.5 percent," the question remains: What is the remaining 0.5 percent? And what effect may it have in a particular experimental system? The purity of even some commonly used reagents is inexplicably low. Over the last two decades, from the shadows of the nitric oxide research, another wave of biological "mediator," hydrogen sulfide (H_2S), emerged. I have been working in this field practically from its beginning, and we have made many exciting discoveries in this area, including applications from blood vessel tone to diabetes, cancer, and Down syndrome. In 2007, when the field was in its formative stages, I wrote a comprehensive review article on H_2S that was only eighteen pages long.[68] While cooped up in the house during the COVID years, two of my colleagues and I wrote a new one. It is more than 250 pages long and has more than two thousand reference citations.[69] A lot of new investigators have entered this growing field of research over the years, and almost all of them, at one point or another, use a salt of H_2S

(the sodium hydrogen salt, NaHS) for their experiments. This is a common way of creating H_2S in the lab: as soon as the investigator puts this white powder into a solution, it gives off H_2S gas, which partially stays dissolved in the solution and partially leaves the solution and exits into the air. A pet peeve of my friend Matt Whiteman from Exeter University—a prominent figure in the H_2S field—is that NaHS, according to its major supplier, Sigma, has a purity of 60 to 70 percent, and we are not told what the other 30 to 40 percent is.

It is also often the case that a smaller company sells the same reagent to several larger ones, which then market it under different names. In the case of antibodies, this can mean that an investigator who wants to "validate" one antibody with another may end up using exactly the same reagent in both experiments—without knowing it. It can also happen that a company switches to sell a slightly different reagent under the same name, perhaps because they have sold out the previous batch and used a different supplier for the repurchase. The newer agent, if it is an antibody, will almost certainly have different biological properties. If it is a small molecule, "only" the impurity profile may be different.

There are also reagent companies that, quite simply, sell the wrong material. In the early 2000s I was leading the research of a start-up (and later mid-size) company called Inotek. We were working on new inhibitors of an enzyme called PARP, which—as discussed in the section on drug repurposing—plays a role in many diseases, from cancer to inflammation and septic shock. In fact, our group was first to advance a potent PARP inhibitor (INO-1001) into clinical trials. (Now several companies market various PARP inhibitors, under various names, for example, Lynparza; they have become widely used in the therapy of certain forms of ovarian and breast cancer.) For reasons that are quite inexplicable, many smaller biotech companies, to this day, sell an old PARP inhibitor compound, called 3-aminobenzamide (3-AB), as "INO-1001." Both compounds inhibit the same enzyme, but INO-1001 is many thousands of times more potent than 3-AB. I have spent quite a bit of time trying to contact these companies to request a correction, but most of them didn't even respond.

Practically no academic laboratory has the time or resources to regularly check whether the material purchased from a supplier is, indeed, what it says on its bottle. This problem can be multiplied when molecular libraries are used. These come in a plate format, for example, in 96-well plates, and each well contains a different drug or compound. These libraries are assembled by purchasing large collections of various compounds (of various purity) from chemical suppliers.[70]

Many standard laboratory techniques, including immunohistochemistry, flow cytometry, and Western blotting, rely on specific proteins called antibodies (also known as immunoglobulins). Antibodies are produced in animals (typically rabbits) that are exposed to ("immunized with") a certain protein. The animal's immune system (specifically, their B cells), in turn, produces antibodies against the injected foreign protein. Theoretically, an antibody is specific, meaning that it only reacts with the intended target. There are two kinds of antibodies, polyclonal (which are produced by multiple B cells, and each could react with your target protein slightly differently) and monoclonal antibodies (which are produced after the isolation of a single B-cell and propagation of this clone in laboratory settings using a cell fusion technique called hybridoma). While polyclonal antibodies are a mixture, monoclonal antibodies are homogeneous. Each type of antibody has advantages and disadvantages. Monoclonal antibodies are supposedly more specific (*cleaner* in laboratory jargon), but some proteins are better recognized and detected with the *dirtier* polyclonal antibody. Some antibodies detect multiple proteins instead of one. This shows up as multiple *bands* on the Western gel. Sometimes there are slightly different versions of the same protein, which is okay. But at other times antibodies can react with completely different proteins too. This is called *cross-reactivity*, and misinterpretation of it leads to faulty conclusions. In the old days, in scientific papers, only the *protein of interest* was shown. In other words, only the part of the blot was shown where the protein was supposed to be, based on its size (*molecular weight*). ("*The dog ate the irrelevant part of the Western blot.*")

Both immunohistochemistry and Western blotting are finicky techniques. They are extremely dependent on the handling of the biological material before the antibodies are added to the system. If the experimental protocol is not followed down to the tiniest detail, the experiment won't work. The results are also dependent on the amount of antibody used in the experiment (scientists call them dilutions, and write them down as, for example, 1:100 or 1:1,000). All of this can be standardized, and when the protocols are followed rigorously, the results can be meaningful and reproducible, if one includes the right kind of controls along with the experimental samples. But when one uses different antibodies from different suppliers to detect the same protein, it is quite common to find different responses. One antibody against Protein X is not always the same as another antibody against Protein X. What's worse, suppliers can run out of batches, at which point they need to produce a new batch of the antibody, which can have different properties than the earlier one.[71]

Many antibodies are not even commercially available; in this case the investigator is at the mercy of a colleague who has produced an antibody, and may spare you a little—or not. In the early days of nitric oxide research, there were no good commercial antibodies for most of the proteins that make nitric oxide. The entire field was at the mercy of colleagues who made it and, who, surely, have received hundreds of requests in addition to ours.

As you see, even if everything is done completely by the book, antibody-based methods can produce unexpected, sometimes inexplicable results.[72] The large "elephant in the room" relates to the provenance of cell lines used in research laboratories all over the world. Even though cell lines are purchased from reputable suppliers, a surprising amount of literature appears to have their cell types mixed up, cross-contaminated, mislabeled, or otherwise misidentified. An enormous amount of incorrect information is present in the literature. One review article talks about hundreds or thousands of papers that claim to use a certain tumor cell line (e.g., various breast cancer cell lines such as the MDA-MB-435 or the MCF-7/ADR), when in fact the cell line is something completely different, such as a melanoma or an ovarian cancer line.[73] This cell lineage problem is so pervasive that many journals have started to demand, or at least strongly recommend, that a submitted paper provide a cell authentication or lineage tracing component. A conservative estimate is that hundreds of millions of dollars of research money is wasted because of cell lineage problems.

When it comes to living animals, many new sources of variability are introduced. The fact that inbred mice or rats are very close to each other genetically (as discussed earlier) is considered a positive aspect because it should help investigators conduct experiments with small animal-to-animal variability. And yet investigators who work with animals know that the exact same strain of a mouse, when it comes from a different commercial supplier, can respond slightly—or, in some cases, not so slightly—different than a mouse from another supplier. In 2016, George Weinstock, associate director for microbial genomics at the Jackson Labs (Farmington, Connecticut), stated that despite their decades of experience and the tightest quality controls imaginable, they find differences between mice bred and maintained at the laboratory's *three different sites*.[74] Apparently not only the supplier but a particular site within the same supplier can introduce some variability.

Many experimental findings are different when mouse strains that have different genetic backgrounds are used. A mouse is always a mouse, but a white mouse (the albino BALB/c) and a black one (the C57BL/6) can respond differently in many experiments. Indeed, many studies show that immune responses, behavior, metabolism, and many other aspects are strain-dependent. In some experiments, investigators are "stuck" with a certain strain: for instance, if they use genetically modified mice that are only available in a certain strain (i.e., on a particular genetic background). (One could change the background using a lengthy process called back-crossing. But most investigators don't do it. They don't even do it when it would be a simple matter of buying a different batch of mice and repeating the experiment.)

The two strains cited here are just examples; there are many common ones. Many of these mice have a baseline immune defect that enables scientists to grow human tumors in them (human cells are rapidly rejected in mice with a normal immune system). In cancer biology, investigators are often between a rock and a hard place. Either they use human cancer cells and grow them in so-called immunocompromised mice, in which case the experiment is closer to human cancer (because the cancer cells are human), but the experiment ignores the immune aspect of cancer biology. Or the investigator could implant mouse cancer cells in a regular mouse that has a healthy immune system. But in this case, the experiment lacks the human aspect. There are more complex—and more expensive, and more time-consuming—models, e.g., humanized mice in which the immune system of a mouse is killed off and replaced with human immune cells. But even this system has disadvantages: the creation of these mice tends to introduce additional sources of variability.

As soon as the mice enter your animal facility, further variability is introduced. How the animals are housed (individually or in groups, with and without various toys provided to them to bring some light into their otherwise, let's admit, pretty boring days), what food they eat, what quality water they drink, and the noise level of the facility could all be important factors that may influence the outcome of an experiment.

Even if all of these factors are standardized, there may be variability from experiment to experiment. For example, there may be a so-called genetic drift in the genome of the mice over time: the genetic material changes due to random mutations, and these mutations pile up and accumulate so the genetic background of a colony becomes different over time. As surprising as it sounds, even inbred laboratory mice that have never been outside and live under artificial light (strictly controlled 12h/12h cycles) can have *seasonal variabilities* in their responses. For example, the level of peritonitis (inflammation of the abdomen) that C57Bl/6 mice develop was found to be more intensive in summer or autumn than the moderate inflammatory reactivity in winter or spring. And there are even strain-dependent differences: the difference is prominent in certain strains of mice but less apparent in others.[75]

There can also be variability in experimental outcomes because of the time of day when the experiment is initiated. This is related to the circadian rhythm of the animals. Unfortunately, even in fields of biology where such details are known to matter a lot, about 70 percent of publications fail to disclose the circadian time when treatment was administered to the animals.[76]

Important differences can be traced back to variabilities in the composition of experimental animals' *intestinal flora* (also known as *microbiome*) as well. Just like humans, laboratory mice have billions of bacteria living in their colon as part of their normal biological environment. The importance of intestinal bacteria in health and disease has been a focus of intensive research in recent years. Mice obtained from different suppliers will have differences in their bacterial flora. Even two "mouse colleagues" of the same strain, age, and sex, who arrived on the same delivery truck, from the same supplier, and were housed together in the same cage, may have different bacterial flora. As they are housed together for a longer time, these differences may diminish.[77]

Each animal facility has its own microbiological environment; none of them are sterile. As a research institution imports mice from different sources (some from collaborators' animal house and not from commercial suppliers), funky mixtures of microorganisms—which are difficult to get rid of—can develop in an animal facility. Many of them don't actually cause diseases, but they are still part of the animal's background and may affect some of the responses. One example is the interleukin 10 (IL-10) knockout mouse (abbreviated as IL-10$^{-/-}$ mice). IL-10 is an anti-inflammatory protein produced by our immune system (immune-cell-derived proteins are called "cytokines"). IL-10 works by inhibiting the production of other inflammation-inducing cytokines.[78] It was exciting to see that mice that lack the ability to produce IL-10 spontaneously develop various inflammatory diseases, for example, inflammation of the gut, and later on even colon cancer on the background of chronic inflammation. (They are also prone to the exacerbation of other diseases. For example, some years ago we found that they develop larger heart infarcts than control mice.[79]) But here is the interesting part. *This only happens in animal facilities that have a certain type of bacterial flora. It does not* happen in others. If the animal house is too clean, the effect is not obvious. When all bacteria are completely absent from their lives, IL-10$^{-/-}$ mice are just fine: they don't develop inflammation. And even in an animal facility where the effect is seen, the degree of the spontaneous inflammation may be wildly variable. For example, one mouse colleague may develop a severe form and the other a very mild form or none at all.[80]

Most of the issues cited so far relate to the baseline condition of the laboratory mice. As soon as the actual experimentation begins, a whole slew of new variabilities is introduced. For instance, when the animals undergo surgeries, various groups may use different anesthesia and analgesia; different labs in different countries may have to follow different sets of guidelines in this respect. Many animal facilities don't let the animal return to the main area once they are removed. They are housed in procedure rooms of different types, often with several different experiments running side by side and investigators coming and going into the room at various times, which may affect the animals' biological rhythms.

A lot of the experimentation involves animal handling, which introduces further variability. Most investigators who work with mice or rats swear that the animals can sense when the investigator is afraid of them. Likewise, an inexperienced investigator may cause more inadvertent stress to the animal than an experienced one. Largely due to the pioneering work of Hans Selye, we know that stress is associated with the release of various hormones, including glucocorticoids, which, in turn, can affect the outcome of a subsequent experiment.[81] Even the way the mouse is picked up or sometimes temporarily restrained can make a difference. There is now a name for these variability factors, chronic unpredictable mild stress (CUMS). By now it should not come as a surprise that various animal strains have different propensities to CUMS.[82]

Laboratory animals can adapt to the daily activity of laboratory personnel, even to individual investigators. Even the sex of the experimenter matters. A study published a decade ago showed that laboratory rats tolerate pain better when it is administered by female experimenters than by their male colleagues; the difference was attributed to the animals' ability to detect male scent.[83] Because of issues of this type, experimenter-free, automatic testing systems are being considered. Such systems may be available, for example, for some behavioral studies—if the group can afford them. However, they do not exist for most experimental protocols. We do not have robots to perform stroke or heart attack surgeries in mice. And even if such systems are eventually invented, they will be out of reach for the budget of most institutions, or they won't be prioritized when it comes to the allocation of a limited research budget.

Most animal studies involve the administration of various drugs (pharmacological agents). Obviously, the route of administration (oral, intravenous, subcutaneous, etc.) matters a lot because the drug will absorb differently and will be metabolized differently. But even small details can matter. For example, if certain drugs are injected into the vein faster or slower, the peak concentration can be significantly different. We have seen this in the field of H_2S research. If a H_2S-generating compound is injected into the vein too fast, the high peak concentration induces a characteristic transient knockdown effect, which could influence subsequent responses.

In other experiments, the housing conditions can matter substantially. For example, in one of our pilot studies, the goal was to measure the blood levels of an experimental compound that was on the way to clinical trials, but we found very strange and variable blood levels; in some animals very long apparent half-lives were measured. Then we realized that the animals were eating some of their own feces lying around in their bedding, which re-exposed them to the drug and to the metabolites of the drug. The solution? Wire floors where the feces fall down and is not available for "secondary murine consumption."[84]

For a long time, most laboratories typically used male animals only.[85] Perhaps this has something to do with the fact that science was mainly done by male principal investigators. Or perhaps there were concerns that female animals may show increased variability if they are studied in a group in which everyone may be in a different part of their menstrual cycle. In 2014, the NIH instituted a new policy mandating that experiments supported by their grants should study both males and females. In the initial years, they gave supplements to laboratories that complained about the costs. Today male/female differences must be included as a factor to be considered in all NIH grants that involve animal studies.

Lots of additional factors can matter in an animal experiment. We are doing our best to minimize pain and suffering of the animals, and we have dedicated committees (called Institutional Animal Care and Use Committees or IACUCs—with dedicated, professional staff, and even with representatives of the community) to do so. One could even argue that this is the most tightly regulated part of laboratory research. However, even this can introduce variability or affect the results obtained. For example, IACUCs do not allow experiments of circulatory shock and systemic inflammation to proceed without analgesia. But the response of the mice—for example, how much cytokines and other factors they produce—is affected by the anesthesia used.[86] One can look at this as something that reflects real life; anesthesia also affects the postoperative inflammation in humans.[87] But one can also look at it as another confounding variable: depending on the type of anesthetic and the dose used (which can be investigator-dependent), another factor is introduced into the experiments that may affect the results. The amount of anesthesia needed to make sure the animal is asleep and won't feel pain during an experiment can also be variable.[88] This way one ends up with animals that sleep equally well, but they may have wildly different amounts of anesthetic in their system. Also, animal committees can (and often do) change the way the animals are sacrificed at the end of the experiment. Whether the termination happens through carbon dioxide exposure or through an overdose of an anesthetic can affect the samples obtained from the animal.

These are examples only and are based on a few selected—and rather basic—laboratory methods. For example, I only discussed a basic cell culture method and did not get into complex cell systems. Various stem cell–derived human cell models can be wildly variable not only between laboratories (due to the lab-to-lab variability of the complex series of steps used to induce the differentiation of the stem cell into a human cell) but also within cell culture dishes in the same culture hood (likely due to differences in the genetics of the individual stem cell donor).[89] In my laboratory, we have had our fair share of problems with this topic. We have obtained normal control stem cells and stem cells from individuals with Down syndrome and tried to differentiate them into neurons using a standard laboratory method. The control cells would comply and nice neurons were obtained, but the Down cells simply refused to do so. We were not surprised by this; we know that Down syndrome cells, just like people with Down syndrome, have many maladjusted genes due to the triplication of chromosome 21. But we could not continue the experiments if we only have control neurons and no Down syndrome neurons. So the new dilemma became the following: shall we modify the differentiation conditions so they are specifically tailored to Down syndrome cells? If we do that, the controls and the Down syndrome cells will have seen different conditions before we started experimenting on them. Would this difference affect the results?

According to the published literature, investigators all over the world seem to have similar problems with stem cell–derived cell projects, especially if the stem cells are not from control donors. In my lab, we investigate the roles of an enzyme called cystathionine beta-synthase (CBS) and its product, H_2S. In Down syndrome, there seems to be too much CBS, and the H_2S it produces inhibits the metabolism of Down syndrome cells.[90] We hope that one day CBS-inhibitor-based therapies can be developed based on this concept. Indeed, in primary cells obtained from many Down syndrome donors, we can see consistently higher CBS expression and beneficial effects when the enzyme activity is normalized.[91] A recent paper using human stem cells reported that they created neurons from Down syndrome stem cells. They, too, concluded that CBS levels in Down syndrome neurons are higher than control levels. But when one looks at the supplemental data in the paper, only two donors were used. In one of them CBS was high, and in the other it remained at the level of control cells.[92] If this is the typical degree of variability found in stem cell–derived neuron experiments, this is likely to bring about some serious reproducibility problems in the future.

As the complexity of the experimentation increases, more and more variables and potential confounding factors are introduced. Very large datasets are now often created: these are best analyzed by specialist statisticians. Collaborators—often in distant laboratories—are called upon to perform certain specialized aspects of the work. The number of methods and the amount of information contained in a single paper—and, consequently, authors on a paper—keeps increasing. As you will see later, this introduces two important factors that are relevant to the reproducibility problems of the biomedical scientific literature. First, the senior author will probably no longer have detailed knowledge of all the methodological details of the project for which the author is formally responsible—including scientific integrity. Second, with such complexity and massive amounts of data, it will be increasingly difficult to include all the relevant methodological details in a scientific publication.

VARIABILITY DUE TO SUBSTANDARD EXPERIMENTATION PRACTICES AND HONEST ERRORS

In the previous sections I dealt with intrinsic variabilities that are baked into the system. It was assumed that experimenters do their very best to do everything perfectly—by following the generally accepted laboratory practices and their supervisors' instructions. Unfortunately, in real life, all kinds of errors and mistakes can find their way into the experiments unbeknown to the experimenter.

These errors can be systematic—for example, if a wrong cell line is used systematically. Somebody, perhaps many months or even a few years ago, might have mixed it up or mislabeled it and may have left the laboratory; the next crew assumes they are working with something, but in fact they are working with something else. There may be errors in a computer script or Excel file created some time ago that is inherited and used by a new group of investigators. There might have been a breakdown of a freezer, and proteins or antibodies may have degraded; unbeknown to the group, they are using an inactive substance (and then wonder why the experiments don't work anymore). There may be contamination of a cell line, and a pesky bacterium called *Mycoplasma* is a common culprit. There may be a constant baseline contamination of the laboratory water, or low levels of LPS may be stuck to laboratory glassware everywhere. Somebody may have made a calculation error and made up a faulty *stock solution* that is later used by everybody in the lab. A commonly used pH meter may become faulty, and every solution in the lab might have the wrong level of acidity. In a laboratory where many people share reagents, somebody might have taken out a reagent from a common bottle with a contaminated spatula. Now you have a new contaminant—possibly an active one—in every experiment that uses that reagent. Bottles may become mislabeled, and labels may come off. A reagent might be kept under the wrong storage conditions for a while before somebody else puts it back into the correct conditions, but by then it is already partially degraded. Many investigators try to save time and money by making concentrated stock solutions from various reagents, which are frozen and used and reused after freeze-thawing. Some compounds don't withstand such torture, and the investigator will end up using increasingly decomposed and eventually perhaps completely inactive reagents.

Cell-based experiments are often done in plates, where each well has cells on the bottom. A typical format is a 96-well plate, which consists of eight rows and twelve columns. Nice and convenient. They can be combined with methods in which a color change happens, which can be measured by specialized machines called *plate readers*. It can even be combined with robotic arms and automated for screening purposes. Very useful, very nice way of doing experiments—except there is something called *edge effect*. When cells are grown in these plates, the outside of the plate receives more oxygen from the cell culture incubator than the middle. Depending on its position, the liquid in the plate evaporates at a different rate as well. Cells at the edges might grow faster or might respond slightly differently than a well in the middle of the plate. Amazingly, even different plates from different vendors behave differently in this regard.[93] If somebody systematically uses the same plate design in which the control cells are always in Row 1 (i.e., near the edge), it can lead to misleading results.[94]

Some problems can also develop with instruments. For example, plate readers sometimes break, and the plate reader machines are also known to produce *row errors*, i.e., depending on the position of a well in a plate, the value it measures may be *off*.

Every machine can break, every machine can lose its calibration, and every machine is prone to user error. Computers can also break and data can be lost. Quality control (QC) processes theoretically can minimize these errors. Lab machines must be calibrated, and good laboratory practice (GLP) can be followed. In pharma laboratories, these processes are typically well controlled. But many academic labs do not have the resources or personnel for all of that. Also, remember this: rules are only as good as the people implementing them.

In addition to systematic problems, there may be lots of random, occasional problems. All reagents need to be kept in specific storage conditions. Proteins tend to degrade upon storage, but small molecule compounds can be oxidized or inactivated by light. A lot of materials are best kept in desiccators or under a so-called inert atmosphere, such as nitrogen, which stops other more reactive gases (e.g., oxygen in the air) from entering the system and potentially reacting with and spoiling the stored material. Some laboratories take better care of these details than others. It is not difficult to inadvertently cause the degradation or contamination of a reagent.

Then there are simple mathematical calculation errors when investigators figure out how much of something must be weighed and mixed with something else. There could be weighing errors on a laboratory balance, or pipetting errors. Typos may occur in a laboratory protocol, or word processor errors may occur when text written in one font is converted to another.[95] Simple human errors may occur by misreading a written protocol or omitting a particular step, by mistake or by judging it as unimportant.

In animal experiments, a lot can go wrong. Not everybody who works in the lab has good hands or works with the highest diligence. Laboratory work does require dexterity, stamina, and the ability to concentrate. If some of these attributes are missing, the experimenter is (and the experiment will be) error prone. Some students learn how to inject drugs into the tail vein of mice easily, but it presents a difficult challenge for others and veins are broken and the drug does not get into the circulation. One common way to administer drugs to mice is by injecting them into their belly (called intraperitoneal injection). If done incorrectly, it can go into the bladder instead or into their colon. Either way it won't be absorbed into the blood of the animal, and it will be less effective or not effective at all. Subcutaneous injections (under the skin) can turn intramuscular (into the underlying muscle) in the hands of an inexperienced investigator. As a result, the drug will absorb differently and have a different effect. When implanting cancer cells under the skin of a mouse, there is variability depending on where exactly the tumor cells go, as well as how the cells were prepared and how many cells are injected each time. Differences can cause tumors that grow to different sizes. In ischemia-reperfusion experiments, a lot depends on where the blood vessel is occluded. In addition to certain anatomical variations in the blood vessel structure, this can cause a lot of variability in an experiment.

Human factors should not be overlooked. In many laboratories, PhD students come and go and postdocs come and go. Research associates and technicians may stay longer. Knowledge generated by one individual may not be fully transferred to the next person. When a postdoc has limited time to wrap things up, it may not be a priority. Or it may be by design if the person who leaves wishes to continue a project somewhere else and does not want to create his own competition in the former lab. If a paper is not completed by the time the postdoc leaves, some supervisors have a policy that the postdoc will lose first authorship, and it will go to the person who completes the project. In other cases, there may be direct or indirect competition between various lab personnel, which certainly does not produce a high level of collegiality and method transfer.[96] I have heard rumors of sadistic principal investigators who put two postdocs, working independently and in parallel, *on the exact same project*. Whoever gets to the finish line first gets to publish. For the loser—tough luck. Imagine the pressure on both postdocs in a situation like this.

To lighten the mood a little, let's go back to the subject of pipetting for a minute. Serial dilutions are made when one takes out a small amount of a solution from one tube and mixes it with a given amount of solvent. Then one takes out some amount from the new solution and mixes it again. If the mixing is done incorrectly, if the tip of the pipet (which enters the solution) is not changed every time and carries over a small amount of drug into the next tube, an experiment can have the wrong concentration of the test substance. It is most likely that errors of this type, coupled with an unreliable system in which false positive results are created, are behind the famous "Water Memory" debacle. In 1988 the renowned French immunologist Jacques Benveniste published a paper in *Nature* titled "Human Basophil Degranulation Triggered by Very Dilute Antiserum Against IgE." He claimed that his solutions maintained biological activity even at very high dilutions in which no active substance was present anymore, which was based on basic calculation principles.[97] Ergo, Benveniste argued, the water must have retained some biological imprint like the snow that maintains your footprints after you walk there. Benveniste claimed that the effect was present "only when the solution was shaken violently during dilution." This was not a small claim because it would have provided experimental proof for a form of pseudo-science called homeopathy, where impossibly high dilutions of substances (in other words, pure water) are sold as a treatment.[98]

The paper was published together with an Editorial Reservation declaring that "there is no physical basis for such activity" and that "an independent investigation is forthcoming." Indeed, a team—consisting, among others, of John Maddox, *Nature*'s editor, as well as the magician and paranormal debunker James Randi—supervised a study in which Benveniste's team repeated the design of the original study's procedure. The team's results were similar to those reported in the original paper. However, in these experiments the experimenters knew which test tubes originally contained an active substance and which did not. As soon as the experiments were repeated in a double-blind manner—notebooks photographed, the entire laboratory videotaped, vials juggled, secret codes assigned, wrapped in newspaper, and attached to the ceiling of the laboratory—no water memory effect was found. A paper with the negative findings was published in the same year, but a bitter battle ensued, and Benveniste publicly called the process a "Salem witch-hunt" and Randi and Maddox "McCarthy-like prosecutors." Another paper, with a title that had one extra word in it—"Human Basophil Degranulation Is *Not* Triggered by Very Dilute Antiserum Against Human IgE"—was also published in *Nature* five years later.[99] But the debate never really stopped. As far as I know, Benveniste and another French scientist, the Nobelist Luc Montagnier, maintained until their death that the "water memory" effect was correct.

THE ABUSE OF STATISTICS

If you have gotten this far in the book, you have read about many factors that contribute to variability in biological experiments and, consequently, to the topic of the reproducibility crisis. Methodological complexities, errors, and mistakes have been covered, as well as some of the tricks the human mind can play on us, including the occasional unrealistic, unshakable belief that scientists are right about something even in light of contrary experimental data. But we haven't addressed the topic of the intentional, unacceptable, massaging of data, let alone outright fraud. So far we have viewed irreproducibility as a result of the intrinsic nature of experimentation itself, coupled with well-intended investigators' actions, including honest errors.

When talking about statistics, issues related to complexity, limitations, and human nature/human errors must also be addressed. Statistics can be extremely complex, and there are inherent limitations in what statistics can and cannot do. And, of course, since it is human beings who do the statistical analysis, human errors can also happen. I don't have the space to explain all aspects of biostatistics, but there are excellent books on this subject. Suffice to say that statistics, by definition, cannot and will not tell you if something is real or not or true or not. Statistics can only tell you what is the chance that the difference you see (or think you see) between two groups of mice in a certain parameter is likely created by pure chance. Also, statistics can tell you if two things correlate, but statistics will not tell you if the correlation is due to causation, or whether it happened due to some completely different outside factor—one that the investigator did not even consider. There are inherent limitations in what statistics can do, and most investigators realize it. Although I will not spend much space on this here, it is likely that overreliance or overinterpretation of otherwise fully valid statistical methods is another part of the reason biomedical literature is unreliable.

Without trying to sound like a grumpy old man, let me mention one more thing. In the 1950s and 1960s, many papers did not contain any standard errors or statistical calculations. In some cases, publications showed a series of original tracings of individual experimental runs. Watson and Crick's seminal paper on the structure of DNA did not have any statistics or graphs with error bars. Even without the power of statistics, many of those old papers contain observations that remain valid to the present day. They could be replicated by other investigators then, and they can be replicated today as well. Some may even demonstrate (of course, using statistical methods, what else?) that many old papers have contained more truth, or produced greater leaps of understanding, than a typical paper today.

Let's begin this section with the uncomfortable truth that even with perfectly honest and well-intended analytical intentions, different groups of scientists, analyzing the same sets of data, can come to completely different conclusions.

One of many examples is a paper published in 2013 in the prestigious journal *PNAS* that compared the changes in inflammatory gene expression in human and mouse experimental systems and came to this conclusion (which also happens to be the title of the paper): "Genomic responses in mouse models *poorly* mimic human inflammatory diseases."[100] The paper was the product of an impressive group of leading scientists working in the field of inflammation. It immediately caused havoc in the eyes of the public, and journalists started writing about wasted public funds for misdirected research. In addition, granting agencies have started to implement new policies to de-emphasize funding of proposals that utilized mouse models of inflammation (never mind that these mouse models have been extremely useful for research and the development of many new drugs, including several classes of anti-inflammatories, for many decades).

But here is the kicker: a reanalysis of the same dataset by a different group of authors came to the opposite conclusion. Their conclusion (and once, again the title of their paper published two years later in the same journal) was this: "Genomic responses in mouse models *greatly* mimic human inflammatory diseases."[101] Soon after this paper was published, an extensive correction had to be published, which wasn't a very good look.[102] But the conundrum persists: depending on how one looks at it, the mouse models are either good or bad at predicting human responses. The confusion on the matter, and the various ripples that originated from this controversy, are ongoing to the present day. Some of the follow-up correspondence has had enlightened titles: "Mice are not men" and "Apples and oranges comparisons won't do." This controversy continues to affect an entire field of science—including granting decisions and even drug development directions.

How different can the conclusions be when the same dataset is analyzed and interpreted by different groups of investigators? This topic had not been systemically investigated until recently. However, in 2023 a group of international authors attempted to conduct a reproducibility study to address this question. Even though the field of the study was ecology and not biomedical science, the conclusions of the project are probably relevant.[103] Different groups, when analyzing the same set of data, can come to divergent conclusions. To add insult to injury, the project also simulated the journal peer review process of the resulting (divergent) papers, and by now it is probably not a shock to learn that the peer review outcomes were wildly divergent.[104]

These examples discussed well-intended interpretations of the same datasets, which produced wildly different conclusions. The next stage of the problem comes when investigators—still acting honestly and with their best intentions—make accidental errors in their choice of statistics, or when entering numbers onto spreadsheets, or when accidentally copying and pasting something wrong. Once again, I will not spend any time on these issues; to err is human.

Earlier I mentioned that a well-designed experiment is blinded (i.e., one creates a separation between who does the experiment and who analyzes the data) and randomized (i.e., controls and experimental groups are done in random order). These are simple concepts, at least in principle. And yet, especially in smaller laboratories and when resources are strained, even these principles are not applied regularly. But I won't spend a lot of time on these issues. After discussing some common issues related to "n-numbers," my main interest in this section is on the place where human subjectivity (bias) and statistics (mis)use clash. This topic is exemplified by three topics: (1) outlier exclusion, (2) p-hacking, and (3) HARking.

Let's start with n-numbers (i.e., the number of experimental repeats). In an animal experiment, where the investigator has nine control mice and nine treated mice, it is easy to figure out what the n-number is: N=9 in each experimental group. But what about experiments in cells or in biochemical experiments on enzymes? Is the n-number the number of tubes or plates used or the number of wells in a 96-well plate? Many investigators cut corners and call it n=6 or n=9 but "forget" to define it in the paper. Some journals have started to crack down on this matter. For example, when Amrita Ahluwalia was the editor-in-chief of the *British Journal of Pharmacology*, she and her statistics colleagues came up with a strict guideline for what is considered a technical replicate (e.g., two wells on the same 96-well plate) and what is a so-called experimental day (when everything should be repeated, from scratch, including growing new cells, splitting them, plating them, making up fresh reagents from stock, etc.) and how many of these (e.g., five) one needs to call it a proper experiment to be considered in the journal. The guidelines were published in 2015 and were updated in 2018.[105] They have been criticized quite a bit by some statisticians, including folks who work at Graphpad, the software company that many biomedical investigators use daily.[106] In my view, nothing can be perfect, but it is a good idea to draw the line somewhere when it comes to n-numbers. Today most journals have a set of statistical guidelines and instructions on their website, and referees are supposed to confirm that the submission complies with them.

The first major area where statistics and subjectivity clash is the area of *outlier exclusion*—or, in a broader sense, *redaction bias*.[107] After the experiment is completed, data are entered in various tables, and statistical analysis is performed. Experimental measurements tend to have a certain range, i.e., the distance between the smallest and the largest value measured. For example, if I were to measure the height of my students, I might find that the range is between five and six feet. But what if I have a PhD student do it for me? When I get the results and look at the table, I see the following: 5.22, 5.42, 5.61, 570, 5.90. I can be fairly certain that there are no 570-foot tall students in Switzerland, so I can safely assume that the number 570 was the consequence of a typing error (*data entry error*): most likely that particular value was 5.70. The measurements can be repeated and the situation can be easily rectified.

So far, so good. But biological variables can be trickier. Sometimes the numbers are clustered around two values. For example, the number of people in a restaurant cluster together around lunchtime and dinnertime. Biological values can have a so-called *bimodal* (or in some cases polymodal) distribution too. Some values have what we call *normal distribution*, and others don't—depending on the distribution of the data, the correct statistical analysis will be different.

In biology, some of the measured values "look like outliers." Let's say that we have two groups of mice with tumors. The tumor sizes for most animals are in the 700 to 1,000 mm^3 range in the control group, but the range is 500 to 700 mm^3 in the group of mice that were treated with an anticancer drug.[108] And then there is one animal in the treated group who has a tumor size of 1200 mm^3. It was part of the treated group; it looked the same as all the other mice; and supposedly it was injected with the same amount of drug. What is the chance that this is a real finding and a certain fraction of mice just don't respond to the drug? Alternatively, what is the chance that the technician who treated the mice did not inject the drug properly? Perhaps the animal was mistakenly injected with a higher number of tumor cells than what was in the protocol. Or what if somebody made a mistake and mixed up two animals, and this mouse is not even a treated one but is an untreated control mouse?[109] Perhaps there is some underlying biological difference in this one animal that accounts for the different response. Remember my earlier description that different laboratory mice are not even like siblings, they are almost like all of them are the *exact same individual*, genetically. At the same time, some of them might have different gut microbiomes than others, or some cages might have had a clogged-up water supply for a few hours and are a little dehydrated. There are literally hundreds of reasons one mouse could respond differently from another.

But why is this one *outlier* such a problem? Why are many investigators reluctant to simply report the data as they are—including the outliers? The reason is *statistical convention*. Once the data are entered into a statistics program, this one little mouse might be the difference between an effect that can be declared *statistically significant* versus *statistically nonsignificant*. In extreme cases, this may be the difference between a publishable versus a nonpublishable paper, or a funded versus a rejected grant application. If one goes back to the laboratory notebooks and finds something "weird" in this one animal, could that be a reason to exclude it from the analysis? And what if nothing looks different? One may repeat the entire experiment, but then one falls into another sin, p-hacking (see below).

Some laboratory adages regarding data outliers are transferred from generation to generation. Without getting into the detailed statistical analysis of it, many lab folks strongly believe that if the outlier is *outside a certain range* it is okay to exclude it even if there is no biological reason to do so.[110] For example, in a dataset that consists of the numbers 47,48,51,53,49,45,52,46,48, and 150, the last number would be considered excludable—even without any biological reasoning. Based on these laboratory adages, even if the last number in this example was 60 (instead of 150), this number would be "excludable" even though the number 60 does not seem to be all that far away from the rest of the numbers in the group.

Some years ago, a few papers appeared in high-profile journals claiming that hydrogen gas (H_2) inhalation can have protective effects, for example, in animal models of stroke. These claims sounded strange because hydrogen is a fairly inert substance, and the claims of being an antioxidant did not make much sense from a biochemical point of view. My group has repeated some of these experiments and found no significant differences when the entire control and treated groups were analyzed as a whole—the way they should be. But it certainly looked like the control values (untreated mice) were clustered together, whereas most of the values in the hydrogen-treated mice seemed lower (suggesting a possible protective effect). But a few other values in this group were super high. Taken together as a group there was no effect of hydrogen. After the outliers were excluded there was a nice effect. Go figure.

The results of this experiment reminded me of my student years in Hungary. We sometimes had visiting speakers from a well-known physiology laboratory in Russia. Typically, they worked on large animals, such as dogs. When a response to some intervention was variable from animal to animal, they did not view it as an annoying problem that messes up their statistics. They tended to ask this question: Was there a *baseline difference* between those dogs who responded a certain way versus the other dogs who responded some other way? We consider ourselves to be modern scientists and that way of thinking is all but forgotten. We don't do this anymore.[111] Why would we? In the old Russian experiments, those dogs were heterogeneous and our laboratory mice are not. They might have been different breeds, ages and sex. Now in Western studies we use inbred rodents. They are, effectively, the same individual, with the *exact same* genetic material.

It is, nevertheless, a possibility that baseline differences may exist between various members of the group. For example, some mice may have had a little fight before an experiment started. Some mice are more difficult to catch in a cage; they may even bite the finger of the experimenter. Such events may increase the catecholamine (e.g., adrenalin, noradrenalin) and corticosteroid levels of the animal, sometimes at a critical time when a treatment injection takes place. Adrenalin and noradrenalin and steroids are powerful hormones that can affect thousands of responses in an animal. Steroid treatment has been shown to regulate up to 10 percent of all genes in our cells; in effect, it reprograms the entire organism. About ten years ago while at the University of Texas my group completed a cell-based screen to look for compounds and compound combinations with powerful anti-inflammatory effects. The two classes cited here came out as strong hits, and in fact they acted in a synergistic manner, i.e., they enhanced the effect of each other.[112] Is it, then, possible that certain treatments and interventions magnify various—already existing—physiological or biological differences in an organism in a way that the "spread" of the responses will be wider? Some responses will be below the average of the control, and a few responses may be widely above it. In this case, if the "high outliers" are "thrown out" from the statistics for whatever (made-up) reason, the conclusion will be that the intervention was effective in reducing whatever response one measures.

The next subject of statistical shenaniganry is the famous p-hacking. This term has moved from scientific language into common hipster language in recent years. In essence, it means that if it looks like the *"experiment will be working"* the investigator does not stop the experiment at a previously determined element size but continues to add more and more experiments until the magical p-value of 0.05 is reached. Prior to the start of a well-designed experiment, one must perform a *power analysis*. One must estimate the expected effect size and the variability of the method and come up with a number of experiments to be performed. Then, and only then, should the experiment be performed and then evaluated. When someone stops and peeks at the data during an ongoing experiment, this should be taken into account in the subsequent statistical analysis, which must be adjusted to take this into account—this is a so-called *statistical penalty*. This type of consideration is well regulated in clinical studies, but in laboratory experiment many investigators—due to lack of training or lack of integrity—don't use the same stringency.

In reality, a lot of experiments start on some usual n value, for example, one row (or part of it) on one 96-well plate may contain three or six wells, and this is a group size (technical replicate). Then the experiment is repeated on different experimental days. Many investigators do this for a while, without any respect for any prior power analysis. Often the statistical method is also tested and retested, without any real basis for doing so, just to see if another method would give the desired significance (figure 3.3). If the difference is "almost significant," many experimenters continue the repeats until they get to the magic $p<0.05$.

3.3 "Even with some fraud thrown in, it didn't work."
Source: Sidney Harris / CartoonStock.com, reproduced by permission.

In reality, there is not much difference between a p<0.05 and a p<0.06. In the former case, the chance that the effect is due to accidental variation of the data is 5 percent, in the latter, the chance is 6 percent. It is a tiny difference. But lab folks quickly learn that with p<0.05 the results will be considered "real" effects (by a referee of a manuscript, for example), whereas with a p<0.06 the referee will likely scold the authors for overinterpreting what the referee will call a statistically nonsignificant finding.

P-hacking is happening all over the world, and one does not need to visit any laboratories or look at any lab books to know this for certain. Statistical methods prove this beyond doubt. When one examines large sets of publications—and this has been done time and time again in various disciplines of biology and medicine—data that are just slightly below the magic 0.05 limit (i.e., almost significant, but not quite) are *very rare* in the literature even though significance values, in real life, should follow a predictable normal distribution. As Masicampo and Lalande stated in the title of their 2012 paper, there is a "peculiar prevalence of p values just below 0.05." In theory, p-values between 0.040 and 0.045 should be about as common as p-values between 0.045 and 0.050 in the literature. But in reality, in biomedical science literature, the latter instances are represented 20 percent more often than expected.[113] It is curious, isn't it? It is as if a magic hand pushes some of these data past a magic finish line.

The next common method of statistical massaging is HARKing, "Hypothesizing After the Results are Known."[114] It is defined as the highly questionable practice of formulating a hypothesis *after* the analysis of the data rather than before conducting the study. An analogy might be when one throws the dart onto a blackboard and then pulls out a white marker pen and draws concentric circles around it.

Imagine that the investigator runs an "omics" study, in which tens of thousands of genes or proteins are measured. Almost invariably they will find some that are affected by whatever drug or intervention was being examined in the project. After the results are in, the investigator picks out some interesting changes and writes a paper with these data in mind. The investigator thus pretends that the hypothesis was already there before the data became available. It is a bad, misleading practice. HARKed "results" are also difficult to reproduce because the HARKed findings often turn out to be due to chance rather than to real effects.

A few years ago, the well-known nutrition scientist Brian Wansink experienced the penalties that come when HARKing is being alleged in an investigator's research work. Wansink (Cornell University, Food and Brand Laboratory) had risen to the status of a celebrity scientist, and in the late 2000s he advised—among others—the White House, the U.S. Army, and Google on nutritional matters. He has written several books for the general public, with titles such as *Slim by Design*, and has published several hundreds of papers that were highly cited in the literature. He was even one of the recipients of the 2007 IgNobel Prize (a satirical offshoot of Nobels, awarded for investigators who study weird or useless things) for "investigating people's appetite for mindless eating by secretly feeding them a self-refilling bowl of soup." Everything seemed to go fabulously in Wansink's career until he published an entry in his blog page titled "The Grad Student Who Never Said No," in which he appeared to embrace the idea of inappropriate mining and HARKing of large datasets and encouraged his students to employ the same problematic approach.[115] Fellow scientists became worried and requested original datasets from Wansink and started to critically examine the data and the statistical practices used by his group. Multiple examples of serious statistics abuses were discovered, and dozens of Wansink's papers had to be retracted. In 2018, Cornell University concluded its internal investigation: Wansink was removed from all of his teaching and research positions. The investigation determined that there was evidence of "data falsification, a failure to assure data accuracy and integrity, inappropriate attribution of authorship of research publications, inappropriate research methods, failure to obtain necessary research approvals, and dual publication or submission of research findings."[116] A comprehensive collection of various flavors of scientific misconduct indeed.[117] One may also wonder where Cornell University was and what it was (or wasn't) doing that allowed all of these activities to go on uninterrupted for a decade.

Entire books focus on statistical and biostatistical issues in general, and specifically in the context of scientific misconduct. Now that you have an idea of what the main problems are, I recommend that you read a recent article by Stefan and Schönbrodt.[118] Just like many statisticians working in the field, they apply the term p-hacking more broadly, and in this broader definition they include the subjects I have discussed separately—e.g., data exclusion and HARKing—as well as a large number of additional problematic practices, such as selective reporting, optional stopping, mishandling of covariates, deceptions related to the graphical presentation of data (e.g., scale redefinition, variable transformation, discretization of variables), and favorable imputation (in essence, "creation" of missing

3.4 "We don't like the term 'data manipulation.' We prefer to think we are curating artisan data."
Source: Courtesy of Eoin O'Sullivan.

data out of thin air to fill in missing values).[119] Others use the word "dredging" as a collective term for various types of statistical abuses, including the practice of building a hypothesis based on nonrepresentative data, and a variety of other sometimes overlapping terms are used for inappropriate uses of statistics (unethical data fishing / data mining, systematic bias, multiple modeling, etc.) that do not need to be fully explained in this book. The important point is this: when applied as a workflow, datasets that show no significance initially can become more and more significant as the consequence of various types of statistical manipulations (figure 3.4). And, of course, the conclusions presented will be further and further away from the actual biological truth. In other words, they will become more and more *unreliable*.

CHAPTER 4

SCIENTIFIC FRAUD—AND THE FRAUDULENT FRAUDSTERS

HIGH-PROFILE FRAUD CASES

In previous chapters, we entered the muddy waters of science. Some of the data massaging approaches discussed there—for example, the outlier exclusion based on purely "statistical" considerations—is viewed as "borderline acceptable" by many investigators. Based on anonymous survey data, we can assume that some of the worse offenses discussed here are also applied, even though investigators know they are not supposed to do it. The majority of what was discussed in previous chapters assumes that *some* laboratory experiments were at least *conducted* in the first place. It assumes that the investigator started with actual laboratory experimentation and followed it up with some data massaging or beautification.

Unfortunately, there is a whole other level of scientific misconduct. It happens when an investigator has a strong, preconceived notion of how the results are "supposed" to turn out and alters the data, or, in some cases, *conjures* data—literally from thin air—to support this notion. This preconceived notion could come from the investigator who is supposed to perform the experiments, or it could come from the head of the laboratory.[1]

Naturally, "results" that are "generated" this way will be completely unreliable. Papers based on results like that also will be irreproducible.[2]

In the following sections, I introduce some high-profile cases of scientific fraud. When a celebrated, decorated investigator gets caught, it is newsworthy. The higher the position of the person is who has been caught in the scandal and the more famous the university is where the person works, the more newsworthy the story becomes. ("The higher you climb, the harder you fall.") Stories like this tend to cause temporary public outrage. Then everything settles down until the next big story comes along.

116 SCIENTIFIC FRAUD—AND THE FRAUDULENT FRAUDSTERS

I am not surprised that everybody—scientists and nonscientists alike—enjoys reading news stories in which high-profile scientific fraud is exposed. Some of these stories are truly mind-blowing: their magnitude, the audacity of perpetrators, and the details of how the fraud was exposed. People love a good horror story. Honest scientists read them too, with a mixture of relief (One down, how many others to go?), hope (Bit by bit, maybe the field is being cleaned up?), resentment (Imagine people doing this in the name of science and rising in rank and reputation!), bitterness (Think about all the wasted grant money that could have gone to support proper science!), and disillusionment (Even science cannot be trusted anymore . . .). Books that focus on stories like this are published regularly. The names of the exposed investigators change and the methods used to perpetrate the fraud change with time, but the basic patterns are similar.

Table 4.1 lists a selection of high-profile cases.[3] The Anil Potti case was designated by the TV program *60 Minutes* as "one of the biggest research frauds ever." But that was ten years ago. Since then quite a bit of competition has come to light, and it is difficult to say which story should be considered the biggest.

Quite a few common themes can be distilled from the cases listed in this table. Let's take them one by one.

(1) Significant fields of inquiry. Most of these high-profile fraudsters have selected medically significant, groundbreaking areas in which to work (better said: "which to infect" or "which to mislead"). It would be a great advance if one could transplant tissues without the need to do immunosuppression in the recipient (Summerlin case). It would be amazing if we could have a genetic test of cancer that would tell us which therapy to use for a particular patient (Potti case). It would be amazing if we could repair a failing heart by injecting some stem cells (Anversa case) (figure 4.1). To have an effective vaccine against HIV/AIDS would be an enormous achievement (Han case). Big breakthroughs like these would be remembered forever. They would change how we practice medicine. The biggest scientific prizes would be awarded for them. And, of course, they would also bring publications in top journals and guarantee grant success (figure 4.2).

(2) Institutional inertia. The institutions where the fraud occurred tended to exhibit some resistance or inertia in investigating the case. In several cases, years and years went by between the first suspicion of potential fraud and the beginning of an institutional investigation. In the Summerlin case, his supervisor,

TABLE 4.1 Selected, high-profile cases of academic misconduct in biomedical research

William Summerlin—Stanford University, Stanford, CA; University of Minnesota, Minneapolis, MN; Memorial Sloan-Kettering Cancer Center, New York City, NY (1967–1974)

Case overview	Transplanting skin (or other organs) from another genetically unrelated organism has been one of the greatest challenges in immunology. Summerlin claimed he had found a simple solution that does not require immunosuppressive therapy of the recipient: a special solution in which the transplantable pieces of skin are soaked prior to transplantation. At Stanford and Minnesota, he already claimed successful transplantations, including in humans: transplantation of a Black donor skin to a White recipient and transplantation of a human cornea to rabbits. At Sloan-Kettering, he focused on transplanting skin pieces from a mouse of one particular strain (a black mouse strain) to another strain (a white mouse strain). It turned out that he was faking his results by coloring the skin of the "transplanted" animals with a felt tip pen.
Noteworthy details	His fraudulent paper (*Transplantation Proceedings*, 1973) has never been retracted. Even after the magnitude of fraud and deception was clear, some public statements tried to minimize the magnitude of the problem. At the conclusion of the affair, Lewis Thomas, president of Sloan-Kettering, stated: "He has not been fully responsible for the actions he has taken nor the representations he has made." He also stated: "If he [Summerlin] recovers from his illness, I hope he will return to medicine and science." The Summerlin affair has also caused sizable damage to the reputation of his supervisor, Robert Good, who has done first-rate scientific work in the field of bone marrow transplantation and was nominated for a Nobel several times.
How were the problems detected?	Summerlin's superior at Sloan-Kettering asked another laboratory member to independently replicate the mouse skin transplantation experiments. They could not be replicated. Animal technicians working for Summerlin discovered the "painting of mice" and reported him. The subsequent investigation, conducted by his institution, confirmed the allegations and proved the fraud. The committee also found his supervisor responsible for lax scientific supervision and a slow response to act in the face of mounting evidence for likely misconduct.
Investigator's fate	In 1974, Summerlin was granted a one-year medical leave of absence while undergoing "psychiatric care for an emotional illness." Subsequently, he was dismissed from his position. He returned to clinical practice and runs his own dermatology clinic in Arkansas.

John Darsee—Brigham and Women's Hospital and Harvard University, Boston, MA (1979–1983)

Case overview	Darsee was hired by Harvard in 1979, and it looked like he hit the ground running. His work was conducted on dog models and focused on the mechanism of damage that the heart muscle suffers after heart attacks. In fifteen months he produced five papers—all published in leading journals. After the investigation was concluded, and large-scale fabrication of data was proven, several papers have been retracted.

(Continued)

TABLE 4.1 (*Continued*)

Noteworthy details	The NIH prohibited Darsee from applying for grant support for ten years.
How were the problems detected?	Two other researchers working in the same department as Darsee suspected foul play and alerted the laboratory director Robert Kloner. Examination of the laboratory notebooks revealed that Darsee had been writing incorrect dates "to make a few hours' work look like two weeks worth of data." The NIH investigation revealed extensive data fraud and manipulation.
Investigator's fate	His Harvard position was terminated. He went into a clinical fellowship in an intensive care department in a county hospital in New York state. In 1983 the New York State Board of Regents revoked his medical license. He is now a medical writer in New York, and he writes under the name JR Hughes-Darsee.
Milena Penkowa—Copenhagen University, Denmark (2000–2017)	
Case overview	Penkowa worked in several areas including brain inflammation and muscle function and exercise. When systematic fraud and data manipulation was detected, more than ten retractions were completed, and many additional "expressions of concern" have been attached to her publications.
Noteworthy details	As the revelations of misconduct started to come to light, fifty-eight Danish scientists signed an open petition demanding a full review of Penkowa's activities. The Danish court concluded that not only did Penkowa commit continuous fraud during her time as an active researcher, but she also committed subsequent additional fraud to cover up the original fraud.
How were the problems detected?	Penkowa's graduate students were unable to repeat her published experiments, and they raised doubts that the prior experiments had, in fact, been performed. When the investigation found evidence of serious and serial misconduct, Penkowa received a three-month suspended prison sentence for data fabrication, embezzlement, and document forgery.
Investigator's fate	Penkowa was dismissed from her university. She has given some lectures at various meetings, including a museum exhibition affiliated with scientology. Her current activities are unknown.
Piero Anversa—Brigham and Women's Hospital and Harvard University, Boston, MA (2001–2017)	
Case overview	Starting with a paper in *Nature* in 2001, Anversa published some highly exciting—but as it later turned out, fraudulent and irreproducible—findings about how stem cells can turn into heart cells and heal the heart after a heart attack. In 2013, the NIH was notified about suspected misconduct. In 2017, the U.S. Justice Department found that the laboratory relied on fabricated data to obtain grants. More than thirty papers were identified as containing fake or questionable data. Most of them have been retracted now.

Noteworthy details	A lengthy and bitter back-and-forth of various lawsuits took place between the NIH, Harvard, and Anversa dragged on. The lawsuits were eventually settled, but the details were not made public.
How were the problems detected?	A group of whistleblowers who were working in Anversa's laboratory became concerned that prior findings reported by the laboratory could not be replicated. They also experienced a pervasive atmosphere in the laboratory that encouraged fraudulent behavior. They took the first step by reporting Anversa to the NIH.
Investigator's fate	In 2015, he was forced out of his position at Harvard. He tried to continue his career in Europe, but was unsuccessful. As a result, he has gone into retirement.

Woo-Suk Hwang—Seoul National University, South Korea (2001–2007)

Case overview	Hwang reported in a 2004 and a 2005 article that he was able to create human embryonic stem cells through cloning. When his findings were found to be fraudulent, his papers were retracted and he was charged with embezzlement and violation of bioethics laws. He received a two-year suspended prison sentence.
Noteworthy details	To produce his fake data, Hwang used eggs from graduate students and from the Korean black market.
How were the problems detected?	The initial charges against Hwang, raised by a former investigator in his laboratory, centered around laboratory fraud as well as around violations of human ethical conduct. Examination of his published papers found irregularities and fraudulent image duplications. One of his coauthors voluntarily withdrew his name from a *Science* paper. An investigation of Seoul National University found that both *Science* articles were based on fabricated data.
Investigator's fate	After dismissal from his university, Hwang moved to Sooam Biotech Research Foundation in 2006 and uses private funding for his work. He continues to publish and has completed cloning projects of various pets for paying customers as well as various rare animal species.

Marc Hauser—Harvard University, Boston, MA (2007–2012)

Case overview	A professor of psychology, Hauser was a prolific and well-respected author in the field of animal behavior and neuroscience. He also wrote several books and was well-known to the general public. In the early 2000s he published a series of articles focusing on the cognition and moral capacity of primates. Harvard and NIH both found several papers that contained data that were fabricated or misrepresented. Only one paper, published in the journal *Cognition*, was retracted.
Noteworthy details	He continues to publish both books and individual opinion articles in the peer-reviewed literature. His latest book, *Evilicious: Cruelty = Desire + Denial*, focuses on the psychological foundations of malicious human behaviors.
How were the problems detected?	Several graduate students were the initial whistleblowers. A Harvard and an ORI investigation followed.

(Continued)

TABLE 4.1 (*Continued*)

Investigator's fate	After the misconduct investigation was completed, he took a year of absence from Harvard, and then resigned. He became an author of popular science books. He also runs an organization called Risk-Eraser that aims to provide support to schools to help children with disabilities learn using "evidence-based strategies and tools derived from the mind and brain sciences."
Anil Potti—Duke University, Durham, NC (2007–2015)	
Case overview	Potti published several dozen papers focused on genomic "signatures" of cancer cells, which would predict favorable responses to a particular anticancer therapy. The papers turned out to contain fake and manipulated data and were irreproducible. The fraudulent research was funded by the NIH, the Howard Hughes Institute, and many other organizations. When all was said and done, dozens of papers had to be retracted or corrected.
Noteworthy details	Based on the fraudulent data, clinical trials were started but had to be stopped when the fraud came to light.
How were the problems detected?	Bioinformatic statisticians at other institutions read Potti's publications, found irregularities in the published datasets, and suspected foul play. An ORI investigation followed and found fraud at a massive level. Interestingly, an earlier whistleblowing attempt (where students in Potti's lab withdrew their names from papers and submitted complaints to Duke) was not followed up.
Investigator's fate	He stopped laboratory research, went into clinical medicine, and currently practices clinical oncology in North Dakota.
Dipak Das—University of Connecticut, Farmington, CT (2008–2013)	
Case overview	As director of the University of Connecticut's Cardiovascular Health Center, Das published more than five hundred papers and was the editor-in-chief of the journal *Antioxidants and Redox Signaling*. Much of his work centered around the beneficial effects of the red wine component resveratrol. The 2005 university committee has determined manipulation or fraud in at least 145 (!!!) instances in his papers. So far, at least twenty-two papers have been retracted.
Noteworthy details	In addition to confirming the serial data manipulation, UConn's investigation revealed that Das has maintained some questionable relationship with several California companies who were producing and marketing resveratrol and other plant extracts with purported health benefits. In addition to UConn, the retractions have involved the University of Debrecen, in Hungary, where Das had collaborative and cosupervisory roles with a junior investigator. Five papers published between 2006 and 2008 were retracted in which this investigator was involved, in most of them as first author.

How were the problems detected?	The initial initiative to investigate Das likely came from a whistleblower from within his laboratory. Multiple investigations, by UConn and by ORI, followed. The investigation found not only that Das instructed his team members to inappropriately manipulate images but that he had done such manipulation himself. Throughout the investigation, he insisted that this is a common practice.
Investigator's fate	During the university's misconduct investigation Das suffered a stroke, which he blamed on the stress his university exerted on him. He passed away in 2013.

Erin Potts-Kant—Duke University, Durham, NC (2006–2013)

Case overview	In 2006, Potts-Kant joined the laboratory of the pulmonologist William Michael Foster and performed a large series of studies related to the reaction of the lungs of mice to various environmental pollutants. The investigation revealed multiple counts of data fraud. Potts-Kant was arrested in 2013 (for credit card fraud) and fired from Duke University. Foster voluntarily retired in 2015. Multiple retractions and corrections followed.
Noteworthy details	Potts-Kant had an undergraduate degree, but no higher degree. Yet she coauthored thirty-eight papers with her supervisor, and her doctored data helped Duke University secure more than $200 million in grant support from the NIH and the U.S. Environmental Protection Agency.
How were the problems detected?	A new investigator, Joe Thomas, joining the laboratory became suspicious of Potts-Kant's data, which appeared to be "too good to be true." He reported the situation, and investigations by Duke and ORI and various lawsuits ensued. Ultimately, the ORI found that Potts-Kant has falsified and fabricated data, and Duke had to pay $150 million in fines, out of which $33 million went to the whistleblower.
Investigator's fate	There is no publicly available information available about Potts-Kant after her dismissal.

Dong-Pyou Han—Iowa State University, Ames (2012–2014)

Case overview	While working on an HIV/AIDS vaccine, Han spiked rabbit blood samples with human HIV antibodies to fake the appearance of immunity development in the animals.
Noteworthy details	For a while, it looked like the case would end with the usual—relatively mild—ORI "voluntary agreement" (his exclusion from federal research for three years), coupled with Han resigning from his current position. However, the case drew the attention of Senator Charles Grassley, who raised public awareness of the fraud, and the federal prosecution was put in motion. A federal prosecutor brought charges against Han. He was arrested, prosecuted, and pled guilty to two felony charges of making false statements to obtain NIH grants. He was fined $7.2 million to be paid to the NIH and sentenced to fifty-seven months in prison.

(Continued)

TABLE 4.1 *(Continued)*

How were the problems detected?	After the purported breakthrough results were presented at various scientific meetings, other groups tried to repeat them—unsuccessfully. They started to publicly complain about the problem. The university started an investigation, in response to which Han confessed his fraud in a two-page letter and called himself a foolish coward.
Investigator's fate	An appeals court in 2016 refused to diminish his sentence. According to the Federal Bureau of Prisons, he was released in September 2019. There is no publicly available information about his current situation.
Haruko Obokata—Riken Center for Developmental Biology, Kobe, Japan (2014–2015)	
Case overview	Obokata published two *Nature* papers claiming to have found a simple way to turn normal cells into stem cells. The procedure was called STAP (stimulus-triggered acquisition of pluripotency) and consisted, essentially, of washing the cells in mild citric acid. It sounded straightforward—but it turned out to be irreproducible. The reason being—as Riken's internal investigation concluded—that it was all the result of a switcheroo: the cells that the group called STAP cells were, in fact, regular embryonic stem cells that someone took from the freezer and relabeled. The STAP papers were retracted and Obokata resigned. In 2015 her alma mater, Waseda University, revoked her PhD degree.
Noteworthy details	Yoshiki Sasai, deputy director of Riken and Obokata's supervisor, became "overwhelmed with shame" and committed suicide in a stairwell of the research building.
How were the problems detected?	Science blogs and Twitter posts exposed doctored images and plagiarism. Soon, it also became clear that other investigators were unable to reproduce the results. When an internal investigation found that the lineage of the original cells and the purported stem cells did not match, demands for a retraction intensified.
Investigator's fate	She wrote a book and published some newspaper articles about the incident. Her current employment is unclear.

4.1 "You are a selfish bastard, Lewis! Those stem cell lines were meant for people who have LOST an organ!"

Source: Nick Kim / scienceandink.com, reproduced by permission.

4.2 "We're pretty sure he falsifies the results, but he does it beautifully."

Source: Sidney Harris / CartoonStock.com, reproduced by permission.

4.3 On the subject of the institutions' response to scientific misconduct...
Source: Leonid Schneider / ForBetterScience.com, reproduced by permission.

Robert Good, was criticized by his institution for being too slow to act in the face of a mounting problem. In the Darsee case, his chairman, Eugene Braunwald, was willing at the initial discovery of cheating to accept his claim that it was an isolated case; Braunwald did not notify the NIH and allowed Darsee to continue under supervision. Only after some time delay—and as further evidence accumulated—did Harvard and the NIH initiate a full investigation. Even in the recent Tessier-Lavigne case—which is described in detail later—seven years before the official inquiry began serious issues were raised in PubPeer and in scientific blogs about some of the papers in question (figure 4.3).

(3) **Prior unethical behavior**. Subsequent digging tends to turn up disturbing patterns of questionable behaviors and associations by many of the fraudulent investigators. For example, after the Darsee case was settled in the early 1980s, it came to light that he had already engaged in fraudulent research in the 1970s while at Emory University. Papers in top journals, such as the *New England Journal of Medicine*, describing various clinical studies were discovered (and subsequently retracted) in which the patients could not be identified; it is likely that some of the patients did not exist and were completely made up. It turns out that already in the late 1960s, while Darsee was a student at the University of Notre Dame, he published a couple of papers in the university's undergraduate science magazine that claimed successful "rejuvenation" of aging rats with various hormones; these papers were likely also based on nonexistent research.[4] In the case of Penkowa, following the fraud detected in her papers, an investigation discovered that her prior PhD work was also based, at least in part, on fraudulent data. As a result, her doctoral degree was revoked in 2017. In the case of Potti, it came to light that he has taken liberties with his own CV, such as listing awards and fellowships that he never received, when he applied for a position at Duke University. In the case of Han, although the "spiked" rabbit plasma samples were only discovered in 2012 at the University of Iowa, subsequent investigation revealed misconduct dating back to 2008 at his previous place of employment, Case Western Reserve University. And it turned out that Anversa, in his early years of research, was mentored by Bernardo Nadal-Ginard, who was characterized by a U.S. criminal court as a "common and notorious thief" who had previously misappropriated millions of dollars from his former employer, Boston Children's Hospital. Nadal-Ginard was a coauthor on the notorious 2001 *Nature* paper—which started the field of stem cell therapy, and which, as we know, is not reproducible. (And which, by the way, still has not been retracted to this day.)

(4) **The wasted money cannot be fully recovered**. The institutions are sometimes induced to return *some* of the money that was used to support fraudulent research. For example, Brigham and Harvard had to pay $10 million to the NIH. This sounds like a lot of money, but it is, in fact, pocket change when we consider that Anversa had received about five times more in NIH grant funding over his career. Duke University had to pay hefty fines to the NIH after the Potti fiasco was settled, and again after the Potts-Kant case, which was the largest False Claims Act settlement against any university. And even this settlement cost the university less in fines (less than $150 million) than the grants

that they received that were based on doctored results (about $200 million). The whopping $7.2 million fine Han must pay personally—if he can—is only a fraction of the $19 million grant support NIH had provided for the development of the fraudulent AIDS vaccine.

These calculations only include the "direct grant money" wasted. If we add the costs related to the unsuccessful replication attempts that were paid by institutions and grant-giving bodies all over the world, the cost of a field of science going in a false direction, and the opportunity costs (money that could have gone to support real science), the amount of wasted money multiplies.

(5) **The personal penalties are relatively mild**. The investigator has to abort the grants and return the leftover money, retract (some of) the publications, and in some cases stop doing laboratory research (figure 4.4). Then the investigator may move into another related career, perhaps in clinical medicine. For more senior investigators, early retirement may also be an option. Sometimes PR

4.4 On the subject of rehabilitation after scientific misconduct . . .
Source: Leonid Schneider / ForBetterScience.com, reproduced by permission.

companies or "online reputation managers" are contracted to repair the online image of the investigator. It is interesting to note that quite a bit of research fraud comes to light precisely around the time when the responsible investigator has just retired.[5] Jail is not an option as a penalty. Even in the highest-profile cases shown in table 4, there was only one prison sentence—although that was a hefty one.[6]

(6) **The misconduct does not happen in isolation**. Often the disgraced investigator utilized or even cultivated a network of like-minded investigators: sometimes former mentors, other times collaborators; often coauthors, other times only fellow "believers." These people may sit on grants committees or journal editorial boards and may help with the acceptance of papers, favorable evaluations when it comes to funding requests, and public dissemination of the results. For example, another stem cell human trial, called SCIPO trial, conducted by close Anversa collaborator Roberto Bolli, was also retracted (Bolli claims that he had no knowledge of Anversa's fabrications and was a victim of his fraud). The 2022 *Reuters* article on Anversa provides deep insights into the circles of friends, supporters, and believers.[7] In the case of the RIKEN/STAP fiasco, Obokata's former mentor, a high-powered Harvard professor, Charles Vacanti, was a coauthor on her *Nature* papers and acted as a high-profile supporter of the concept even later on when incredulity and criticism began to intensify around the paper.

(7) **Admission of personal responsibility is rare**. The vast majority of investigators caught cheating maintain their innocence or minimize the affair. According to one common laboratory adage, if caught, the first author should blame the last author; the last author should blame the first author who is typically a junior investigator (figure 4.5), and middle authors should claim that they had nothing to do with the study in question. On his current LinkedIn page, Summerlin characterizes his leaving of academia as a result of "falling out with his research project mentor." Penkowa's response to the verdict was that " 'deliberate malpractice' is another matter and something I have never done." Potti and Hauser, in their respective ORI settlements, "neither admitted nor denied" research misconduct. In various interviews, Anversa placed the blame on various middle managers who worked in his laboratory, and he continues to maintain his innocence.

In the case of the RIKEN/STAP affair, in her 2016 book *That Day*, Obokata places the responsibility on her coauthor Teruhiro Wakayama, saying that "crucial parts of the experiments were handled by him and him alone."[8] University of Connecticut's Dipak Das not only rejected the accusations of fraud, but in a long

4.5 On the topic of penalizing those who were found responsible for scientific misconduct: "Student meets bus."
Source: Leonid Schneider / ForBetterScience.com, reproduced by permission.

rebuttal letter he accused his university of attacks motivated by "anti-Indian racism." The John Darsee case is complicated: at one point he claimed he had "no recollection" of committing the abuses, but he later wrote an apology in the *New England Journal of Medicine* saying, "I am deeply sorry for allowing these inaccuracies and falsehoods to be published in the Journal and apologize to the editorial board and readers."

 (8) *Once a believer, (almost) always a believer*. Many of the disgraced investigators—just like Benveniste with his "water memory" findings mentioned earlier—bitterly maintain that fundamentally they were right about the concept. (Some of the figures or data in those papers may have been wrong, but the underlying fundamental concept is right.) (figure 4.6) Obokata's mentor and coauthor, Charles Vacanti, continued fighting for the issuance of a STAP-related U.S. patent for several years after the STAP paper was retracted and he had closed down his lab and retired. In the Hauser case, a 2007 paper—which has been criticized by the Harvard and ORI investigations but not retracted—was supplemented

4.6 "I already wrote the paper. That's why it's so hard to get the right data."
Source: Benita Epstein / CartoonStock.com, reproduced by permission.

with a 2011 addendum, in which the authors of the original paper "returned to the laboratory to re-run the three main experimental conditions reported in the paper, videotaping every trial and accompanying the video records with field notes."[9] The addendum concludes that, upon replication, the authors found the exact same results as stated in the original article. Hwang has successfully fought for, and eventually obtained, a patent on a stem cell line originally reported in his now retracted papers in *Science*.

(9) **Litigation, murkiness**. Quite a bit of litigation happens in many of these cases. For instance, the Potti affair involved many lawsuits. The Anversa affair involved a complicated and acrimonious legal back-and-forth: while Harvard was suing Anversa for fraud, Anversa was suing Harvard because Harvard's investigation damaged his reputation and cost him job offers and destroyed their chances to sell their spin-off biotech company. After Das was found to have committed data and image manipulation on an industrial scale and was dismissed from his position at the University of Connecticut, he intended to file a $35 million

defamation suit against the university. It is difficult to figure out the final result of the lawsuits; the settlement terms are often kept secret. Thus even years after the misconduct affair is supposedly settled, many details remain murky or are hidden behind confidential legal settlements. For example, Harvard University still has not produced a full public accounting (let alone an apology) of the Anversa affair.

(10) **"Runaway trains" and slowly responding scientific systems.** Once a field of biology or medicine is in motion, it is hard to stop it—even if fraud has been the basis of an entire misguided subfield of research and the fundamentals of the field are called into question. For example, it is now clear that Anversa's papers on how stem cells turn into heart cells are based on made-up and doctored data. Yet a recent *Reuters* investigation finds that the area of repairing a damaged heart with stem cells lives on, with hundreds of millions of dollars of grant support being devoted to it.[10]

Another investigator whose name has been in the public eye recently is Marc Tessier-Lavigne, who was the president of Stanford University from 2016 until his resignation in 2023. Tessier-Lavigne works in the field of neurodegeneration and Alzheimer's disease. Some of the problematic papers go back to 1999, when Tessier-Lavigne was an executive at the biotech firm Genentech. Many of his high-profile papers—including papers in *Nature*, *Science*, and *Cell*—came under scrutiny by a committee specially assembled at Stanford, consisting of highly respected scientists, including the Nobelist Randy Schekman.[11]

Doubts about Tessier-Lavigne's papers go back quite some time, beginning with comments on PubPeer in 2015.[12] Not much happened for seven years. Then, in 2023, Stanford's student newspaper, the *Stanford Daily*, published a series of articles about these publications and raised some questions related to potential scientific misconduct.[13] One of the juiciest allegations is that, apparently, Genentech has developed some doubts about the validity of the data generated by the Tessier-Lavigne laboratory, and several complaints were filed internally claiming misconduct. *The Stanford Daily* article also mentions that at some point the validity of the work of an investigator named Anatoly Nikolaev, who worked in the Tessier-Lavigne group, came into question. To see if something fishy was going on, Nikolaev was asked to repeat his studies—but, first, his reagents were switched over without his knowledge. There is no way the experiment could have worked under these conditions. Yet miraculously Nikolaev still delivered the "correct" results.[14] At this point, it should have been clear that he was falsifying data, and the paper should have been retracted right away. But it wasn't.

Stanford's investigation was completed in the summer of 2023. The findings have been made public fairly quickly—which is an exception rather than the rule in cases like this. As a result, Tessier-Lavigne resigned from Stanford's presidency but maintained his position as Professor of Biology at the same university. The committee concluded that "Dr. Tessier-Lavigne's labs engaged in inappropriate manipulation of research data or deficient scientific practices, resulting in significant flaws in five papers that listed Dr. Tessier-Lavigne as the principal author." It also stated that his research work had "multiple problems" and "fell below customary standards of scientific rigor." Subsequently, several high-profile papers had to be retracted, two from *Science* and one—which had undergone extensive corrections in 2015—from *Cell*.

But the findings were presented to the public with wildly divergent spins on them. The *New York Times* chose the title "Stanford President Will Resign After Report Found Flaws in His Research," and *Science* went with "Stanford President to Step Down Despite Probe Exonerating Him of Research Misconduct." The key is the following sentence: the panel found that "Dr. Tessier-Lavigne had not participated in the manipulation, was not aware of them at the time, and had not been reckless in failing to detect them." Some of the public's reaction was rather savage. One reader's comment after the *New York Times* article stated: "So the former business exec is forced out of his position as an administrator for suppressing falsified scientific data but gets to keep his job as an academic scientist. Strange world we live in."[15]

I will not spend much space on recent high-profile articles for which potential problems have been raised but are not yet settled. In articles published in *Science* or *Nature*, attention has been drawn to potential data irregularities, duplications, and retractions or corrections in many high-profile laboratories.[16] But one high-profile case should be mentioned because as it has already made the pages of many mainstream papers. This case focuses on problematic images in the papers of the hypoxia signaling pioneer (and recent Nobelist) Gregg Semenza of Johns Hopkins University.[17] The increased scrutiny has resulted in at least eleven retractions and ten corrections or expressions of concern. According to a recent article in the *Baltimore Sun*, it is unclear whether Johns Hopkins is looking into this matter further. Data detective and microbiologist Elisabeth Bik aptly commented: "They're not going to kill the chicken that lays the golden eggs." "So, the institutions usually drag their feet and nothing happens for years and years."[18]

Another series of recent scientific misconduct allegations concerns Orazio Schillaci, the Italian Minister of Health, who published a series of papers while working at the University of Rome Tor Vergata between 2018 and 2022. During this time, he published a new paper, on average, every twelve days. Several of these publications appear to have multiple image duplication "irregularities" (in which the exact same image is claimed to represent completely different experiments or conditions in different papers or different panels of the same paper). Schilacci is "not worried," however. He issued the following statement: "I have not manipulated anything. The images do not come from my laboratory, but from other colleagues that have not done anything wrong."[19]

In the science blogs of the data detectives (more on them later), one can find thousands of cases where questionable-looking images and figures are being scrutinized—in some cases involving high-profile investigators. But one must not jump to conclusions. In science, just like in any other walk of life, everybody is innocent, of course—until proven guilty.

EASILY DETECTABLE METHODS OF MANIPULATION

Focusing on high-profile stories can be misleading. It may give the impression that scientific misconduct is a rare, isolated event, perpetrated by a tiny fraction of people who work in science. Focusing on high-profile cases may also leave one with the impression that misconduct is something that gets exposed and will be swiftly dealt with. The "only a few bad apples" concept—a few who are quickly weeded out by the self-cleaning nature of the scientific process (nothing to see here, please move on . . .)—was the prevalent view until the late 1980s.[20]

Later on, the analogy changed and the situation was compared to the tip of the iceberg. High-profile cases that come to light may represent a fraction of the actual cases. However, the prevailing view became that science, overall, is pretty clean. Yes, sometimes a fraudulent investigator crops up and is exposed, and it is likely that a bunch of other fraudsters also exist who remain under water—unexposed. But overall, we have beautiful, clean blue water surrounding these icebergs. Imagine a pristine fjord in Norway.

In my view, even this analogy is severely misleading. If we want to stay with nature analogies, my feeling is that we are dealing with a big scientific swamp, with various swamp creatures of different sizes and shapes living in it. There are some relatively clean areas of water, too, and there are regular life forms as well. But there are an awful lot of swamp creatures who happily coexist in their natural environment, taking away food and resources from the regular life forms. In addition, a whole ecosystem built around the swamp is benefiting from it. The people who are supposed to manage the swamp, or perhaps drain it, are nowhere to be found. And, of course, good luck trying to build a house on the swamp.

In the following sections, I describe some of the common techniques that are used to drastically alter and fake results. I would like to stress, in advance, image fraud is not the only way—and most likely not even the most common way—to create fake figures for publications. Image fraud came into focus simply because it is the easiest to detect. It is certain that a lot more fakery is going on—the true scale of which is anybody's guess at this point. And most of it is impossible to detect just by examining the figures in a publication.

One picture says a thousand words, the saying goes. In many scientific articles, pictures of cells or clusters of cells or histological slides, which contain magnified microscopic images of a certain organ or tissue, are shown. In some cases, these are simply native images, and the cells or tissues are shown as they look under the microscope. In other cases, they are labeled with a dye to produce a colorful image and to emphasize certain cellular constituents, for example, the nucleus of each cell. In yet other cases, a specific protein (antibody) is used to detect the presence of a certain protein in the cell or the tissue. I have already explained that this method has some inherent variability due to the antibody used and due to various steps in the immunohistochemical process.

What is important for the topic of image fraud is that every cluster of cells, or every histological tissue section, is *unique*. It is like a fingerprint. No two sections look exactly the same. And here is where an examiner with a sharp eye—or, more recently, computer programs that can scan and analyze multiple images—comes in handy. They can detect duplicated images that were shifted, rotated, or otherwise manipulated. In the context of a scientific publication, one can, then, prove with high certainty that some sort of manipulation took place. For example, if one takes a picture of a neuron, and then zooms in to the exact same picture of the neuron, and in the figure legend claims that the neuron has grown in size in a certain time period, by examining the characteristics of the neuron in the two pictures and examining various so-called background artifacts that are parts of all photomicrographs, one can show that the two pictures are the same.[21] Likewise, if in the same multipanel picture of images, the exact same picture, or part of the exact same picture—or the same version of the same picture after some rotation or cropping or shrinking or expansion—appears twice, or multiple times, and the legend to the figure claims that each panel represents different treatments or conditions, the examiner can determine that this simply cannot be correct. When one examines multiple papers, the exact same immunohistochemical picture cannot claim, in one instance, to depict the presence of protein A and in another paper protein B. Instances like these must be due either to a mistake or—especially if they happen serially, in multiple papers, over several years—to suspected intentional image fraud. One can be sloppy once or twice, but when an investigator has dozens and dozens of papers flagged in PubPeer, it becomes much more difficult to believe we are dealing with an unfortunate long series of honest mistakes.

But here is the worst part. Just because a paper does not show evidence of image manipulation does not mean that the presented results are correct. It could also simply mean that the investigators did not make obvious "mistakes" in assembling fraudulent figures. Histological figures and Western blots are supposed to show representative findings. Each part of a multipanel picture is supposed to be one example—a *representative* example, i.e., a typical example—for each group. An investigator who wants to fake a difference where there is, in fact, no difference could simply choose various pictures that fulfill this pretense. Many histological slides are produced in a typical experiment. Some of them can show, for example, a slight degree of damage or inflammation, others can show more. If one has a protective effect of an intervention or a drug, the slides in the control group typically show no damage, the slides in the disease group show some degree of injury (which may be variable from animal to animal), and the slides of the treated group may show less injury than the control group. But if the treatment is ineffective, an unscrupulous fraudster could show a normal tissue and call it a treated one. Or the person can pick the slide that shows the least amount of damage and call it a representative example. This is why it is important to examine and grade histological sections in a blinded manner, by an investigator who has no information about which slide belongs to which group. But if the goal of the fraudster is intentional—if the goal is to produce the illusion of an effect—this can be achieved easily. From a sufficient number of pictures, virtually any result or any effect may be "created" on the computer screen of a scientific fraudster. As long as each of the pictures presented in a multipanel figure is a *different picture*, no image fraud will be detected. But, of course, this does not necessarily mean that the figure represents the biological truth. When a sufficiently crafty fraudster puts together a series of images and calls them "representative," the image fraud will not be detectable by simply looking at the paper.

If an unscrupulous fraudster who is hell-bent on faking some sort of imaginary "result" is worried about detection of fraud, that person could also "generate" the primary samples in such a manner that the fakery could *not* be detected by any observation or analysis of the resulting paper. For example, let's say that a cheater wants to "prove" that an experimental drug works in a model of inflammation. If the cheater slips a known anti-inflammatory drug, like a steroid, into the treatment group during the animal experimentation, the resulting histological images will turn out beautifully: no subsequent image fraud will be "needed."

There are also ways to manipulate the various stages of the immunohistochemical staining process to "create" realistic-looking effects. Of course, the risk of this can be reduced by blinding and randomization the treatment phase and separation of the various phases of the experiment to multiple investigators. But this requires additional planning, coordination, and more research personnel. In reality, a lot of academic laboratories rely on small groups; often a single investigator performs all stages of the research. If this investigator is hell-bent on intentionally faking the results from the onset, and carefully covers his tracks, the fraud will be very hard to detect.

With the invention of Adobe Photoshop in 1990, it became easier to scan histological pictures into the computer, label them with arrows or scale bars, and assemble various composite images. At the same time, this led to the rise of fraudulent "Photoshop Artists" in science. With Photoshop, it became a matter of a few simple clicks for a cheater to increase the contrast of a picture or even to copy and paste various clusters of cells to the background of another picture, for example, to create the illusion of an inflammatory focus, or to erase some cell clusters to "clean up" an unwanted part of the picture. There are thousands of flagged images in PubPeer where this type of blatant manipulation is completely apparent.

While I was writing this book, a PubPeer comment appeared in relation to one of my own papers, which was published in a minor pharmacology journal twenty-five years ago. One figure in this paper had a three-panel histological picture. The top panel was the control, the middle panel was an example of an inflamed tissue, and the bottom subpanel was supposed to be an example of a tissue from an animal that showed lesser inflammation because it was treated with a pharmacological agent. The PubPeer image analysis showed that the third panel, was, actually, a higher magnification and 180-degree rotated part of the first panel. The histological work and the resulting images were assembled by the first author of a paper, a PhD student who worked in my laboratory whom I will call Antonio. I did not notice the problem, nor did anyone else in the laboratory, nor the referees of the journal, nor any of the readers. The information contained in the picture was consistent with the rest of the information in the paper, where various measurements of tissue inflammation took place that did not rely on histology. Back then pictures of histological slides were taken by a camera attached to a microscope. The film was mailed to a photo-lab off-site, which developed the film, made shiny color pictures, and sent them back to the laboratory, where multipanel images intended for publications were assembled by hand. Actual shiny paper copies of these color figures were sent, in the mail, to the journal for publication; and these pictures were then sent further, by mail, to the reviewers. We thought we were running a tight ship in the laboratory: primary data were checked regularly; laboratory notebooks were maintained according to the best standards at the time; results were written and inserted; and notebooks were checked and countersigned and dated. But we are all humans. At some point, a mix-up with these pictures could have happened. An honest mistake. First, I considered doing what most people do on PubPeer comments like this, which is to apologize and claim an honest mistake—because, obviously, after twenty-five years, we cannot go back to check the original data (no laboratory can keep them for that long).[22] But when I looked at PubPeer, it turned out that in the subsequent career of Antonio—after he left my laboratory and worked in various academic centers—more than one hundred of his subsequent papers received PubPeer comments for obvious image duplications. This either points to an incredible degree of sloppiness or to something much worse.[23] In light of this, I decided to initiate retraction of the entire paper. Another lesson learned.[24]

The second commonly detected source of image fraud relates to Western blots. As explained in chapter 3, this method is used to measure the amount of various proteins in a biological system (cell or tissue), and—just like the immunohistochemistry—it relies on the binding of an antibody to a certain protein. In practice, a mash-up (homogenate) of cells and tissues is loaded into small pockets (wells) at the top of a flat gel (imagine a thin layer of yellow) and exposed to an electrical current running through the gel from its top to the bottom. Each protein has a certain size and is electrically charged and will move with the current and spread out in the gel. Then the proteins are transferred from the gel to the thin

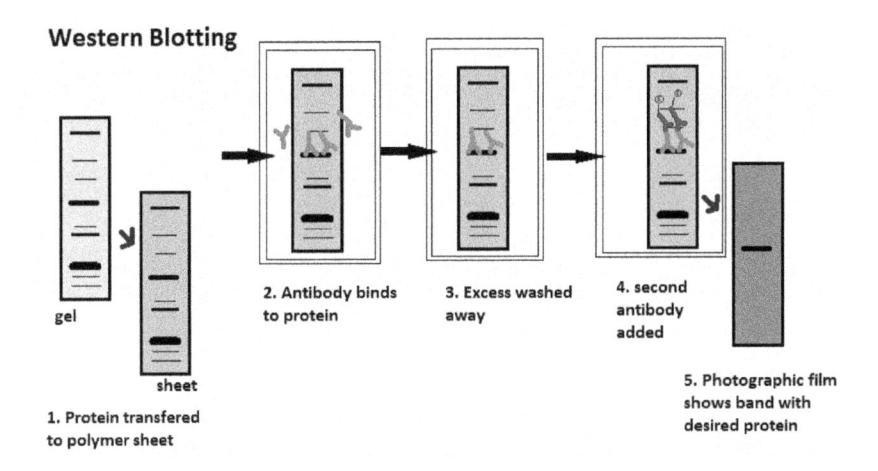

SCIENTIFIC FRAUD—AND THE FRAUDULENT FRAUDSTERS 137

4.7 The principle of Western blotting. The method relies on the binding of antibodies (depicted as Y shaped proteins) to proteins that are laid out on a gel. In the end, a "blob" is produced; its location on the gel corresponds to its size. *Each blob*, in each experiment, *has an individual character*, similar to human fingerprints.
Source: Courtesy of Cawang / Wikimedia Commons.

sheet of special fabric (membrane), again using electrical current. Last, the membrane is exposed to an antibody of choice, which ideally binds only to the selected protein of interest. The reaction is developed and detected using a machine called a gel box. Ultimately the proteins are visualized as dark blobs of various sizes. The position of the blob on the gel corresponds to the size of the protein. The intensity of the blob corresponds to the amount of the protein (figure 4.7).

With a method like this, every step of the experiment is crucial, and one wrong step in the process will ruin the whole experiment. Even in a "perfect" experiment, there may be some small errors related to the gel, which can contain small bubbles or inhomogeneities; the side of the gel runs slightly differently from the middle, the concentration of the antibody used and the time of exposure with the antibodies affects the staining intensity, and so on. Because of these factors, ideally, an entire experiment (e.g., controls and treated groups) must be run on the exact same gel, with lanes next to each other representing different groups of replicates. (Visualize it as a 100-meter sprint competition, with eight sprinters running side by side, each in their own lane.) Some of the lanes are also used for control proteins (so-called standards). It is important that into each experimental lane the investigator puts the exact same amount of protein homogenate.

Another important thing—in addition to detection of the protein of interest—is that the *exact same* Western blot run must be used for loading control to detect a so-called housekeeping protein that is *not expected to change* with the experimental intervention. For example, this protein could be actin: a common, structural protein that is constantly present in all cells and tissues at a predictable amount. The change in the target protein must be viewed in the context of this loading control. If the loading controls are variable, the experiment was not performed very well. Obviously, there is no such thing as a separate loading control experiment; loading controls must be done together with the actual experiment that seeks to quantify the target protein.

Western blots are extremely useful tools, but they have to be performed and interpreted with utmost care. Even when everything is done by the book, Westerns can be misleading due to nonspecificity of the antibody used, interpreting the incorrect lane that may be a product of an experimental artifact, or hiding true differences due to incorrect overexposure of the gel. Even with the best intentions, one can sometimes find misleading results and publish findings that may not be reproducible by another investigator. But, of course, if a fraudulent investigator is hell-bent on cheating, various steps in the process can be manipulated (from the loading of the gel to the analysis) to distort the results and "create" various "desired" findings.

Adobe Photoshop is useful when assembling multipanel Western blot figures that contain examples of various experiments within the same figure. But Western blotting is also a method in which the fraudulent Photoshop artists can truly shine. Unscrupulous fraudsters can relabel each lane of a Western blot, and a whole different experiment could be claimed. (The runner analogy would be that after the 100-meter sprint the name tags of the runners are switched: so in the end not the winners but some other runners would step up to the podium—perhaps those who did not even take part in the race.) Individual blots, or chunks of blots, can be copied and pasted. This incorrect practice is called "splicing." The weight of the protein could be intentionally misrepresented. Or a cheater could take the loading controls from a completely different experiment or reuse it repeatedly for different experiments, and different figures, to create the illusion that equal amounts of proteins were loaded into each lane of the gel (figure 4.8). (In reality, these loading controls, of course, belong strictly to each individual experimental run and must not be used interchangeably or for unrelated experiments.) If a

4.8 On the subject of loading controls for Western blots...
Source: Leonid Schneider / ForBetterScience.com, reproduced by permission.

fraudster loads different amounts of a protein extract on a gel or spikes an extract with some pure protein and then splices an equal loading control under it, the appearance of any effect can be created.

At the top of Western Blot Fraud Olympics are the so-called Frankenblots where every single "blob" in every single lane is copied and pasted together, from original gels of questionable origin—often, simply from other Western blots that are "lifted" from completely unrelated, previously published papers (figure 4.9).

Thankfully, each blot contains its own pattern, its own "fingerprint." This means that the fakery can be detected visually by an eagle-eyed examiner—or even better, by automated programs that can enhance the contrast and create pseudo-colors

4.9 On the subject of Western blot manipulation...
Source: Leonid Schneider / ForBetterScience.com, reproduced by permission.

to check for background irregularities or other signs of manipulation. The situation has gotten to the point that journals that care about fraud have started using software to try to detect Western fakeries, as a matter of course, as the first step of manuscript evaluation.[25] In fact, Western blot fakery became so pervasive that journals have instituted a policy that *full gels* (i.e., uncropped gels) that supposedly represent the entire experiment must be provided to the reviewers at the submission of the papers (figure 4.10). In response, many fraudsters have started to *fake entire gels* by photoshopping the "desired" blot patterns onto the background of a regular gel.[26] A regular war of escalation is going on, and both sides—the cheaters and the checkers—are mobilizing more and more sophisticated tools.

Flow cytometry is another technology that relies on the detection of proteins through various antibodies. The method itself is the analysis of single cells in solution—as they flow past various lasers. The measured parameters (light scatter, fluorescence parameters) often rely on the binding of an antibody to a protein (that is typically on the surface of the cells). In the end, the results are often presented on a figure that has two axes; each axis may depict the intensity of binding of a certain antibody to a certain protein. Typically, the lower left quadrant represents a negative cell cluster, the upper left quadrant shows the cell population that contains the protein indicated on the vertical axis, the lower right quadrant depicts cells that contain the protein indicated on the horizontal axis, and the upper right quadrant shows those cells that are positive for both measured proteins. In the end, cells end up in one or more clusters of dots, each dot represents one cell. One can imagine how easy it is for a fraudster to manipulate a picture

SCIENTIFIC FRAUD—AND THE FRAUDULENT FRAUDSTERS 141

4.10 Further on the subject of Western blot manipulation . . .
Source: Leonid Schneider / ForBetterScience.com, reproduced by permission.

like this—for instance, using Photoshop or any type of graphics program—to shift cell clusters, or to add or remove additional dots/clusters of cells. This type of fraud became so prevalent that it now has several different names: instead of flow cytometry it is called *flaw* cytometry (the linguistic invention of the Novartis data scientist Kevin Vervier) or *faux* cytometry (the linguistic creation of the Roche software engineer Konrad Rudolph).

Some of the high-level scientific fraud cases discussed earlier in this chapter involve histological, photomicrograph-based, or Western blot "irregularities" of the type shown here (figure 4.11). A 2016 analysis that examined more than twenty thousand publications from 1995 to 2014 concluded that about 4 percent of the examined papers contain problematic figures that show signs of deliberate

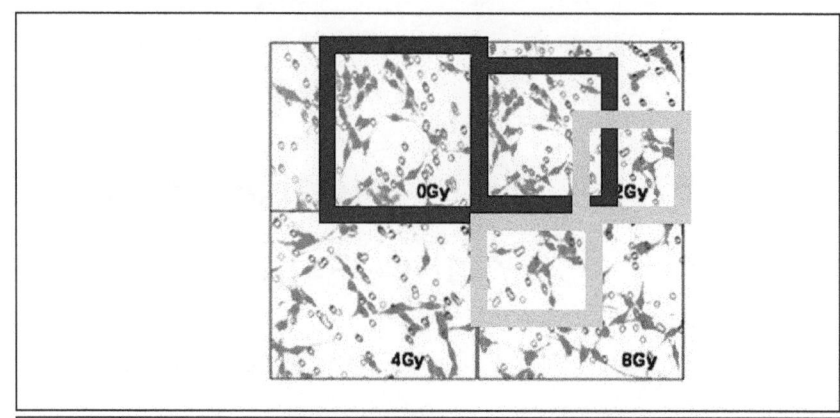

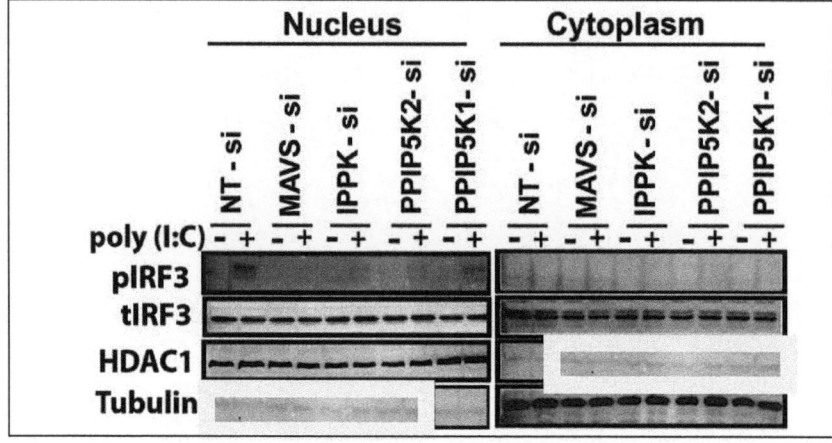

4.11 Common examples of image fraud. *Top:* The four subpanels are supposed to represent groups of cells from four different experimental conditions: control cells and cells exposed to increasing doses of radiation at 2, 4 and 8 Gray. However, the cluster of cells in the black bracket is the exact same in the 0 Gy and the 2 Gray conditions, which cannot happen. The same cell cluster cannot be at two places at once. Likewise, the cluster of cells in the 2Gray and 8 Gray panels is also the exact same. This also cannot happen. This paper was retracted in 2015. *Bottom:* example of another very common type of image fraud, when the exact same "loading controls" for Western blots are reused and claim to represent different conditions. In this case the area labeled with white brackets is the exact same in the left and right blot—i.e., both contain a particular sequence of lanes and gel artifacts that identifies a "fingerprint" of a particular experiment—even though, according to the claim of the publication, the two blots supposedly represent controls for completely different experiments. Once again, this is physically impossible. This paper was retracted later, in the same year.

Source: Figures reproduced under the Creative Commons (CC BY) license. The same examples were also highlighted in E. Bik's excellent paper on the subject of image fraud. See Yang Liu et al., "Carbon Ion Radiation Inhibits Glioma and Endothelial Cell Migration Induced by Secreted VEGF," *PLOS One* 9, no. 6 (June 2014): e98448, https://doi.org/10.1371/journal.pone.0098448; Niyas Kudukkil Pulloor et al., "Human Genome-Wide RNAi Screen Identifies an Essential Role for Inositol Pyrophosphates in Type-I Interferon Response," *PLOS Pathog* 10, no 2 (February 2014): e1003981, https://doi.org/10.1371/journal.ppat.1003981; and Elisabeth M. Bik, Arturo Casadevall, and Ferric C. Fang, "The Prevalence of Inappropriate Image Duplication in Biomedical Research Publications," *mBio* 7, no. 3 (June 2016): e00809-16, https://doi.org/10.1128/mBio.00809-16.

manipulation, with the incidence of the problem increasing fourfold (!!!) over the examined ten-year time period.[27] In another survey, Sholto David has examined 715 papers published in the journal *Toxicology Reports* that included various images. He found that 115 contained inappropriate duplications—a 16 percent rate.[28] Let me stress once again that this survey only focused on *one type of potential misconduct* simply because this type is easily detectable.

Image fraud is not the only type of fraud that can happen in a paper; far from it. It is simply the type of fakery that is *most easily detectable*. If the publication does not contain images, but instead has a series of numbers in a table or just a bunch of bar graphs, some bars higher and some being lower, there is no certain method to determine, simply by looking at a paper, whether the presented figures are based on real experiments or not. Some statistical methods can be used to determine if the reported numbers in a table are likely to be based on real data or not. It is more difficult to do it with summary data and bar graphs and easier if an entire dataset (so-called raw data) is also available, for example, in the form of a supplemental table. In this case, if clusters of numbers repeat themselves, it is likely that some copy-paste shenaniganry was involved. If certain figures repeat more than others, it is more likely that the authors have "created" data by entering random numbers, because it turns out that the human mind is not completely random. Or, if a certain sequence of numbers repeats in the exact order, chances are that somebody made a mistake—for instance a sloppy copy-paste error in an Excel sheet. Using these mathematical/statistical procedures we can detect potential or likely fraud.[29] But we are dealing with statistical likelihoods and not—as in the case of image fraud—irrefutable certainty.[30]

But here is the *real dilemma*. If we see a paper that contains obvious fakery in Western blots or histological images, what are we supposed to think about the rest of the paper? What is the chance that the authors took the same liberty with the data in the rest of the paper? On one hand, "where there is smoke, there is fire." On the other hand, as stated previously, everybody is "innocent until proven guilty."

THE EXTENT OF INTENTIONAL FRAUD IN THE SCIENTIFIC LITERATURE

To what extent is blatant, intentional fraud responsible for the reproducibility crisis? How much of the problem is related to the inherent variability and complexity of the methods used versus various so-called borderline approaches of beautification, versus blatant, intentional falsification? As you can imagine, most fraudsters will not confess to their fraudulent ways—not even in anonymous surveys. Nevertheless, we can still learn quite a bit from these surveys and make some deductions. One of the widely publicized surveys was done by *Nature* in 2016. More than 1,500 researchers were surveyed, and about 40 percent of the respondents believe that fraud often contributes to the irreproducibility and a further 30 percent believed that it sometimes contributes to the irreproducibility. In a recent anonymous survey of Dutch scientists, 8 percent of the respondents who work in the field of biomedical research stated that they have falsified or fabricated data at least once over the past three years. In addition, more than 50 percent of the respondents stated that they regularly engage in questionable research practices.[31] A meta-analysis conducted in the United States a decade earlier reported a 2 percent figure for investigators who admitted data fabrication/falsification, but when the question was asked about the respondents' colleagues the number increased to 14 percent. When it came to lesser offenses (e.g., data exclusion based on a "gut feeling" or changing the design, methodology, or results of a study in response to pressures from a funding source), up to 30 percent (!!!) of the responders admitted to such behaviors.[32]

The suspect behavior seems to start early in the career of biomedical scientists. In a survey published in 1996, from a group of postdoctoral fellows at the University of California San Francisco, 3 percent admitted to "modification" of their data, but 17 percent said they were "willing to select or omit data to improve their results."[33] From biomedical trainees at the University of California, San Diego, 5 percent admitted that they have "modified research results in the past," but 81 percent (!!!) were "willing to select, omit or fabricate data to win a grant or publish a paper."[34] In a metasurvey published in 2021 in the journal *Science and Engineering Ethics*, it was reported that 15 percent of researchers witnessed others commit falsification, fabrication, or plagiarism; and 40 percent were *aware of others* who engage in questionable research practices.[35] (This survey was compiled from many different types of surveys and aggregated many fields of science—not only biomedical.)

Many smaller surveys produced numbers in similar ranges. If we consider the well-known psychological phenomenon of deflection (or the Muhammad Ali effect)—i.e., when the respondent claims not to engage in a certain type of wrong behavior but believes that almost everyone else does—then probably the higher numbers in the survey (i.e., those that relate to the behavior of others) are closer to the truth.[36] In that case, we will end up with the depressing conclusion that one in five people working in biomedical science have engaged or are engaging regularly in fraud and that almost half of them engage in questionable practices. We should also consider that groups that cut corners or make up data are probably more productive in terms of their number of publications than are their colleagues who behave ethically.[37]

Considering all of the above, it is not surprising that some observers of the situation simply conclude that it is "time to assume that health research is fraudulent, unless proven otherwise." This is the conclusion that Richard Smith, a former editor of the *British Medical Journal*, reached in 2021:

> We are realizing that the problem is huge, the system encourages fraud, and we have no adequate way to respond. It may be time to move from assuming that research has been honestly conducted and reported to assuming it to be untrustworthy until there is some evidence to the contrary.[38]

PSYCHOLOGICAL ASPECTS OF SCIENTIFIC MISCONDUCT

A brief discussion should also be devoted to the personal motives and personality issues associated with scientific misconduct. Actually, it is difficult to find solid information on this topic beyond the usual factors of hypercompetition and external and internal pressures, in part because very few of the scientists who have committed misconduct talk about their motivations and thought processes. Their typical response is to maintain innocence, blame others, or plead the fifth. In the social sciences one investigator, Diederick Stapel, has come clean. Stapel is a Dutch social psychologist, who, in his heydays, regularly published in top journals such as *Science*, and who was caught cheating and had to retract many of his papers—but at least he fessed up to it. In fact, he wrote a whole book about it titled "*Derailed*." He described, in excruciating detail, how he faked data entries in his Excel sheets, and he revealed his motivations—ambition, careerism, and self-deception—behind these actions.[39] C. K. Gunsalus came up with the acronym "TRAGEDIES," which highlights some of the inner workings of scientists who commit misconduct (see table 4.2).[40]

A lot has been written about the sociological and psychological—and even the evolutionary—aspects of cheating in a more general sense. Clearly, cheating is a biological phenomenon that is not unique to human beings and will emerge wherever and whenever there is *competition for limited resources*. In fact, there

TABLE 4.2 Various pitfalls of research misconduct

Initial	Meaning	Example
T	Temptation	"Getting my name on this article would look really good on my CV."
R	Rationalization	"It's only a few data points, and those runs were flawed anyway."
A	Ambition	"The better the story we can tell, the better a journal we can go for."
G	Group and authority pressure	"The PI's instructions don't exactly match the protocol approved by the ethics review board, but she is the senior researcher."
E	Entitlement	"I've worked so hard on this, and I know this works, and I need to get this publication."
D	Deception	"I'm sure it would have turned out this way (if I had done it)."
I	Incrementalism	"It's only a single data point I'm excluding, and just this once."
E	Embarrassment	"I don't want to look foolish for not knowing how to do this."
S	Stupid systems	"It counts more if we divide this manuscript into three submissions instead of just one."

Source: C. K. Gunsalus and Aaron D. Robinson, "Nine Pitfalls of Research Misconduct," *Nature* 557, no. 7705 (May 2018): 297–99, https://doi.org/10.1038/d41586-018-05145-6.

are interesting studies on the existence of bacterial cheaters or animal cheaters (e.g., a portion of organisms who do not contribute to the community and skim off others' "work"). There are even studies and calculations on what percentage of a cheater population is tolerated in a group of animals.[41]

I don't think I will surprise anyone if I say that many individuals cut corners and play various angles in all walks of life. On occasion, used car dealers, lawyers, and even some politicians have done it. So nobody should be surprised that some portion of individuals working in science are cheating too. What makes the case of scientific fraud infinitely more disgusting is that the very essence of our vocation is (supposed to be) the search for truth.

A large body of literature is available about sociopathic/psychopathic and pathologically narcissistic behavior in society. One does not need to have a degree in psychology or behavioral science to suspect a connection between these pathological character traits and blatant scientific fraud (figure 4.12). Even though

4.12 Further on the subject of external pressures . . .
Source: Leonid Schneider / ForBetterScience.com, reproduced by permission.

individuals who exhibit these traits are few in a society, the end result of their behavior will be careerism and an excessive number of publications: after all, fake papers are faster and easier to produce than real ones.

A triangle of fraud is also often mentioned. It consists of (a) motivation (in the case of science, existential and financial pressure, as examined in this book); (b) justification and rationalization (in the case of science, a lot of the times pressure from the supervisor and other outside factors are being blamed); and (c) opportunity (situations and circumstances that permit fraud).[42]

From the various factors that facilitate misconduct that are listed above, clearly, some of the root causes (supply/demand imbalance, fundamental aspects of human nature) will never be eliminated in the world of science. Nevertheless, with better checks and balances, and with increased attention and vigilance to the problem, I have no doubt that we can do better than the current situation. Various suggestions in this regard are discussed in chapter 6 of this book.

CHAPTER 5

A BROKEN SCIENTIFIC PUBLISHING SYSTEM

FROM MANUSCRIPT TO PUBLICATION

In principle, the role of scientific journals is simple. Their role is to provide a conduit through which new scientific information can reach the public. Perhaps in the so-called old days this was, indeed, their primary role. Over time, however, journals have multiplied and a lucrative for-profit industry has sprung up. Moreover, a certain hierarchy of journals emerged, to which the various participants of the research infrastructure began to attach increasing importance. As discussed in chapter 1, publications became increasingly intertwined with scientific career building and grant funding as well as with fraud—at a shockingly high level. There are many problematic aspects of the entire publication game. All of them contribute to the unreliability of the scientific literature, some to a larger degree than others. Let's take them one by one.

Once a research project is completed, or once the investigators decide that there is sufficient material for a new publication, a scientific manuscript is prepared. They choose a journal that publishes on their particular topic and try to match the quality of the material to be published with the reputation of the selected journal. Each manuscript must follow different guidelines because each journal has a slightly different format, which is specified in the journal's guidelines. Typically, an original paper that contains novel laboratory research data is structured in a series of sections: it begins with the title, list of authors, abstract (summary), introduction, methods, results, discussion, conclusions, and finally the references. In the published version, figures and tables are interspersed within the results section. Supplemental information in the form of additional figures, datasets, and other large files that don't fit into the main body of the paper are also provided.

Once a manuscript is written and all of its figures and supplements are assembled, the material is submitted to a journal. It is either rejected immediately by the editor or the editor's staff, or it is sent out for peer review. Peer reviewers are scientists who are expected to be experts in the field, and they are asked to provide a fair and impartial judgment of the paper, including both its strengths and weaknesses. Based on these comments, the editorial office makes a recommendation. This can range from immediate acceptance (a rare event), acceptance with small modifications, a request for minor or major revisions, or the complete rejection of the paper. The authors are provided with comments from the referees and from the editorial office, and they can revise the paper if the additional work requested is not more extensive than what the authors are willing to do. Otherwise, they can reformat the paper (with a lot of time and attention) to the style of another journal and resubmit the manuscript to a journal that typically is less prominent than the originally selected one.[1]

The submission/evaluation process continues—often over many months and many journals—until the paper is accepted by a journal. Accepted papers are turned into galley proofs, which the authors can read and make small corrections on. Many journals make these uncorrected proofs available for others to read on their website. Corrected proofs are published as the final paper.

In the old days, journals were published on paper, in several volumes (i.e., physical copies of an issue consisting of many articles placed after each other and bound in a book format) each year. Then came the hybrid publication model, in which paper and electronic forms of a journal were published in parallel. Many journals, over the years, have stopped printing on paper. Most scientific journals only publish in electronic form today—but the format and conventions of the bygone era of paper-based publishing remain.

The classical format of the scientific paper itself has been repeatedly criticized, and sometimes even called a fraud.[2] It is based on the artificial separation of the introduction (where the authors pretend that the reader is an empty vessel to whom the background of the field is presented), followed by the hypothesis (which seemingly comes from pure deduction as opposed to actual experiments, some of which will be presented later in the paper), followed by the discussion, where the newly gained knowledge is communicated in a pure form. All of this sounds highly unrealistic and artificial—we can all agree. At the same time, to my knowledge nobody has presented a better format or a better solution to the problem.

The order of authors is a very important in a scientific paper. The last author (typically the corresponding author as well as the person who takes overall responsibility for the material in the paper) is generally the head of the group. Typically, this person supervises and advises everybody else and had a major role in setting the direction of the research, securing the funding for the work, usually writing the paper—or at least the introduction and the discussion sections—and assembling the final version, submitting it to the journal, and corresponding with the editor and referees. The last author is also supposed to guarantee the overall integrity of the paper, from design to data collection, including analysis and preparation of the figures.

The first author is another important position; in most research studies this is the person who has "run" the study in the laboratory and usually has conducted most of the hands-on investigation. As projects get larger and larger, the author lists get longer, and the rest of the authors include personnel working on the project in the last author's laboratory or in collaborating laboratories. Sometimes there are shared first authors (i.e., the first two authors) or multiple corresponding authors (i.e., the last two authors). Although there are well-established rules regarding what role justifies authorship and what does not, these rules are often stretched or violated, so papers may have additional authors who did not really contribute (called honorary authorship). A big name coauthor may be included to help with publication in a top journal, which is not easy at all.

Most journals don't publish rejection rates, but leading journals such as *Nature* or *Science* reject 97 to 99 percent of the submitted manuscripts, most of them without any external review. *Science Advances* has an impact factor of around thirteen, which is considered quite good (but not anywhere near the level of *Science* or *Nature*), and receives twenty thousand manuscripts a year and rejects 90 percent of them—again, mostly without peer review. The Nobel laureate Randy Schekman has called *Nature* and *Science* "luxury journals," and ten years ago he criticized them for seeking out "sexy subjects or make challenging claims" that, in turn, "builds bubbles in fashionable fields where researchers can make the bold claims these journals want, while discouraging other important work, such as replication studies." He also warned that "the lure of the luxury journal can encourage the cutting of corners and contribute to the escalating number of papers that are retracted as flawed or fraudulent."[3] He was not wrong. A recent analysis demonstrates that the papers published in the so-called top journals tend to overestimate their likely true effect *more* than papers published in average

5.1 "I didn't exactly write the article, but . . . well, I didn't exactly do the research, either."
Source: Sydney Harris / CartoonStock.com, reproduced by permission.

journals, and the material published in them does not seem much more reliable or reproducible than the literature average.[4]

As demonstrated in the section in chapter 4 focusing on high-profile scientific fraud, authors who are more "remote" from the actual work may not have direct access to the primary data and may get "bamboozled" into the role of an uncritical supporter, even if the primary data are problematic (figure 5.1). There may be unjustified extra authors on one hand, and perhaps authors that have been unjustifiably left off the paper, on the other hand (figure 5.2). One can read a lot about authorship issues and problems. Strictly from the standpoint of *reproducibility*, however, authorship issues are not considered a major problem. If the material presented in the paper was based on solid research, authorship issues do not contribute to the unreliability issue that is the central focus of this book.

The refereeing game is a more problematic aspect of the publication process, however. The way this process is set up provides plenty of opportunity for inappropriate behavior and is viewed by many investigators as murky and unreliable. Even at the first stage of the process, when the paper gets the initial review in the

5.2 On the subject of honorary authorships...
Source: Leonid Schneider / ForBetterScience.com, reproduced by permission.

editorial office, a level of subjectivity is introduced. Journals—especially the top ones, but also lesser ones—receive many more submissions than they can publish. The editorial office must make hard decisions, early on, about which submission should be rejected right away (this decision, called desk rejection, is the fate of the majority of submissions) and which can go out to peer review. The editor in chief is usually a well-established scientist who has a certain field of expertise, but the journal's scope is always broader than any single person's expertise. There are editorial board members who bring different expertise. But many journals make the initial decision based on parameters such as novelty and potential impact, which, in reality, are not far away from newsworthiness. Although the desk rejections are typically signed by the editor in chief, they are often made

5.3 On the subject of independent peer review. . . .
Source: Leonid Schneider / ForBetterScience.com, reproduced by permission.

by less experienced staffers who work in the journal office. There are plenty of examples of papers rejected by top journals that later turn out to be seminal.[5] Clearly, randomness and potential bias are already in the system at this stage.

The next stage, when selected referees evaluate the manuscript, is also subject to heavy criticism (figure 5.3). After all, the selected referees are supposed to be experts in the field of the paper. On one hand, they may be active in the field and may be supporters of a concept, including potential former colleagues or friends (in which case the paper may be viewed in a more favorable light than it deserves), but on the other hand, they may by direct competitors (in which case the paper may be rated less favorably than it deserves).

It is difficult for an editor to judge the potential level of conflict. The reviewers are anonymous, and their comments can range from unrealistically supportive to unrealistically harsh, unreasonable, or even rude (figure 5.4). Estimating the percentage of submissions in which the referees get an idea for an experiment in their own field or when referees may reject a paper in an attempt to hold back a competing study until their own paper comes out on the same topic is difficult to predict.[6]

It is often the case that reviewers of manuscripts suggest further specific experiments, and it is often the case that the acceptance or rejection of a manuscript

5.4 "It is a bad paper and, as a reviewer, I should reject it, but it cites five of my own papers..."

Source: Hagen / CartoonStock.com, reproduced by permission.

depends on the outcome of these experiments. Interestingly, in my thirty years as a referee, I have never seen a revision in which these additional experiments—requested by me or by the other reviewers—did not turn out exactly as expected, i.e., consistent with the main message of the paper and according to the expectations of the referees. I know that referees are smart people, but how can they be *that smart*? I don't have any proof of improper behavior. I just know that when authors feel that they have a foot in the door with a major journal, they will have extra motivation to pass the final step of the evaluation process.

All of the refereeing work is done pro bono, i.e., as part of academic life, as a service to the scientific community. Early-career scientists consider it somewhat of an honor to be invited to referee papers. However, as scientists get further ahead in their career, this tends to turn into a source of distraction and frustration. Balázs Aczél and colleagues from the Eötvös University in Budapest calculated that academic referees are, essentially, donating the equivalent of $1.5 billion per year to the (for-profit) publishing industry worldwide.[7] They calculated that six work hours are spent on refereeing each paper, which seems realistic.

Refereeing is not only time-consuming but is also hard work. The rewards are minimal; one can get some experience from it, and investigators can put it on their CV—which impresses exactly nobody. In the minisurvey I conducted with fifteen of my full professor colleagues, they estimated that, on average, 16 percent of their workday is spent reviewing papers. (Add to this the 30 percent that is spent on grant writing, and a scary number is produced. Almost half of the workday of many academics is spent on activities that are not related to doing actual research.)

I receive requests to referee several papers each day—every single day, including weekends. If I accepted all of these requests, I would instantly become a full-time referee. Refereeing is frustrating and is not a very rewarding task. Authors can argue back and forth. They can refuse to present original datasets. Editors can reverse reviewers' decisions or ignore various signs of manipulation of fraud.

A lot of the submissions contain obvious fraud and should not be published, anywhere, ever. Paul Brookes at the University of Rochester estimates that about 50 percent of the manuscripts he receives for review contain manipulated data or other signs of misconduct, and my experience is similar.[8] But, still, even the weak or problematic papers and those where the misconduct is plain to see will eventually get published *somewhere*. In these papers, even simple, helpful suggestions for improvements—made in previous review cycles for other journals—are regularly ignored. In cases when the referee identifies a potentially fraudulent component in the manuscript, the authors may "correct" the fraudulent part by replacing it with other fraudulent content that is less detectable; in this case the referee, in essence, has been turned into an involuntary accomplice to scientific misconduct.

It is the refereeing and editorial decision stage where reputable journals are engaged in a fight against an ever-increasing wave of manipulation and fraud. As discussed earlier, journals are instituting guidelines and recommendations for study design, randomization, presentation of data, and, in recent years, the inclusion of raw data with their submissions. As discussed in chapter 4 regarding Western blots, journals now demand full uncropped blots—to which some fraudulent groups respond by faking the entire uncropped blots. Journals also rely on various fraud detection software of increasing sophistication to identify image fraud in submissions. In a recent paper in the *Journal of Clinical Epidemiology*, a group of scientists published a seventeen-page (!!!) treatise with a set of recommendations for journal editors and referees on how to detect warning signs of a paper mill product.[9] So is this what science has become? Is this what journals must do just to avoid getting duped by fraudsters? And can any journal seriously expect volunteer academic referees to act as forensic investigators for every paper they referee?

"REAL JOURNALS" AND THEIR PUBLISHERS

A student who enters the world of biomedical science quickly learns about the hierarchy of journals. For laboratory-based (basic) science, the big ones are *Nature, Science, Cell*, and *Proceedings of the National Academy of Sciences USA* (*PNAS*). Historically, these journals have published the most significant discoveries in the fields of biology and medicine.

These are also journals that historically have had the highest impact factor. This factor—which has, by now, gained a mythical status among scientists—is the number of literature citations made, on average, about each article published in a given journal.[10] Technically, the impact factor reflects the journal in general rather than each article published in the journal. In fact, most of the citations are made for a relatively small number of articles, and the majority of articles receive few citations.[11] Nevertheless, publication in a high-impact journal—regardless of whether it later proves to be significant or not, or whether it is cited in the field or not—has long been considered a badge of honor and is a major positive factor in the eyes of promotional and grant committees (figure 5.5).[12]

What is considered a journal with a good impact factor depends on the particular field of science. For most investigators working in common fields, such as cardiovascular or cancer research, to publish an original article in a journal with an impact factor around ten is considered very good, whereas a paper in a journal with an impact factor below two is considered borderline. *Science* and *Nature*—with their impact factors around fifty—are out of reach for the majority of scientists. Part of the story is that there has been a degree of impact factor inflation over the years, and the impact factor of most journals is slowly creeping upward while, as we will see later, the *actual impact* of the published material is creeping downward.[13]

Even without checking the impact factor of a journal, most investigators have a fairly good idea of what is considered an excellent journal, or a good one, or perhaps a mundane but decent one in a field. But it is difficult to keep up with the ever-increasing flood of new journals, both real ones and problematic ones. For instance, despite its headquarters being in China, its editorial board made up mostly by Chinese researchers, and its published papers' authors predominantly from Chinese institutions, a new journal called *The Innovation* is listed as *Innovation (Cambridge)* in scientific databases, and it has achieved an

5.5 On the subject of publishing in top journals . . .
Source: Leonid Schneider / ForBetterScience.com, reproduced by permission.

astonishingly high impact factor of thirty-two in just three years.[14] One should also keep in mind that there are various ways to artificially inflate a journal's impact factor, for example, by including many review articles or guideline articles instead of original papers in a journal. When it comes to impact factors, the famous Goodhart's Law comes to mind: When a measure becomes a target, it ceases to be a good measure.[15]

In recent years many of the top journals have decided to start new journals to funnel some of their submissions—those that are considered good but still not publishable due to the sheer volume of submissions—into their own next-tier publishing realm (figure 5.6). Springer Nature, the publishing company of *Nature*, has created many baby journals, for example, *Nature Medicine*, *Nature Metabolism*, *Nature Chemical Biology*, and *Nature Immunology*, for potential high-impact articles in specific subfields. They also created *Scientific Reports* and

5.6 "Great news from the science journal. They want us to rethink our methodology, but they love our results."
Source: Robert Leighton / CartoonStock.com, reproduced by permission.

Nature Communications, which publish a large number of articles each year. The general idea is that these journals publish articles that are technically well done but are expected to have a lower impact in the field. The American Association for the Advancement of Science, which publishes *Science*, has created *Science Signaling*, *Science Translational Medicine*, *Science Immunology*, and various "partner journals," including the not too creatively one titled *Research*. Cell Press, which originally published the journal *Cell*, now has about fifty journals, from *Cancer Cell* to *Cell Reports* and *iScience*. The *Journal of Clinical Investigation* now has *JCI Insight*, and *PNAS* now has *PNAS Nexus*.

This journal multiplication is the result of a massive explosion in the number of scientific articles. A 2022 estimate indicates that about *one million* new articles are published in biomedical science each year: that is almost a 50 percent increase over the previous decade. There are more than fifty thousand (!!!) journals in biomedical sciences, and about fourteen thousand of them are indexed by the Institute of Scientific Information.[16] There are biomedical mega journals, such as *PLOS One*, that publish about twenty thousand papers each year. Much of the explosion in the number of articles published can be traced to a relatively few recent publishers, including Frontiers, MDPI, and Hindawi. These publishers also tend to publish disproportionately more so-called Special Issue articles, which are commissioned and edited by so-called Guest Editors, and in which the process of article evaluation is less transparent and possibly more biased—as indicated by the often suspiciously rapid turnaround times between submission and acceptance, which tend to have drastically shortened in these journals over recent years—than the regular process typically followed with regular articles.[17]

At the same time, the actual impact of scientific publications has been steadily declining over time. Although the number of papers is exploding, each paper is becoming less and less groundbreaking—to use a fashionable word, less *disruptive*—in recent years. This has been clearly demonstrated by a new metric called the CD index, which, for life sciences and biomedicine, has *decreased by more than 99 percent* over the last fifty years, from 0.2 to 0.0006 (figure 5.7).[18] This could be visualized as an elephant shrinking to the size of a cat.

The trend goes across all journals; even the "Big 3" (*Nature, Science, PNAS*), the most prestigious journals in which a publication can significantly boost the authors' careers (figure 5.8) are affected by it. To translate this into plain language, a lot of material is being published these days, but most of it makes no or very little real-world difference. The Nobelist Sydney Brenner talked about "high throughput, no output science" in 2008. A recent editorial in *Nature* declares: "Disruptive science has declined."[19]

All of this is counterintuitive. With all the cool new methods and fancy model systems, not to mention the immense increase in data-processing power, one would think that the impact of new discoveries should *increase*, not decrease.

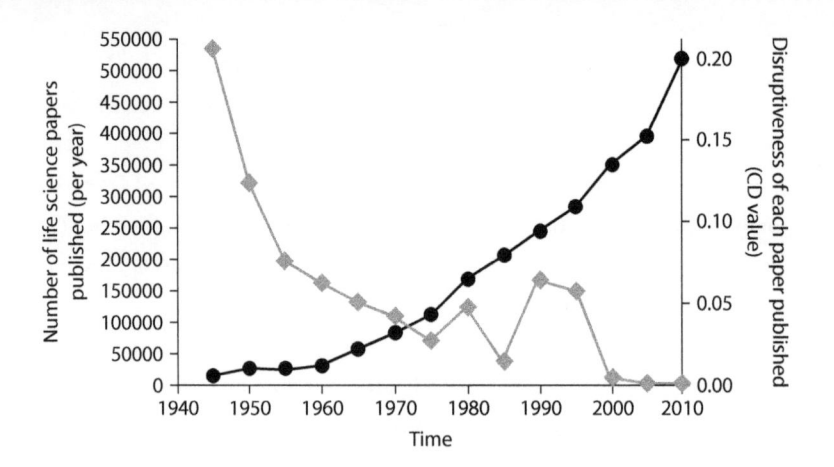

5.7 Explosion of the number of life science publications published each year (black circles, *left axis*) and the level of disruptiveness of each publication (gray squares, *right axis*) over the years.

Source: M. Park, E. Leahey, and R. J. Funk, "Papers and Patents Are Becoming Less Disruptive Over Time," *Nature* 613, no. 7942 (January 2023): 138–144, https://doi.org/10.1038/s41586-022-05543-x.

5.8 On the subject of publishing in a major journal . . .
Source: Leonid Schneider / ForBetterScience.com, reproduced by permission.

The most pessimistic interpretation is, of course, that much of what is being published is honestly conducted but UNRELIABLE work at best—and fake garbage at worst. A less pessimistic interpretation is that we see a lot of "salami science" in which information that belongs in a single paper is sliced up and sold as a series of separate articles.[20] Yet another interpretation—which would be tragic if it was the case—is that all the important things have already been discovered. But this is unlikely. One alternative hypothesis is that a substantial portion of the new information coming out in recent years is still meaningful (at least the fraction of the publications that contain results that are, indeed, reproducible), but there is not enough time and not enough resources to unpack all of it—i.e., to follow them up and translate them into something medically useful.

Disruptive content or not, the scientific publishing industry is a *very big business*. The entire scientific publishing industry is estimated to be a $19 billion business globally and annually, with about half of the business held between five companies—Elsevier, John Wiley & Sons, Taylor & Francis, Springer Nature, and SAGE—who together publish about ten thousand journals.[21] Biomedical science is estimated to be at least half of the above figures, with double-digit profit margins. The massive pro bono work that scientists provide to this industry should be considered in that context.

One of the greatest revolutions in the field has been the introduction of "open access" journals, which has been driven in large part by the switch to online publishing. In the old system, legacy journals were published on paper, and university libraries would subscribe to them; these subscription fees made up the majority of the publishers' income. Articles published in these journals were accessible for scientists working at these institutions—but for nobody else. (Many investigators also personally subscribe to *Science* or *Nature*, which publish scientific news articles in addition to regular scientific content.)

The publication cost of the open access journals is *paid by the authors themselves*.[22] But, in return, everybody can read them, even lay audiences: no library and no subscription is needed. The hope was that this new publishing system would be more transparent, more cost-effective, and everybody would benefit. Over time, however, the open access fees kept creeping up, and the open access model became a big moneymaker for publishers. The average open access fee is currently a few thousand dollars per paper, but some of the top journals have begun to charge $10,000 or more. Many new publishers have jumped onto this gravy train after seeing the kind of profits that can be made in this area. After all, the new model didn't need a printing press, and it didn't need a distribution network; the only "product" of this new kind of publisher was a pdf file made accessible for everyone to download. Many publishers even force the authors to do the professional-looking formatting of this file. And—as stated earlier—the refereeing of manuscripts is principally done by scientists who work for free. This development was certainly a factor in the creation of many new low-quality open access journals where the search for profit takes precedence over scientific quality.

One time, as I sat through a seemingly endless editorial board meeting of a pharmacology journal, the publisher's office presenter focused on new ways to market the journal through social media. In the presentation, the speaker kept referring to the articles published in the journal as "content." One of my friends and colleagues poked my side and whispered, "and you thought we were scientists? We are content creators!"

I have served several years on various editorial boards. Most of these journals had a long history of existence, and many have long associations with various professional scientific societies. The scientists I met on these board meetings, for most part, had good intentions and worked hard—and, of course, without any financial compensation. They all believed that they were doing an important public service. Occasionally, some borderline efforts were introduced to artificially boost the journal's impact factor. For instance, one idea was to make sure all papers included the guidelines published in the same journal, and in prior (relevant) papers that happened to have been published in the same journal.

In some instances I encountered unacceptable practices. For a few years I served on the editorial board of two journals that belong to an empire of open access journals that are not predatory (more on predatory journals later), but as I slowly realized, they did not deserve any stellar reputation either. The two journals with which I was involved were, supposedly, the most highly regarded ones in the journal family. One day I received a submission from a handling editor and was asked if the paper should be sent out for peer review. I noticed three images that supposedly represented histological pictures from different patients were versions of the exact same image. Slight changes in the visual fields had been applied, but they were essentially the exact same clusters of cells (meaning that they were misrepresented; all came from a single individual). I told this to the handling editor and recommended immediate rejection and suggested that they

5.9 On the topic of journal specialization...
Source: Sydney Harris / CartoonStock.com, reproduced by permission.

contact the ethics committee of the authors' university. To my surprise, the handling editor went into a back-and-forth with the authors, who offered to change these pictures for different ones and came up with some sort of excuse for why the initial submission had this problem. The handling editor kept pushing for a peer review of the manuscript. I told them that I had zero trust in this submission. Also, the next day I resigned from the editorial board of both of these journals. (I am curious about whether this particular paper will eventually be published or not, and in which journal.)

I have not lost all of my trust in the scientific publishing process yet. Yes, we may be looked upon as content creators. Yes, there is a high degree of specialization in the scientific journals, and one could even argue that there are too many journals and too many papers. (figure 5.9).Yes, there are problems with the way the journals operate and how the refereeing is set up. Yes, there is randomness in the entire process. Yes, we all work for free for for-profit publishers. But the biggest problem remains that the published material has a high level of unreliability.

The quality journals—even if the majority of the high citations they generate come from a surprisingly small subset of papers they publish—continues to command a certain degree of respect. *Nature* and *Science* are at the forefront of the movement that raises everyone's attention to the reproducibility crisis and to various emerging problems that the entire scientific community faces. But, as you will see in the next section, there is another level of shadow industry in which content creation reaches the level of brazen insolence and serious criminal activity.

PREDATORY AND FAKE JOURNALS AND THEIR PUBLISHERS

The legacy publishing industry (let's say the old-school publishers) is clearly a for-profit enterprise, but they try to maintain a level of standards and quality control—or, at the very least, the *appearance* of these things. There is a whole other level of the publishing industry, however, that is strictly and unashamedly in it—only for the money. Some in-between journals (also called gray journals, run by gray publishers) are balancing at the edge of legitimacy; they maintain a real editorial board, but many problematic articles still slip through and end up in print.

As we go even deeper down into scientific publication hell, we will encounter the growing land of predatory journals and publishers whose only motivation is money.[23] They may have a cursory peer review evaluation, or better yet pretend to have one. (In many instances, they don't even pretend to have one.) They hunt for submissions, and publish them for money. Employees (I will not use the word "editor" to flatter them) who work at these journals seek out unsuspecting authors and present themselves as a legitimate journal, but their sole aim is to get researchers to send them their publications, which they publish regardless of the scientific value of the material. The word "predatory" implies that the prey is unsuspecting, but many of the authors who publish in these journals are acutely aware of the low quality and sketchy reputation of these journals. They publish in them anyway because it helps their career advancement.

The famous "Beall list" of predatory journals was created and maintained as a blog in Jeffrey Beall's spare time; Beall is an academic librarian at the University of Colorado. The list contains thousands of journals and is a handy resource for

5.10 On the topic of open access predatory journals . . .
Source: Leonid Schneider / ForBetterScience.com, reproduced by permission.

academic scientists. As you might expect, Beall's list was not well-received by the publishers on the list (figure 5.10). The OMICS group, based in Hyderabad, India, threatened to sue Beall. In response to other threats of lawsuits, *MDPI*—a publisher officially registered in Switzerland, but in reality run from China—was taken off the list. So was *Frontiers*—another publisher registered in Switzerland. In 2017, Beall was "forced to shut down the blog due to threats & politics."[24] But it is active once again on beallslist.net.

I get at least half a dozen email invitations *every day* from predatory journals. They come from completely unknown journals, often written in exuberantly flowery or broken English. When I examine the "editorial board" of these journals on their website, I see imaginary or nonexistent scientists, scientists who are recently deceased, or names that resemble a real scientist's name and affiliation but with a few letters altered. I contacted a few of the scientists listed on these websites (with the typos corrected). The typical response was that they had no idea they were listed and didn't have anything to do with these journals—and, of course, they don't want to have anything to do with them either.

A predatory journal is only marketable if it is listed in PubMed and if it has an impact factor. Although there are safeguards and policies against it, many predatory journals do find their way onto respectable journal databases.[25] There are several ways to achieve this feat. For example, if the editor in chief has NIH grant funding, the journal can be listed in PubMed. Or real journals can get hijacked: they can be purchased by an unscrupulous group of fraudsters and turned into a vehicle to publish indiscriminate material, including completely fake papers produced by "paper mills" (much more on these later). Another form of journal hijacking occurs when a journal assumes a name that is very close but not an exact copy of the name of a real journal. Retraction Watch maintains a list of hijacked journals, and more than two hundred journals are listed.[26] Some real academic databases, such as Scopus (a major literature database maintained by Elsevier), have also been infiltrated; a recent survey found that more than three hundred predatory journals were listed on the Scopus database.[27]

There are ways to artificially inflate the impact factor of a journal, and predatory journals have found "creative" ways of doing it. For example, they use papers published in the same predatory journal (or in other predatory journals that may be part of the same predatory empire) and cite various articles, often "in bulk," that have absolutely nothing to do with the subject matter of the citing article.

To illustrate the degree of fraud that is embedded in the predatory publishing industry, some outraged scientists resorted to comedy tactics. In one such attempt, a Polish scientist came up with a completely fictitious CV with the name Anna O. Szust ("oszust" means "fraud" in Polish). The CV consisted of a list of nonexistent positions and publications. The CV was sent to 120 journals on Beall's list, and forty journals quicky accepted "Dr. Szuszt" as an editor, many within a few hours of the submission of her CV. Some predatory journals don't even try to hide what they are doing. I recently received a bulk email from the journal *Pharmaceutics and Pharmacology Research* (published by "Auctores Publishing LLC, USA") that proudly boasts an impact factor of 3.02 and states that if I "aspire to become an Editorial Board member, the minimum qualification is a Ph.D. or M.D. degree."

In 2005, David Mazieres and Eddie Kohler (New York University and University of California, Los Angeles) became so annoyed with the constant bombardment of requests to submit an article to a sketchy journal called *International*

Journal of Advanced Computer Technology that they wrote a paper titled "Get Me Off Your Fucking Mailing List." The abstract and the body of the paper contained the same sentence, repeated over ten pages. The paper also contained two figures that were made up of repeats of the exact same words. The paper, in its full glory, is available to be admired on Mazieres' website at Stanford University.[28] But here is the scary thing. They actually submitted this manuscript to the above-mentioned predatory journal, and the journal accepted it—for a reasonable publishing fee. An anonymous reviewer even rated the manuscript as "excellent."

In another sting operation, jointly executed by John Bohannon and *Science* over a period of six months in 2013, Bohannon submitted about three hundred nonsensical papers to various suspected predatory journals that listed completely fictional authors and described the effects of nonexisting agents, including the painful-sounding compound "constipatic acid" that apparently "inhibits the growth of murine polyploid epithelial carcinoma cells." More than half of the journals accepted the bogus submissions, and most of them did so without any sign of peer review.[29]

I don't want to give potential fraudsters any new ideas, but with the emergence of artificial intelligence (AI) it will be even easier to create fictional meeting abstracts and papers. There are, in fact, embarrassing telltale signs of indiscriminate AI usage. Sometimes the AI-using authors forget to remove the last two words ChatGPT spits out ("regenerate response") from their copy-pasted ChatGPT product, and these two words end up as part of their paper—either as part of the actual text or as a curious subheading of a section. Then laughable explanations and corrections follow.

Using ChatGPT I tried my hand at creating a fictional meeting abstract related to the anticancer effects of an imaginary substance I decided to call "nanocapsulated zumbroid." ChatGPT had no problem with the fact that the substance in question does not exist. In response to my brief prompts, in a matter of seconds AI created a wonderful abstract that can be admired in Appendix 2 of this book. There are even several Free Academic Paper Generators online. Just type in the author names, add some key information on what you want your paper to say, and—hey presto—a manuscript is produced.[30] In this brave new world of AI, we all must learn a new word: "aigiarism."[31] This word means "plagiarizing using AI."

Nobody has any idea how to fight predatory publishers. The Federal Trade Commission tried the legal route in 2016. They sued an organization called OMICS group in India, and won. The group was supposed to "halt deceiving practices," was ordered to pay a fine of $50 million, and was told "to cease its deceptive business practices, including failing to disclose fees, misleading authors about the legitimacy of its journals and marketing conferences with star speakers who never agreed to participate."[32] But last time I checked they were still going strong: they still maintain an active network of more than five hundred journals and organize an impressive number of fraudulent international conferences—better known as "scamferences" (more on them later). And this is only one group; there are hundreds, perhaps thousands, of them operating in hard to reach/hard to sue jurisdictions and in dark corners of the internet.

"WE ARE THE MILLERS": CRIMINAL PUBLISHING GANGS

In the 2013 comedy *We're the Millers*, Jennifer Aniston and Jason Sudeikis depict the hilarious misadventures of a family of scamsters in the world of drugs and deception. But now meet a completely different group of millers: the science millers also live in the world of drugs and deception, but they operate in the "scientific paper mills." These criminal fake scientific paper generating organizations have become so prolific that they are responsible for a significant portion of the published literature. Paper mills employ ex-scientists who "went rogue." The workshops are, apparently, mostly based in China and Russia, although this is the type of activity that can be easily done by remote work from any corner of the planet. Paper mills are notoriously hard to track down, and often we only see their "products" and cannot expose the individuals working in them or running them. According to the best information available, paper mills work for paying "customers" who wish to have a paper produced for them in a certain field.

Let's say a customer "needs" a paper that will "show" the effect of some factor, e.g., a natural compound or a micro-RNA in some model of cancer. (This example is not random: paper mills often focus on somewhat peripheral and exotic areas of biology, for instance, miRNAs, lncRNAs, circRNAs, and noncoding RNAs.[33]) The customer pays, and the millers produce and deliver a paper with the requested findings.

The millers have amassed a database of Western blots, photomicrographs, pictures of various mice with various sizes of tumors, etc. Using collages of these datasets, they create totally fake results and write them up in the usual format to produce a complete paper—with Abstract, Introduction, Discussion, and References—*literally out of thin air*. Part of the service also includes publication of this paper in a peer reviewed journal with a certain previously agreed impact

5.11 "Publication Workshop": On the subject of paper mills and ready-to-order scientific articles . . .
Source: Leonid Schneider / ForBetterScience.com, reproduced by permission.

factor. No actual laboratory is involved, no actual experiments are needed. This is why everything in the paper can be *beautifully consistent*: the effects can be statistically significant, and the standard error bars look nice and tidy. If the paper mill does not use blatant image manipulation, does not reuse figures or photomicrographs from prior papers, and does not use plagiarized (copy-paste) text that can easily be detected by plagiarism software, referees, and editors may have a hard time telling the difference between a real paper and a paper-mill-generated fake one. A "high-quality" paper mill product can have the appearance of a solid real study, whereas a low-quality one can be visually detected (or at least suspected) by someone with a trained eye (figure 5.11).

The millers are paid in full only when the paper is published. If they can dupe a journal with a higher impact factor into publishing it, they can charge a higher price for the paper. This is why the "quality products" from the better paper mills can end up in real journals—not only in borderline or predatory journals.

To ensure smooth publication of the "product," the paper mill operators suggest "friendly" reviewers who are connected to the paper mill, or more often simply use fraudulent email addresses—ones that resemble real people and real institutions, perhaps with a single letter variation. This way the manuscript will not go to a real expert in the particular field of who may have the "Joe.Schmoe@harvard.edu" email address but ends up in the miller's own email box at Joe.Schmoe.harvard@gmail.com—this was the email address that the miller recommended as a "suggested referee" during the submission process. In other words, the millers themselves "take care" of the peer review evaluation of their own paper. Guess what? The miller/referee will rapidly accept his own fake creation; perhaps even congratulate himself for a job well done.[34]

Why then would any editor invite any suggested referee? Many of them, in fact, are not doing this anymore—for this reason.[35] Some editors will still invite them because they are inundated with submissions, and when dozens of expert referees all reject the task of looking at the manuscript, not many reviewer alternatives are left except the author-suggested ones.[36] In other cases, such as in the aforementioned hijacked journals or rogue guest editors, the whole publication process is rigged, with the editor colluding with the authors and referees to make it appear as if a legitimate peer review process had taken place.

Another approach to getting paper mill products published is the partial, temporary invasion of a (previously) legitimate journal in the form of "fake guest editors." Many journals publish focused series of papers that are organized and handled by an external guest editor, typically a significant expert in a particular field. As revealed by *Nature* in 2021, fraudsters have been impersonating real experts, and the temporarily hijacked fake guest editors, in turn, facilitate the publication of fake paper mill products—by the hundreds. Elsevier has recently retracted 165 "in press" articles and has three hundred additional ones in the works—all published in various special journal issues. Springer Nature has recently retracted sixty-two similar articles.[37]

Last year Roland Seifert, editor in chief of the well-respected pharmacology journal *Naunyn-Schmiedeberg's Archives of Pharmacology*, discovered that his journal fell victim to a "massive attack" of paper mill products.[38] The journal retracted ten papers that were already published and stopped the publication of another twenty. Seifert also published a twenty-point list of "warning signs" that should help others identify paper mill products at the time of submission.[39] These include some of the patterns already mentioned (e.g., fake referees, cut-and-paste images, and the inability to provide raw data when requested).

If a paper mill product is rejected by a journal, the exact same manuscript will be submitted to a different journal—but often with a *completely new set of authors.* Various websites and databases that are posted and maintained on data detective websites (see below) list various sets of retracted paper mill products—often several hundred at a time. One further indication of a paper mill product is when different papers, from seemingly different authors and institutions, have a similar-looking multipanel figure style. Unfortunately, this usually only becomes apparent after the discovery and takedown of an entire batch of paper mill products.

Clearly, there is a significant demand for fraudulent journal authorship—and consequently, there is a black market to meet the demand. In one version of this marketplace, an already accepted (most likely completely fake) paper is being shopped around on Facebook or on some sketchy website where the name of the journal and the topic of the "study" is announced: interested potential authors can place their bids. Package deals are offered: if a group of interested authors buys both a first and a last authorship, the price is reduced. Some paper mills quite brazenly advertise their products in the open. For example, a Russian paper mill called 123mi.ru proudly announces that they have published more than five thousand articles for more than twenty thousand satisfied customers over the last five years. They run many mirror websites, including a website called buy-sell-article.com for, well, buying and selling articles. Some of these websites look very neat, highly user-friendly, and, of course, very businesslike. They contain multiple papers in various topics that are being sold: each entry starts with the title of the paper, followed by an abstract, keywords, and the journal's impact factor (the journal title is grayed out—this is only available for those who pay). Underneath, neat green and red boxes show potential customers which authorship position is available and which one is already taken, and, of course, how much it would cost to buy a certain authorship position. Then you click on the red button that says "buy" and—hey presto—for the reasonable cost of $500 to $1,000 you have just become a coauthor on a soon to be published paper in a journal that will have a decent impact factor, will be published by an internationally known publisher, and will be indexed in PubMed. (Here, here, everybody come here! Papers for sale!! They are almost as good as the "real" ones! And for only a fraction of the price! And without any complicating factors, such as laboratory experimentation!)

Anna Abalkina, a sociologist from the Free University of Berlin, has analyzed more than one thousand journal authorship "advertisements" posted on the 123. mi.ru site and demonstrated that about half of them, as promised, produced papers—many of them published by major scientific publishers (figure 5.12).

5.12 For sale papers. The screenshots are from "science-publisher.org." The left-side example was subsequently published in December 2021 in *Frontiers Genetics*. It was retracted on September 4, 2023. The retraction notice states that "concerns were raised regarding the contributions of the authors of the article," which is quite an understatement. It also specifies that "the authors do not agree to this retraction," presumably because they have paid good money for the authorship positions to the miller, and they will never get their money back.

Source: Science-Publisher, "Co-Authorship in Scopus, Web of Science Research Papers." Website where authorships on papers were offered for sale, September 7, 2021 (but already removed and archived), https://web.archive.org/web/20210907141327/https://science-publisher.org/coauthorship/articles/; and Elham Zeinalzadeh et al., "The Role of Janus Kinase/STAT3 Pathway in Hematologic Malignancies with an Emphasis on Epigenetics," *Frontiers in Genetics* 12 (December 2021): 703883, https://doi.org/10.3389/fgene.2021.703883. Retracted article.

If one understands how the illegal marketplace works, one also understands why these papers contain very strange (i.e., suspicious) constellations of authors, often from diverse geographical locations, often with very different backgrounds, and without any evidence of prior interactions or collaborations.[40]

Recently, I received an email titled "Academy Collaboration Invitation from China" from an organization called "Alliance Academy of Science." After a few brief email exchanges, it became clear that this organization attempts to find willing scientists who can serve as facilitators to satisfy the authorship needs of paying customers. Two paths were offered: "Project Collaboration" and "Journal Article Publishing." I was provided with a handy series of tables that outlined the amount of money offered and the payment terms if I manage to publish something for their authors. Articles in journals with an impact factor of two to three would be rewarded by $3,000 to $4,000; but a paper published in a journal with an impact factor of sixteen or higher would pay as much as $24,000. In light of these numbers, it can be estimated that a single campaign, when a Special Issue in a midtier proper journal is guest edited and filled up with dozens and dozens of suspect articles, may be rewarded to the tune of $50,000 to $100,000—not a bad renumeration for a couple of weeks of . . . hmm . . . let's say "creative work."

Bulk retractions—when paper mill clusters are being simultaneously exposed and subsequently retracted—make the news with alarming regularity. The data detectives Smut Clyde and Elizabeth Bik discovered one that they affectionately call tadpole mill, with more than four hundred fake papers. In 2021, the Royal Society of Chemistry retracted seventy paper mill articles. In the same year, the editor in chief of *DNA and Cell Biology* noticed a series of paper withdrawals that came from the same IP address and used the same language. Sometimes entire journals are being shut down when it comes to light that they are heavily compromised by paper mills. Hindawi—which did not have a stellar reputation in the publishing industry to start with—had to close down four of its journals in 2023 (*Computational and Mathematical Methods in Medicine, Computational Intelligence and Neuroscience,* the *Journal of Healthcare Engineering,* and the *Journal of Environmental and Public Health*).[41] The Cambridge Crystallographic Data Centre had to flag nearly one thousand protein crystal structures because they were likely products of a paper mill.

Most recently, the data detective Smut Clyde has exposed a cluster of paper mill products showing various tumors, placed next to each other, and in each paper *the same ruler* was placed next to them for size reference. Not the same type of ruler—*the exact same ruler* with the exact same small indentations, smudges, and irregularities, in dozens of different papers, from different authors, from different laboratories, from different countries. This has to be the story of a magical space- and time-traveling ruler. Of course, the reality is less mystical: out of sheer laziness, the paper miller copy-pasted the image of the exact same ruler into an entire series of fake papers that were then sold to various customers.

In another instance, a one thousand plus article paper mill was discovered by Smut Clyde that was churning out fake papers on the imaginary effect of a class of molecules called metal–organic framework compounds. As he puts it:

> The mill's outputs are a curious hybrid of crystallography and medicine, where these fascinating meta-crystalline structures acquire therapeutic applications,

so that they meet the requirements of the mill's clinician customers. These compounds [depending on the paper milled paper in question] gain anesthetic properties, or kill cancer cells or bacteria, or stop inflammation.

In this particular case, the method of discovering the paper mill products followed an unusual path. Smut Clyde discovered that the paper mill uses a "stock" series of references, which have nothing to do with the actual topic of the papers. (This is done presumably to save time for the paper mill producer, but it is also possible that some of these are "sponsored" references that will boost the citations of some paying customer.[42]) By simply identifying those papers that cited a certain set of articles, Smut Clyde identified the whole cluster of the paper mill products. All of this makes one wonder if these papers—even though many of them were published by journals from reputable, major publishers—have gone through *any* real refereeing. Any referee who took even a cursory look at the list of references would have noticed the problem. It is, therefore, more likely that the paper mill products were not only written by fraudsters but that the refereeing process must also have been diverted to fraudsters—most likely the exact same ones who wrote the articles in the first place.

The neuropsychologist Bernhard Sabel and colleagues have come up with an ingenious method to identify potential fake paper milled papers (figure 5.13). Their method (called red-flagging) identified potentially fake publications based on common themes, such as the authors having a private rather than an institutional email address. Other potential identifiers of problems included "international coauthors" and "hospital affiliations." (Hospitals, i.e., not a university or a research institute. The presumption here is that hospitals, in many cases, do not have a research laboratory where the work described in the paper could have been carried out.) If all three of these themes were applied, the method had a 90 percent sensitivity to detect fake papers—although it also had a 37 percent false alarm rate. Sabel concludes that more than 350,000 fake papers are being published each year (!!!). He estimates that 34 percent (!!!) of neuroscience papers published in 2020 were likely fake or plagiarized; in medicine, he estimates a 24 percent incidence.[43] Assuming that each fake publication generates a $5,000 to $10,000 profit, the size of the worldwide paper mill business is at least $1 billion, but possibly as much as $3 billion.[44] In an interview for National Public Radio (May 14, 2023), Sabel calls this "probably the biggest science scam ever."

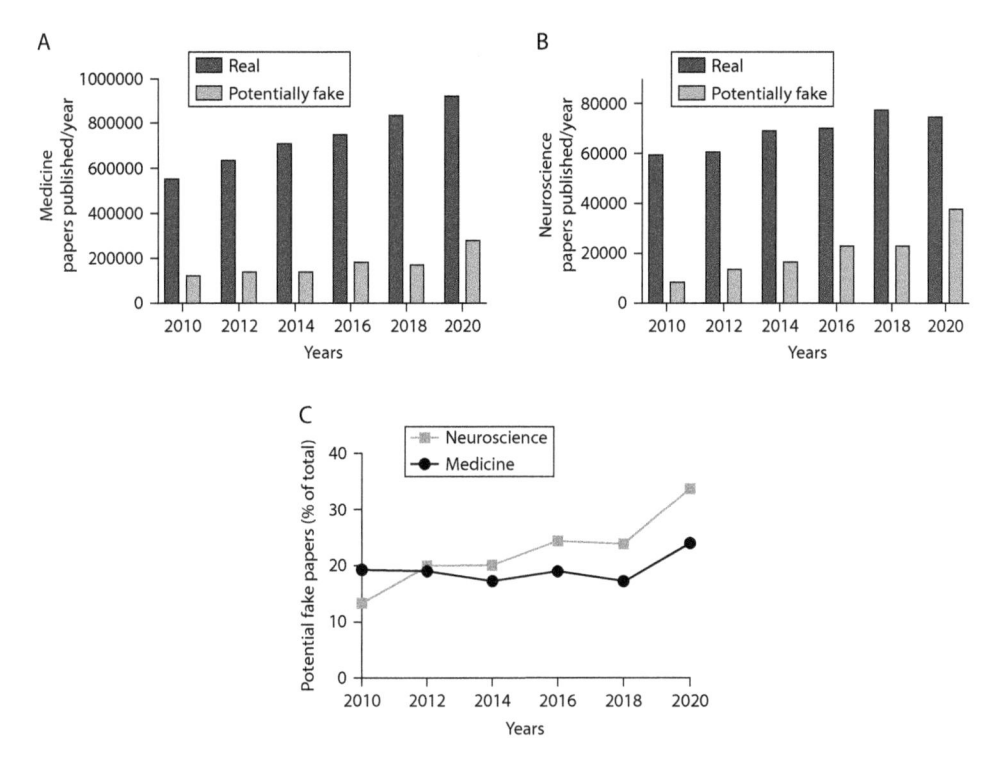

5.13 Potentially fake papers identified by the "red-flag method." (A) Real and potentially fake papers in medicine published per year. (B) Real and potentially fake papers in neuroscience published per year. (C) Percent of potentially fake papers in medicine and neuroscience.

Source: Bernhard A. Sabel, Emely Knaack, Gerd Gigerenzer, and Mirela Bilc, "Fake Publications in Biomedical Science: Red-Flagging Method Indicates Mass Production," MedRxiv, October 18, 2023, https://www.medrxiv.org/content/10.1101/2023.05.06.23289563v2.

No paper mill industry would exist if there was no demand for it. Its proliferation clearly shows that there is a massive worldwide demand for published papers. And, apparently, almost *any* paper will do. It can be a real one, or it can be a fake one—for some people, it seems, any publication is okay, as long as it has their name on the front page and is published in a journal with an impact factor. Because the demand exists and because the business is lucrative, scientific fraudsters keep coming up with new innovative ways to infiltrate the scientific publishing process.

In my estimation, the sole purpose of most paper mill products is to satisfy a certain advancement in an unscrupulous person's career. Most paper mill products do not rise to a level that can really influence the direction of science—although I am fairly certain that many realistic-looking mill products remain undetected. Thankfully, most mill products don't directly threaten human lives. Most of these papers, after all, are only read by the authors and the referees—and, as we have seen, these two can be one and the same—and never receive any citations or follow-up.[45]

But none of this changes the fact that these papers are blatantly fraudulent, pollute the scientific literature, and cost a lot of time and effort to detect and correct. And, of course, they represent the worst kind of farce (or perhaps tragicomedy): they fly in the face of everything we value and respect in science.

PLAGIARISM, CRYPTOMNESIA, CITATION BIAS

In addition to the inclusion of doctored or nonexistent data into a scientific paper, there are many other ways to distort a paper. The most common one is plagiarism, where the authors of a paper lift various parts of someone else's writing and present it as their own. There is also self-plagiarism, in which authors are reusing the same text in multiple papers. The standards of plagiarism have changed over time. For instance, twenty to thirty years ago it was acceptable—in fact, it was preferred—for authors to use the exact same text in the Methods section of subsequent papers, if, indeed, the exact same methods were used. This way—everybody thought—readers will have an easier time identifying the method. These days, even this is considered to be plagiarism; the way to avoid it is just to cite a previous paper that contains the method rather than pasting it into the Methods section again.

One day in 2023 I saw, to my great surprise, that a paper from my own group, originally published in the *British Journal of Pharmacology* in 1998, seemed to have gotten a new life.[46] It was republished, word for word, in 2022 in a journal called *National Journal of Pharmaceutical Sciences*, and again in the same year in a journal called *Research & Reviews: Journal of Pharmacology and Toxicological Studies*, and a third time again 2023, in a journal called *Indian Journal of Pharmacy and Drug Studies*. All of these papers were "written" by a set of four authors from India whom I had never heard of before. There were some slight variations in the titles (for some titles, the sequence of the words in our original title were mixed up

A BROKEN SCIENTIFIC PUBLISHING SYSTEM   177

5.14 Examples of "carbon copy" papers. Our original paper was published in 1998 (top left panel shows part of the first page). Several plagiarized versions of the same paper were published in 2022 and 2023 (part of each paper's first page is shown).
Source: Author's own data.

almost beyond comprehension), but the abstract and text, and the figures, and the list of references is exactly the same as our original paper (figure 5.14). The only difference is that the last figure we had in our 1998 paper, which were some histological and immunohistochemical pictures, was omitted from the copycat papers, probably because the plagiarists did not want to pay the extra fee that usually comes with publication of such illustrations in most journals.

Our attempts to contact the original journal or the various "new" journals did not result in any response. This is truly a remarkable case. The four "authors" from India managed to commit blatant plagiarism and repeated self-plagiarism *at the same time.*

The percentage of plagiarized text in the biomedical literature is difficult to determine, but it is likely to be substantial. Of course, the term "plagiarism" encompasses a whole range of activities, from mild to severe. It can range from copy-pasting sentences from others' work, or paragraphs, or sections of reviews into dissertations, and even republishing entire papers under someone else's name. According to one survey focusing on papers published by authors from Dutch universities, the extent of self-plagiarism (recycling of some of the authors' own work in subsequent papers) is about 3 percent in biochemistry (which is much better than the 14 percent found by the same authors in papers in economics).[47] Other surveys and estimations are somewhat higher and believe that about 20 percent of the published biomedical literature may contain at least some degree of plagiarism or self-plagiarism.[48] In a meta-analysis of multiple surveys—which mainly included biomedical scientist cohorts but also incorporated a few surveys from other disciplines, such as economics—only about 2 percent of the respondents admitted to such behavior, but the same respondents state that they have witnessed about 20 percent of their colleagues of doing it.[49]

From time to time, high-profile cases appear when political figures are busted for plagiarism or other forms of scientific misconduct, often in their doctoral theses, which were often prepared decades earlier. Several German politicians had to resign when their copy-paste jobs came to light, including the former German defense minister Karl-Theodor zu Guttenberg and the former vice president of the European Parliament Silvana Koch-Mehrin (both of them lost their PhDs and resigned in 2011), the former education minister Annette Schavan (her PhD was rescinded and she was forced to resign in 2013), and Franziska Giffey, Germany's former minister for Family, Senior Citizens, Women, and Youth, who resigned in 2021. Even Ursula von der Leyen, the former minister of defense for Germany, and the current president of the European Commission has been criticized for "mistakes" and "inconsistencies" in her doctoral thesis written in the early 1990s. There is a website and computer system set up to detect duplications and plagiarism for German authors (vroniplag.fandom.com) where hundreds of plagiarism cases are painstakingly documented. The Hungarian president Pál Schmitt had his PhD revoked for plagiarism and had to resign in 2012. In Russia, thousands of politicians have plagiarized their diploma theses, including Vladimir Putin, who appears to have plagiarized several sections of his 1996 thesis. In early 2024, the

Harvard president Claudine Gay was forced to resign after multiple instances of plagiarism were revealed in her academic papers. Many of these plagiarism cases are not in biological or medical sciences, but are in law, economy, sociology, or sports policy. But some of them are, and these include Ursula von der Leyen's thesis (medicine).

How much does plagiarism contribute to the unreliability of the scientific literature? It depends on the type of plagiarism involved. If the plagiarized material consists of copy-pasted (but fundamentally correct) statements, although unseemly and improper, it is less corrosive than a completely fictitious paper mill product. Even a complete republication of somebody else's article won't do much harm to the overall body of the literature—if the original article happened to be correct.[50]

Plagiarism is annoying, of course, especially from the standpoint of the original author. It can also end the career of individuals, as we have seen. Naturally, we must fight it. But in the grand scheme of things, I believe that it is not plagiarism but the many other problems discussed in this book that lie at the heart of the current reproducibility crisis.

I have already discussed how the current, anonymous peer review system (of publications and grants) can create problems when a dishonest reviewer picks up a new, emerging concept or idea from the reviewed material and pursues it as their own. A related phenomenon is cryptomnesia (sometimes termed "inadvertent plagiarism"), when a forgotten memory/idea returns later on as one's own. The phenomenon itself is well-documented in the psychology literature.[51] It is very likely to happen as part of the biomedical scientific process, but to my knowledge it has not been systematically studied.

Several other phenomena are somewhat related to the topic of plagiarism and are unbecoming and annoying—and may be potentially viewed as minor contributors to the reproducibility crisis. One of these is *selective citation*, in which an author intentionally and selectively leaves out literature references from the paper if they do not support the overall "story" (figure 5.15). Yet another way—unfortunately, very common these days—is to ignore prior publications on which the article is built to increase the perceived novelty of the current work. A variation on this theme is to cite the prior work, but not in the Introduction section, where it belongs, but hide it somewhere, usually in one of the last paragraphs of the Discussion section, or cite it without fully explaining that it contains relevant *prior art* to the current paper.[52]

5.15 On the subject of literature citations . . .
Source: Hagen / CartoonStock.com, reproduced by permission.

Incorrect literature citations can also come from the peer review process itself. In a recent survey conducted by *Nature*, about two-thirds of responders said that they have been subject to coercive citation practices, where referees or journal editors seem to pressure them to cite unrelated or superfluous work.[53] These requests can stem from the referee's lack of knowledge of the field, or may represent an effort by the referee to boost their citation record, or—if the suggested citation is from the same journal where the article is being considered for publication—it may be an effort by the editor to improve the impact factor of the journal. As long as the peer review process remains anonymous, this type of shenaniganry will persist.[54]

Fake articles, especially paper mill products, can also be detected by looking at their language. There are some very interesting discussions about common *tortured language* usage in paper mill products.[55] The route by which various strange and inappropriate—not to mention, ridiculously comical—words and expressions end up in these papers is through the action of synonym generators. These are used by the millers to avoid detection by plagiarism detector software. In the bizarre world of paper milled articles, "logarithmic growth phase" becomes "logical" or "logistical" growth phase (or even "growth phage"); "final density" becomes "ultimate density"; "serial dilution" becomes "serious dilution"; "successful synthesis" may turn into "triumphant synthesis"; and "procedure" into "preformation."[56]

Yet another form of language-related misconduct is *linguistic obfuscation*, i.e., the use of confounding language, jargon, and unnecessarily complicated ways of writing. It may simply be an author's attempt to make the article sound "more scientific," but it can also be used to hide problems or inconsistencies in a paper. A recent analysis has proven that linguistic obfuscation is more frequent in fraudulent papers than in honest ones.[57]

Thankfully, computer programs can identify similar or identical texts, and all self-respecting journals use these programs right away when a manuscript submission is received. Of course, low-level predatory or gray-zone journals have no problem with publishing, for money, whatever comes their way.

As far as the legitimate portion of the biomedical literature is concerned, due to the plagiarism detection software usage, we can expect that the incidence of the straight-up copy-paste type of plagiarism will decrease in the literature in the future. However, whether this will result in a significant improvement in the reproducibility of the actual material published remains to be seen; I happen to be rather skeptical in this regard.

CONFERENCES, SCAMFERENCES

In the old days, scientific conferences were supposed to serve as vehicles for scientists to inform each other on the progress made in a given field and to foster an exchange of ideas to facilitate progress in the field. Over time, this role has diminished as scientific meetings became places where investigators hunt for ideas and new concepts to be, let's say, *acquired*.

A famous incident, well documented in a BBC documentary in the early 1990s, relates to the identification of the mysterious factor EDRF as nitric oxide.

In chapter 1, I related the story of a major paper published in *Nature* in 1997 demonstrating that EDRF is, in fact, nitric oxide. Salvador Moncada, the lead author of the paper, however, did not receive a Nobel prize, which was puzzling to many people in the field. As it turns out, there was a vascular biology meeting held in July 1996 in Rochester, Minnesota, where Robert Furchgott presented his latest findings, which strongly suggested that EDRF is in fact nitric oxide. Moncada was sitting in the first row at his lecture. He did not ask any questions but went back to his laboratory and conducted his own experiments, which were published in *Nature* in June 1997. The paper does mention Furchgott's suggestion at the beginning of the article. But Furchgott's paper—which was a book chapter of a symposium and not a high-profile paper in a leading peer reviewed scientific journal—was only published one year later, in 1998. Most people reading the literature connected Moncada's group with the discovery and were puzzled when Moncada was not included among the list of nitric oxide Nobelists—only Furchgott.[58]

In the mid-1990s, I learned my own lesson on what scientific meetings are for. At a large meeting in Japan, I proudly presented some recently published data that my group generated on the biological roles of a particular enzyme we had been working on. Toward the end of the talk I included some recent data we had generated in the lab but was not yet published. At the meeting, I talked to an investigator who had a higher-level model of the same disease condition. I suggested a collaboration and even mailed some compounds to him when I returned to my laboratory. The understanding was that our group would publish our paper first, and this would be followed by a joint paper using the higher-level model. Instead, the other investigator went ahead and conducted a series of quick studies and published these results. While our paper was going back and forth with referees in one journal, the other investigator's paper had smoother sailing in another journal—where, at the time, prominent editorial board members could simply "communicate" articles, which would then be rapidly accepted—and was published before ours.

Because of this experience, I decided to *never* disclose any unpublished material at future scientific meetings. In fact, disclosing a new idea or presenting an unpublished set of data at a meeting is something that the vast majority of scientists will not do anymore. Especially now, when everybody has a cell phone, and scientific lectures turn into Taylor Swift concerts, and every second person holds up a cell phone and takes a video or a series of photos of each slide presented—even if the speaker explicitly asks the audience not to do that. *Especially* in cases like that.

When specific questions are asked at the end of talks, most investigators give evasive or low-information answers—*especially* when it comes to a new idea or to work in progress. Instead of a climate of collaboration, we now have a climate of competition and stonewalling.

Why, then, do scientists still attend scientific meetings? To be honest, I am not sure anymore. If one is invited to give a plenary lecture at a significant meeting, it still carries some prestige I guess. There is some value in traveling and being away from the day-to-day hectic laboratory life and organizing your thoughts and ideas. There are also opportunities to meet collaborators from other labs and other countries. Many meetings feature commercial exhibitions where new techniques or new equipment may be viewed. Sometimes these exhibition stands are useful; at other times they are rather overwhelming. At large meetings that both basic and clinical scientists attend, such as the annual meeting of the *American Heart Association* or the *American Diabetes Association*, the difference is truly grotesque. For example, compare a basic science poster section (typically a poorly attended event in which postdocs are standing in front of their average-looking posters hoping a potentially interested colleague passes by to whom they can show their science) with the flashy, professionally designed, castle-like contraptions of commercial exhibitors (plush carpets, professional lighting, sharply dressed, confidently smiling company reps, freebies such as pens, notebooks, laser pointers, drinks and food, and raffles, etc.).

At least those conferences are real ones. In parallel to these, almost like a parallel universe, an entire industry of fake conferences ("scamferences") has sprung up in recent years. Often these are organized by the same groups that are running predatory journals. I receive email invitations to such events, literally *every hour*.[59] In some cases, the bots that push the meeting pick out a publication from the literature, take the title of the paper, find the affiliated email address, and send an "invitation" ("scamvitation"?). Some of these emailers are relentless; they never seem to stop. Many of them are written in atrocious English; many of them are not even addressed to me (because the bot took down the wrong info from the internet), and almost all of them are totally unrelated to my research field. Unsubscribing from the email lists or blocking the spammer does not help.

While writing this book, I examined several scamference websites. They list various so-called invited speakers and conference organizers. I contacted several of these scientists, and it turned out that—in most cases—they had absolutely no idea that their name and face were being used to advertise the meeting. In some cases, the scamference organizers listed completely made-up names and institutions as their organizing committee. In other cases, they posted the names and affiliations of real, but recently deceased, scientists. As discussed earlier, the exact same tricks, with some variations, are used for the so-called editorial boards of most predatory journals. This is no coincidence because the fraudsters who stand behind the scamferences are often the same ones who run the predatory journals.

If a real scientist attends a scamference, that person might find that the program has been completely redone at the last minute and various unrelated sessions have been patched together. The unfortunate attendee is subjected to the mockery of a session in which a series of talks, from marginal-quality speakers representing wildly diverse fields, takes place. Scientific value: zero. For some scamferences, once the registration fee is pocketed no actual meeting is held, and the organizers ride off into the sunset—never to be found.

In 2019, Ruari Mackenzie, a journalist for technologynetworks.com, attended one suspicious conference and wrote an article describing his experiences.[60] The event was organized by "Conference Series," which is part of the Indian OMICS group of $50 million penalty fame discussed earlier. The meeting was called the "23rd International Conference on Neurology & Neurophysiology and 24th International Conference on Neurosurgery & Neuroscience" in Edinburgh, Scotland. Registration and the overall organization were completely disorganized, from the beginning. The keynote speaker, Professor Koji Abe, of Okayama University in Japan, gave an opening talk but slipped out for the rest of the meeting. The sessions consisted of a hodge-podge of topics and speakers, without any organizational principle and without following the originally announced schedule. Many speakers did not turn up, and random attendees were asked to chair various sessions. Mackenzie wasn't able to track down most of the scientists who were, supposedly, the organizing committee of the meeting. Some of them had never even heard of the meeting until they got Mackenzie's call. Other organizers were real; they were incentivized by being offered various positions on journals (predatory, of course) that were also part of the OMICS empire.

The last time I checked, the scamference series was going strong, with the most recent meeting held in London in the spring of 2023, and the next one was scheduled for Zurich in 2024. As you can learn from their website:

> Neurophysiology 2023 participants include Professors, Researchers, Physicians, Business Delegates, Scientists, students all over the world. The goal of

the Neurophysiology Meeting is to provide you with a rare opportunity to build on what you already know in the research and development of new therapies and techniques and accelerate efforts to enhance health and well-being.

This vacuous statement was followed by literally many hundreds of (!!!) professional societies they claim are "Related Societies and Associations," even though, very clearly, they have absolutely nothing to do with the scamference. The (fake) show must go on. In Appendix 3, I include a handy tool prospective scamference organizers can utilize when formulating their next harassing (I mean, invitation) letters.

Who attends scamferences? A few misguided junior scientists may have fallen prey to an invitation—and they will be in for a surprise. But I am fairly certain that most attendees *know perfectly well* what type of meeting they are attending. It is a way for them to have their institution pay for travel, hotel, and meals in a tourist destination city in a far-away country. In other words, the name of the game is embezzlement of grant or institutional money—but with some degree of plausible deniability.

Judging by the number of meeting invitations and meetings listed online, scamferences are a thriving industry. Their scientific value is close to zero. Moreover, abstracts or conference proceedings released by these meetings further pollute the scientific literature. In addition—needless to say—they divert time and money from actual research.[61]

RETRACTIONS AND THEIR CONSEQUENCES

If the material published in a scientific paper proves to be problematic, various courses of action can be taken. If the problem is localized to a certain part of an article, the authors can request a correction, which means deletion of a problematic part of the article (typically, parts of the Results section, e.g., some figures) and replacing it with—what they claim to be—the "correct" material. In some cases, these corrections are so extensive that the word "mega-correction" is used (figure 5.16).

In other cases—typically when the nature of the problems is such that pretty much everything stated in the paper is called into question—the entire paper can be retracted, i.e., pulled out of the scientific literature. This does not mean,

5.16 On the subject of article corrections...
Source: Leonid Schneider / ForBetterScience.com, reproduced by permission.

however, that the paper disappears into thin air: it just means that an article called Retraction Notice is published in the same journal, which notifies readers that the paper in question has been retracted and should not be considered as valid information anymore. In many cases the pdf of the paper is reprinted with a big red RETRACTED watermark pasted diagonally over each page (figure 5.17). Sometimes the retraction is initiated by the senior author and cosigned by some (or, in rare cases all) authors. Other times—e.g., when a series of paper mill products comes to light—the journal's editor initiates the retraction. The main reason for retraction is, of course, scientific misconduct of some type, either research misconduct (such as image fraud) or publishing misconduct (such as plagiarism). It is rare that a retraction comes from an investigator who discovered some systematic problem in the laboratory or evidence of prior misconduct.

For reasons that are not entirely clear, there is also something called an "Editorial Expression of Concern," a sort of scientific no-man's-land category. This is published when the journal discovers problematic material in a publication and wishes to warn the readers but does not wish to proceed with a correction or a retraction for whatever reason.

5.17 The first retraction ever...
Source: Leonid Schneider / ForBetterScience.com, reproduced by permission.

The number of retracted articles has seen an unprecedented increase over the last decades. In 2010, there were about five hundred retractions. In 2020, there were about 3,500.[62] In 2023, the number of retractions increased to a staggering ten thousand. PubMed lists more than thirty thousand articles as retracted. In other words, the vast majority of retractions have appeared in the last few years. In the field of cardiovascular research, the number of retractions increased about tenfold over a ten-year period, with the majority of the retracted papers coming from Chinese and American laboratories.[63] Some clusters of retractions come from repeat offenders, i.e., authors who seem to have built entire careers based on fraudulent/bogus materials but eventually have been caught. Other clusters come from busted paper mills, where hundreds of papers are often retracted at the same time. But retractions affect essentially all disciplines, all journals, and all types of investigators, including Nobel prize winners.[64]

Although there is a remarkable increase in the number of retractions, the process can take many months or even years to complete. I recently saw a paper retracted in which suspect images had been reported already back in 2014.

All too often it can take many years for a journal to retract a paper. Obviously, nobody is fond of retractions: it does not look good for the authors of the paper, and it does not look good for the journal's publisher, editors, and referees, either. It is not surprising, therefore, that retractions can take a long time—often several years. As discussed earlier, for a long time there were no clear guidelines of what constitutes acceptable manipulation of figures, and what constitutes image fraud, and what the consequences should be: correction, mega-correction, or full retraction. Also, in many cases not all authors agree with the retractions; in other cases, the retractions may happen in the context of some data integrity investigation at the authors' workplace—another process that, on its own, can also take quite some time.

Just because a paper is retracted does not mean that it disappears from the literature. Other articles—published prior the retraction—may still cite it. Even after the retraction is published, an undead "zombie paper" may continue to misdirect the field and may even collect further citations. Sometimes retracted papers may even find a new home later on in a different journal. A couple of authors may be missing from the new version, suspicious Western blots may be replaced with better-looking ones, but the material essentially remains the same.

There are many different analyses related to retracted papers. Some observers estimate the money wasted that went into the work that ultimately had to be retracted. (Unsurprisingly, we are talking about a lot of money.[65]) Others look at the career progression of the investigator who had to retract various papers. (Unsurprisingly, the productivity and the extramural funding of the investigator started to decline in the years prior to publication of the retracted work, and even more unsurprising, it continued to decline thereafter.) Yet others look at the careers and career prospects of the investigators who had to retract papers. (Again, unsurprisingly, their careers suffer: their papers receive fewer citations, their career opportunities diminish, and they are subject to different types of social penalties and restrictions.[66])

The analysis that is missing, however, is how to determine the percentage of published material that is questionable, fraudulent, or irreproducible, and yet remains unretracted. When we consider the massive amount of questionable and fraudulent material being released each month, which is discussed elsewhere in this book, retractions—which represent less than 0.1 percent of the published biomedical literature—represent nothing but a *small tip of a large iceberg*.

HOW TO READ A SCIENTIFIC PAPER?

Because of all the problems I have discussed in this book, one has to use a healthy dose of skepticism when assessing any information related to science—from an article presented in a lay journal to a paper presented in the scientific literature. When it comes to articles written by science writers in a daily newspaper, one must immediately assume that the article is overhyping and overinterpreting a finding as the result of a joint effort by the journalist and the scientists involved. It is all too common to talk about new cures of a disease, for example, when the actual article is a laboratory study in mice or in cells in a culture. Journalists tend to make enormous jumps from a cell-based study where some drug mixture, apparently, modifies some hallmark of aging in a cell to touting a new miracle medicine that reverses aging in humans. The best way to go about this is to try to find the actual primary article and take a good look. Of course, this is easier for people with some level of training in biology and science than for people without that background.

When it comes to assessing primary scientific articles—the type that starts with an Introduction and has Methods, Results, and Conclusions—a few things can be immediately assessed. One of them is the list of authors: are they from a well-respected organization, and what sort of reputation do they have? Do they have any past controversies associated with them, or a history of retractions? Do they have any conflict of interest with the subject matter? For instance, if the authors have a spin-off company related to the study subject, they might be more exuberant about the findings than are justified.

Also, one can look at the journal where the findings were published; if one is not familiar with the journal, and if the journal seems to have no impact factor, or a low impact factor, it can be a warning sign. Typically, journals with a longer history, or journals affiliated with a respectable professional organization, are generally accepted as more trustworthy. If the paper was published in an open access journal, or in a journal that has been flagged as predatory or near-predatory (in the so-called gray zone of journals), this should also raise some red flags. One can look up the editorial board of a journal and determine the quality of individuals involved in it.

One can also look at the time that passed between the submission of the paper to the journal and the acceptance of the paper (this information is usually on the cover page). If the time between these two events is very short (e.g., less than a week), it can be suspicious; it may indicate an overly "friendly" refereeing process.

Even if the data published seem convincing and the article seems to check all the right boxes, it must be kept in mind is that this is one single study. If the paper was published earlier, one could look for similar articles to see if the findings have been confirmed by others—for example, by looking up subsequent articles that cite the paper in question. One can also look for comments on data detectives' websites and on PubPeer.

After that one can also look at the actual information contained in the paper. (At this stage, the more biology/statistics/research background one has, the better.) One can look for the key design aspects of the study that were discussed earlier in this book: blinding, randomization, power analysis, etc. One can try to spot any image duplication, suspicious-looking blots, or other potential problems in the figures. One can look at the effect sizes: an effect may be statistically significant (signified by one or more asterisks the usual way) but may not be all that impressive in terms of a percent increase or a percent decrease. And, of course, one can look at the model systems used—have they used the right type of controls and have they been properly validated? Also, how close or far are they from the human relevance of the study? For example, a study in a cluster of cells or in a model organism (worm, zebrafish) is typically further away from a human disease or condition than a mouse model.

And here is one more tip: If it is too good to be true, it is probably not true. Be especially skeptical when it comes to claims around breakthrough papers related to life extension, rejuvenation, antiaging (figure 5.18), downloading a human brain's content to a computer, or pretty much anything that contains the words "cleansing," "toxins," "acidification," or "alkalization."

None of these assessments are 100 percent foolproof, of course. It may well be that a flashy paper in a top journal from a brand-name institution becomes problematic or irreproducible later. In any case, a healthy dose of skepticism is recommended throughout the process.

One must also be cautious with review articles. In theory, papers that aggregate the results of multiple studies (meta-analysis papers) are supposed to be more reliable than any individual study, and as a rule of thumb, when such material is available, it should always be considered. It is still not 100 percent foolproof, of course, because the meta-analysis can only be as reliable as the individual papers

5.18 Scientists devise yet another way of delaying death.
Source: Nick Kim / scienceandink.com, reproduced by permission.

it is built on. A meta-analysis of dozens of biased, flawed papers may not bring much better conclusions than the original papers. One should also be cautious with the so-called state-of-the-art review articles. In these articles, the generally agreed consensus in a certain field is presented—supposedly. But, in many cases, review articles reflect the opinion or bias of the authors who typically work in the subject field, and they are likely to emphasize data that supports the authors' line of thinking and may be critical or dismissive of studies that go against their concepts. In some cases, nonscientific issues—political and geopolitical matters—can color the opinions expressed in a (supposedly) scientific article, which was exemplified in a recent review article published in *Nature Medicine* on the origins (figure 5.19) of COVID-19.[67]

With review articles, one should also keep in mind that there is a lot of ghostwriting going on. Many pharmaceutical companies have sizable ghostwriter units that prepare entire review articles, which then are presented to willing academics, who submit these papers under their own names and affiliations.[68] This type of shenaniganry is mainly related to reviews of marketed drugs, but it can also

5.19 "I've been sent home to lay low. If anyone asks, no one in my virology lab knows anything about a rampaging army of pangolin-bat hybrids."
Source: Nick Kim / scienceandink.com, reproduced by permission.

happen with basic research topics. A pharma company that has a product in a certain area may do some "astroturfing" (or "disease-mongering"), arranging for many articles and papers on that product to come out in the scientific literature (which, then, will seep into the lay literature)—with the goal of drawing emphasis to a particular pathway or disease, and in turn, of course, selling more drugs for its management.

Ghostwriting is almost impossible to detect just by looking at a paper because many authors do not declare their conflict of interest, even though they are supposed to do so when the manuscript is submitted. However, reading an overly enthusiastic or one-sided summary of a brand new drug—perhaps by an academic author with no previous scientific activity on the subject—should always be taken as a warning sign.

CHAPTER 6

THE WAY FORWARD

WHERE IS THE OVERSIGHT?

The day-to-day task of research integrity oversight is placed on the principal investigator, the head of the laboratory, i.e., the person who is running the research operation. This person, typically, is intimately involved with the research and has a key role in planning, designing, supervising, analyzing, and interpreting the findings. These individuals are easy to identify: in the vast majority of the papers they are the authors of correspondence, and they are the last author listed. At the institutional level, institutions may have different "offices of research integrity" that organize integrity training courses and investigate potential issues that are brought to their attention. Grant-giving organizations also have research integrity offices, for example, the NIH maintains the Office of Research Integrity (ORI), which investigates and publishes a small number of cases every year.

Another level of integrity oversight is placed at the level of the scientific journals. It is, of course, the fundamental interest of any self-respecting journal to publish material that is of high scientific standards (or, at the very least, material that is not intentionally fraudulent). As the reproducibility crisis has come to the attention of the scientific world, the journals have begun implementing various checks, including plagiarism detection (which is easy and pretty much standard today) and image fraud detection (e.g., analyzing Western blot irregularities). Nevertheless, many journal editors and most referees don't think it is their job to act as the scientific integrity police. Most referees assume that the material to be evaluated is not intentionally fraudulent, and their job is mainly as assessors of scientific novelty and quality.

It appears that maintaining research integrity is, on one hand, the duty of everybody and, on the other hand, the duty of nobody. It is also clear that a lot of the stringency comes as a reactive process (e.g., in response to whistleblower reports) rather than being a proactive process aimed at improving standards and preventing misconduct.

In recent years, several individuals have set up databases or initiated websites that deal with the topic of scientific misconduct, fraud, and retractions. The first website focusing on fraud in science—at least the first one that I came across—was called science-fraud.org, and its attention was devoted to "Highlighting Misconduct in Life Sciences Research." It began in 2012 and was maintained by an anonymous investigator, who clearly had a background in biological science and who dedicated a significant amount of time to identifying problematic papers and investigators with clusters of problematic papers. The mysterious blogger used the name "Frances deTriusce" (an anagram of "science fraudster") and had a clear talent for science "sleuthing." Whoever was maintaining this site was extremely frustrated with the magnitude of obvious misconduct in biomedical science and had good intentions, and this person was providing an important public service. The website focused on clearly fraudulent image duplications, inappropriate Western blot splicing, unacknowledged recycling of data, and identifying repeat offenders. Some of the people identified were powerful scientists who worked at brand-name institutions, and one was a Nobel prize winner. The author of the blog did not mince words: used the word "fraud" when this appeared to be the case, and sometimes also used satirical and crude language to make a point.

After about six months of operation, the blogger's identity was revealed to be Paul Brookes, an associate professor at the University of Rochester. In a 2014 interview titled "Paul Brookes: Surviving as an Outed Whistleblower," Brookes talked about the legal threats he had received, his need to hire a lawyer to protect him, and the reaction of his university to the entire affair.[1] Clearly, the university was "not pleased" and did not offer any support or protection for him. The blog was the first to call out by name many researchers who had published strings of problematic papers. In the face of various legal challenges and impeding lawsuits from scientists he had written about, after a year of operation Brookes had to shut down his site.[2] A small victory for Brookes is that none of the six scientists who sued him for defamation won their lawsuits, and all of them ended up retracting several papers. However, about 80 percent of the problematic material identified on his site remains uncorrected and unretracted.[3]

The longest-running currently active website that focuses on scientific integrity is Retractionwatch.com, which was founded by Ivan Oransky and Adam Marcus in 2010. Oransky has a background as a medical doctor and held editorial roles in various biomedical journals (*Medscape*, *MedPage Today*, *Scientific American*, and *The Scientist*) throughout his career. Using donations (e.g., from the MacArthur Foundation), they have built an institution called the Center for Scientific Integrity. This center maintains the Retractionwatch blog, and its motto is this understatement: Retractions provide an "insight into the scientific process." Retractionwatch scrutinizes retractions in biomedical science and maintains a searchable database of retracted articles that has grown to more than forty thousand papers. Among those, one can admire the top ten most retracted authors, including the German anesthesiology researcher Joachim Boldt (194 retractions) and the Japanese anesthesiology researcher Yoshitaka Fujii (172 retractions). Retractions can also be analyzed by country. The highest rate of retraction (fourteen out of ten thousand published papers) comes from Iran, followed by Romania, Singapore, India, Malaysia, South Korea, and China. Retractionwatch does not proactively identify problematic articles, instead it focuses on closed cases—i.e., articles that have already been retracted. This policy keeps them out of the trouble (e.g., lawsuits) that Paul Brookes' blog was subjected to. They also publish articles related to various emerging issues in scientific integrity and publish a weekly reading list of articles on that subject.

Pubpeer.com navigates in more controversial waters. The associated entity (PubPeer Foundation) is registered as a nonprofit, public benefit corporation, and it defines itself as an online journal club. (Journal clubs are a regular academic activity at most universities. A group member carefully reads, scrutinizes, and presents a recently published scientific article to their research group, and the group discusses the merits and demerits of the paper.) PubPeer was cofounded in 2012 by Brandon Stell, a neuroscientist working at the French National Center for Scientific Research in Paris, France. The originally stated goal of the site was to "enable scientists to search for their publications or their peers publications and provide feedback and/or start a conversation anonymously." At a hearing of the U.S. House Committee on Science, Space, and Technology in 2022, Stell discussed his experience with the site.[4] In the beginning he "naively thought that comments would be similar to scientific discussions." However, to his surprise, "the overwhelming majority were about more fundamental problems with data. An astonishing number of comments began flooding in that pointed to serious flaws in articles." Stell asserts that "numerous regular PubPeer users are now clearly more expert than journal and institution staff when it comes to the forensic examination of the literature." Anonymity provides protection against lawsuits aimed directly at the commenters, but the site itself has been subject to various threats of lawsuits over the years. PubPeer has flagged many problematic articles that later came into focus for the general public—including articles by Marc Tessier-Lavigne, the embattled ex-Stanford president discussed in chapter 4.

Even a cursory look at the comments on the site can be a sobering and depressing experience: fraudulent pictures, manipulated images, and clearly "doctored" data, in a seemingly endless succession. My conservative estimate is that PubPeer currently contains at least fifty thousand flags (!!!) highlighting some type of "image irregularity" or visible evidence of intentional fraud or inappropriate beautification. Many authors have dozens of flagged problematic articles, some even more than one hundred. Note that many of these problematic papers are *not retracted*: they are flagged, comments made, and the shenaniganry is out there in plain sight for everybody to see. The last sentence of the comments is usually a request that the authors of the flagged papers respond and clarify the matter in question.

David Sabatini, aka "mTORman" (a former professor of biology at the Massachusetts Institute of Technology's Whitehead Institute in Boston and codiscoverer of the mTOR pathway), has issued at least ten corrections to various papers his group published in top journals between 2018 and 2022. Sabatini gave the following useful advice on Twitter on how to respond "when a PubPeer troll starts to harass you on Twitter over a minor mistake in a 10 year old paper":

(1) Block the steaming turd immediately.
(2) Post corrected figure on PubPeer.
(3) Contact journal to ask what they would like you to do.

In 2021, Sabatini resigned from all of his positions at MIT and at the Howard Hughes Institute—but this was *not* in conjunction with the image irregularities discovered by the above-mentioned steaming turds.[5]

In many other cases, the responses/excuses of the authors of the PubPeer-flagged articles can be bizarre, even comical. I have not yet seen "The dog ate the raw data" or "Aliens did it," but some responses are close and can only be classified as dark comedy. The classic all-purpose winner is, of course, this statement: "results and the conclusions drawn from the data were not altered by this correction." An entertaining selection of further excuses can be sampled in Appendix 4.

On occasion even more farcical events take place on PubPeer, for example, when the authors replace one set of clearly fraudulent data with another set of data that turns out to be equally fraudulent. I have not seen anybody, anywhere, honestly fessing up to any intentional manipulation or beautification. *Not once.* If it happened, it was only done to "illustrate the truth better."

All of this brings us to a contentious area: what is considered "acceptable manipulation of images" in a scientific paper? The standard has been shifting over the years; in 1994, the rules were not clearly established, and cutting and pasting various parts of blots was acceptable practice as long as it "illustrated" the actual findings. The bar has been raised over the years to the point where no splicing is allowed, in principle; if for some reason it is unavoidable, then the separation between different parts must be clearly indicated and full gels must be shown to the referees. Many journal guidelines are still nebulous when it comes to manipulation. Elsevier, for example, still maintains that some image manipulation is okay. For example, the Author Guidelines for the *Journal of Hepatology* state: "While it is accepted that authors sometimes need to manipulate images for clarity, manipulation for purposes of deception or fraud will be seen as scientific ethical abuse and will be dealt with accordingly." This has been used to exonerate some investigators in various data integrity investigations. The defense goes like this: the authors were not manipulating the images to *create the impression* of some effect that was not really there. They just changed a few things to *better illustrate* their findings.[6]

And here is the most depressing part: most of the problematic articles are flagged, discussed—*and then not much happens.* No explanation is given by the authors, no article correction happens, and no expression of concern is published, let alone a retraction (figure 6.1).

Another website, ForBetterScience.com, is run by the independent science journalist and occasional cartoonist Leonid Schneider, based in Germany.[7] He has more than a decade of experience with hands-on laboratory research (molecular biology, stem cells, and cancer research), but he is no longer active in the lab and now focuses full-time on scientific integrity problems. His crowd-funded website is well-intended, but some readers may find it blunt and confrontational.

6.1 Further on the general subject of scientific misconduct.
Source: Sidney Harris / CartoonStock.com, reproduced by permission.

His sense of humor is savage, but hilarious. In the "About . . ." section, Schneider sets the tone:

> The climate of fear is omnipresent in science: bad science is very rarely exposed in the open, fraud and ethics breach[es] are routinely covered up, while accusations and criticisms are raised only behind closed doors and are often unfairly or even wrongfully assigned. There are many good and honest scientists out there, but they have no voice. My site is for them.

The website contains a massive amount of information and regularly exposes fraudulent papers, clusters of papers, and groups of researchers responsible for these papers. Schneider is highly critical of the entire scientific infrastructure, which is pictured as being reactive/defensive rather than proactive, and operates with low transparency and massive inertia.[8]

Problematic papers and authors exposed on Schneider's blog have subsequently been picked up by mainstream news organizations—typically without giving any credit to Schneider. But Schneider—along with several of his guest authors—soldiers on. On a daily or weekly basis, they expose problems at all levels of science—from Chinese paper mills to problematic publications of prominent scientists, sometimes even Nobel prize winners. Schneider has been hit with several defamation lawsuits over the years, but came out reasonably okay and—is still standing today. Even though, in the fall of 2024, in an extraordinary move, the country of Italy has blocked all access to his website, in response to Leonid's multiple articles exposing misconduct and fraud committed by various Italian biomedical scientists. One example of a major breaking article on his blog related to the data fabrication and retraction scandal affecting Harvard's Dana Farber Cancer Research Institute (which Schneider, in his inimitable linguistically innovative style, designated as "farberifications"). Another article on his website resulted in the resignation of the University of Kiel president Simone Fulda, when concerns were repeatedly raised about image manipulations in her papers.[9] Surely by the time this book will be published, dozens and dozens of additional new cases will have been added to the lists at ForBetterScience.com.

Numerous individuals or perhaps groups of individuals (data sleuths or data detectives) who work on these issues use pseudonyms (e.g., Clare Francis, Smut Clyde, Aneurus Inconstants, Artemisia Stricta, etc.). They spend significant amounts of time exposing bad science—often working closely with PubPeer or ForBetterScience. Those who work at the cutting edge, discovering and exposing hitherto undetected problems, often remain anonymous. Anonymity protects them from retaliation, but at the same time it can diminish their impact: many journals and journal editors dismiss and do not investigate accusations of fraud that come from anonymous tips.

Other data detectives use their own name. Perhaps the most prominent is Elisabeth Bik, who maintains a blog and has detected countless problematic papers over the years—including whole clusters of paper mill products—resulting in about one thousand retractions and one thousand corrections.[10] Another one is Nick Wise, whose keen eye has resulted in more than eight hundred retractions.[11] Recently, Smut Clyde's identity was revealed through an article he published on the subject of—what else?—paper mills.[12] His real name is David Bimler, and he is a psychologist formerly with Massey University in New Zealand. He has single-handedly taken down several thousand paper mill products and other fraudulent and questionable publications.[13] Another prominent blogger and activist is Dorothy Bishop. Her scientific background is in experimental psychology, but her blog covers a wide range of topics—from paper mills to the emerging practice of fake/fraudulent (joint) academic affiliations—that apply to all areas of biomedical science.[14] I also recommend the blog of a second experimental psychologist, Adam Mastroianni, who covers a wide range of topics around research, including the futility of the peer review process, various problems related to research grant applications, and how the strong-link/weak-link problem hinders scientific progress.[15] There are also several young vloggers on Youtube who are focusing on various aspects of scientific misconduct and fraud. One example of these "young turks" is Pete Judo, of Warwick University in the UK.

There are some rare instances of support and occasional public recognition for prominent data detectives. In 2021 Elisabeth Bik received the John Maddox Prize for "courageously advancing public discourse with sound science." Oransky and Bik are often featured in news articles that point the public's attention toward scientific misconduct and the reproducibility crisis. But the proactive aspects of scientific fraud detection (i.e., when the issue is *not* raised by an institutional whistleblower and the system is forced to react) are primarily based on the initiatives and efforts of concerned and dedicated individuals, who often do this on a shoestring budget as a supplemental activity in addition to their day jobs—with potential defamation lawsuits hanging over their heads.

Let me repeat this because it is mind-blowing: the proactive "cleaning up" of science is largely based on *private initiatives* (!!!). Proactive data sleuthing of the published literature is *not* done by the scientific institutions where the work is being performed. Nor is it done by the grant-giving bodies who pay for all of this work. And it is not done by ministries of science or health or education—although, broadly speaking, this subject belongs in their jurisdiction. And it is definitely not done by scientific publishers or journals—even though they are the ones who publish these irreproducible and often fraudulent materials. The one NIH-initiated site, PubMed Commons, which was supposed to serve as a venue to comment on published papers, began in 2014 with some fanfare but was unceremoniously shut down after five years. Usage was minimal, with comments submitted on only 7,500 of the twenty-eight million articles indexed in PubMed.[16] The likely reason for its failure is that, unlike PubPeer, anonymous comments were not allowed.

Not everybody is happy with the idea of having anonymous, self-appointed data detectives investigating and exposing scientific fraud. It has been suggested that people who act in bad faith—e.g., disgruntled former colleagues or competitors seeking an unfair advantage—may abuse the cover of anonymity and abuse the system for personal attacks.[17] At the same time, if anonymous comments are not allowed, the system does not seem to work either, as seen with PubMed Commons. It seems like a catch-22.

The entire process of detecting problematic publications and cleaning up the literature lacks systematic coordination, does not have a commonly acceptable set of standards (e.g., a clear definition of where mistakes end and where fraud begins), remains severely underfunded, and only scratches the surface of the problem. Also, as discussed earlier, it focuses on easily detectable problems—essentially, image fraud—which is only one of many types of possible research misconduct. It is as if the official stakeholders do not have much of an appetite for getting to the bottom of the problem.

It is safe to say that scientific fraud detection is one of the most thankless jobs imaginable. At the same time, it is also one of the most important ones. The occasional celebration of individual data vigilantes is nice but is misguided. The task is much more important than something that is left to volunteer individuals—no matter how eagle-eyed and dedicated they might be. The various stakeholders identified here need to step up and raise the process to the next level.

CRIMINALIZATION OF SCIENTIFIC MISCONDUCT?

As shown in previous chapters, scientific misconduct that involves basic laboratory research and associated misrepresentation of data or figures in basic science papers or grant applications is rarely penalized in a court of law. In some instances the investigator and his institution have sued each other (Anversa case, see table 4.1), and sometimes the NIH and the institution where the misconduct occurred had a legal battle and, as a result, some of the wasted grant funding had to be returned (e.g., the Potts-Kant case: Duke University vs. NIH, see table 4.1). But rarely has a basic scientist gone to jail for authoring fraudulent publications or paid fines or penalties out of their own pocket. The only exception is the Dong-Pyou Han case (see table 4.1), in which the regular ORI settlement path was elevated to a federal case due to the actions of Senator Charles Grassley.

The way misconduct in basic sciences is treated legally is markedly different from misconduct that affects clinical trials, patient care, or patient lives. In this area—which is beyond the scope of this book—there have been numerous civil and criminal lawsuits and some jail sentences. In fact, even in the

Dong-Pyou Han case, one of the reasons for elevating it to a more stringent level of prosecution was that it related to research on an AIDS vaccine, which could have endangered patient lives further down the road. One other case of research misconduct that did end up with a sentence relates to Eric Poehlman (University of Vermont College of Medicine). In the late 1990s, he conducted basic research on patients in the field of obesity and diabetes, in part in the context of hormone replacement therapy, and was found to have committed fabrication and falsification of patient data. He spent a year in jail as a consequence. When it comes to falsification or fabrication *in clinical trials*, several additional criminal cases are known.

But no criminal cases have been brought based on misconduct perpetrated in "pure" basic research—even though all the elements of crime are present, or can be present. These *perpetrators* create and publish fraudulent data. They *knowingly* commit the crime (as the number of fraudulent papers increases, the common excuse of "inadvertent mistake" is less and less believable). The most obvious cases of *fraud* (e.g., image fraud relating to histological pictures or Western blots) can be *easily proven* (i.e., the exact same picture cannot represent two completely different things), and if a follow-up investigation looks at the laboratory notebooks, it will likely turn up additional misconduct as well. There is also an element of *financial motivation*: clearly, CVs boosted by a long list of publications will lead to academic advancements (i.e., higher salaries), more grant funding (which, in essence, means that the government or some granting agency is defrauded in the process), or the possibility of patents and spin-off companies (i.e., investors are defrauded and more money goes to the fraudulent investigator). Scientific prizes and awards may come to the fraudster (many of which come with cash into the pocket of the fraudulent investigator). And, of course, there are *victims* of these crime as well. These include the general public, i.e., patients. When the advance of science is held back or diverted into unfruitful directions, progress in medicine is losing out—not only in terms of money but also in terms of time. The victims also include fellow scientists who play by the book only to lose out in the competition for funding and career advancement. Young researchers who are unlucky enough to undergo scientific training in a group of scientific fraudsters—instead of receiving real scientific training—are also victims. And granting agencies are victims, too, when their funds are misdirected to support fraudulent groups.

The idea that *all* forms of scientific misconduct—including many forms of misconduct discussed here—should be criminalized is not some sort of crazy overreaction. It has been discussed for several decades, even when the generally believed notion was that misconduct was a relatively rare event. Thirty years ago Susan M. Kuzma (at the time with the U.S. Department of Justice) wrote a lengthy article on the subject and argued for criminalization.[18] In this article she states that "misrepresenting research results in order to obtain federal grant money is essentially theft by false pretenses" and that "the federal interest in and ability to prosecute does not depend on the scientific context" (i.e., it should not applied only to patient-related cases). She also states that "there is no federal statute that specifically and discretely punishes misrepresentation of research results as 'fraud in science' or 'scientific misconduct.'" The two applicable federal statutes she identifies are (a) submitting false statements to the federal government and (b) defrauding someone through the use of a federally controllable means.

Kuzma proposed that a two-tier statute should be implemented. At the lower level, a misdemeanor provision (for the base offense) for "knowingly making a materially false statement or representation about research results or about the method by which they were obtained." At the next level (aggravated form of the crime) she proposed enhanced penalties if "particularly egregious behavior were proved, such as acting with intent to defraud, engaging in multiple and discrete acts of misrepresentation, or causing harm by inducing reliance on the falsified research." As far as I know, in the last thirty years these suggestions have never been considered for implementation.

Discussion on the topic was reignited in 2014, when a pro/con article was published in the *British Medical Journal*.[19] Zulfiqar Bhutta (Hospital for Sick Children, Toronto, Canada) argued that criminal sanctions are, indeed, necessary to deter growing research misconduct. The contrary argument was presented by Julian Crane (University of Otago, New Zealand), who took the position that criminalization would not have any deterrent effect and would undermine public trust in the scientific process. We can expect this discussion to continue in the future. It is likely that the debate will flare up with every new high-profile misconduct case. My own opinion is that Kuzma's suggestions made perfect sense in 1992 and make even more sense now because we can see the magnitude of the problem.

Based on polling data, the general public as well as the community of scientists would both favor harsher penalties for scientific misconduct. According to a survey published in 2018, 96 percent of the American general public sees scientific fraud as morally unacceptable, and more than 50 percent of the respondents believe that jail sentences of one year or more would be justified for those who perpetrate it.[20]

This is a complicated and difficult subject. A paper by a group of bioethicists from Stockholm University has carefully considered various practical problems of implementation—for instance, the question of where to draw the line between honest error and intentional fraud.[21]

TRADITIONAL SUGGESTIONS FOR GRADUAL IMPROVEMENTS

In the 1970s and 1980s, the typical institutional response to anyone who dared to raise the issue of reproducibility or scientific misconduct was an attempt to minimize the problem and talk about the dangers associated with the public losing trust in science.[22] Thankfully, this type of stonewalling has, by and large, stopped. Many news articles now raise this issue for the public's attention; among others are the data detectives Elisabeth Bik and Ivan Oransky, who are often featured in these articles. Top scientific journals such as *Nature* and *Science* regularly publish news articles and opinion pieces on the subject. The NIH and other granting bodies are also openly discussing the problem.

Entire journals are dedicated to the subject. The Taylor and Francis publication *Accountability in Research* investigates the problem down to the finest detail: finding the best methods to identify "questionable" journals,[23] deciding if research data mismanagement (RDMM) should be classified as a questionable research practice (QRP),[24] and determining the difference between "reckless" versus "merely negligent" behavior in the laboratory.[25] We have gotten to the point now that research misconduct itself is a stand-alone academic field of study, with platitudes like "intentionality is irrelevant for the impact on validity" and "we conclude that encapsulated in challenging prerequisites of the national research integrity promotion work, the meaningfulness of such work emerges through integrating research integrity in all aspects of academic life to entail indispensable systemic changes in academia."[26] Other such comments include "a focus on prevention is crucial."[27] And finally, "recalibrated responses are needed to a global research landscape in flux."[28] Some of the discussion is becoming too abstract and too academic, but at least there is a discussion. The elephant in the room is finally being acknowledged. What to do about it?—well, that is another question.

The better journals have fortified their fraud detection efforts right from the beginning of the publication process. In addition, some replication efforts are being funded by granting agencies, although compared to the money that supports regular research projects, these resources are miniscule. Ten years ago the NIH director Francis Collins published an article in *Nature* on the subject of reproducibility titled "NIH Plans to Enhance Reproducibility."[29] However, when we look at the initiatives and changes they proposed, we don't see a lot of successful implementation. PubMed Commons was a failure (discussed earlier). The concept of anonymizing the NIH peer review process (i.e., reviewers should be blinded [i.e. anonymized/masked] to the name of the investigator whose application is being evaluated) was floated but ultimately not implemented.[30]

When I look at the opinion pieces published in prominent journals, I see reiteration of various problems, but I don't see a lot of new ideas on how to improve the system. The suggestions focus on slow, incremental changes: "let's enhance access to primary data," "let's increase the transparency of research," "let's strengthen the journals' fraud detection systems," and "let's improve scientific integrity training at various levels of all institutions."

The most comprehensive and sensible set of guidelines and recommendations on how to improve the reproducibility of preclinical research has been put in place by Leonard Freedman and colleagues at the Global Biological Standards Institute (Washington, D.C.). The Reproducibility 2020 Initiative was launched in 2016, and it includes four core areas where improvements must be made: (1) study design and data analysis (with improved training, increased attention to statistical issues, clear and early disclosure of study design and analysis parameters), (2) reagents and reference materials (with improved cell line verification, antibody, serum, and other common reagent verification, improved standards), (3) laboratory protocols (consensus standards, protocol disclosures, and repositories), and (4) reporting and review (more detailed reporting of the findings in papers, open access to research protocols and raw data, and cooperation with replication efforts).[31]

Another article, in fact, a manifesto—cowritten by several top experts in the reproducibility field including Brian Nosek and John Ioannidis—summarizes the same concepts in table form (see table 6.1). Among the topics in the table, the "reproducibility mantra" seems to have been discussed the most. In a 2013 *Nature* article, Lawrence Tabak, the NIH's principal deputy director, stated: "If the premise isn't validatable, then we're done; it doesn't matter how well you

TABLE 6.1 The current thinking: commonsense approaches to increase reproducibility in biomedical research

Theme	Proposal	Examples of initiatives/potential solutions (extent of current adoption)
Methods	Protecting against cognitive biases	• Blinding, randomization
	Improving methodological training	• Rigorous training in statistics and research methods
	Independent methodological support	• Involvement of methodologists in research • Independent oversight
	Collaboration and team science	• Multisite studies / distributed data collection • Team-science consortia
Reporting and dissemination	Promoting study preregistration	• Registered reports • Open science framework
	Improving the quality of reporting	• Use of reporting checklists • Protocol checklists
	Protecting against conflicts of interest	• Disclosure of conflicts of interest • Exclusion / containment of conflicts of interest
Reproducibility	Encouraging transparency and open science	• Open data, materials, software • Preregistration
Evaluation	Diversifying peer review	• Preprints • Pre- and postpublication peer review
Incentives	Rewarding open and reproducible practices	• Registered reports • Transparency and openness promotion guidelines • Funding replication studies • Open science practices in hiring and promotion

Source: Modified from Marcus R. Munafò et al., "A Manifesto for Reproducible Science," *Nature Human Behaviour* 10, no. 1 (January 2017): 0021, https://doi.org/10.1038/s41562-016-0021.

wrote the grant." The NIH officials were also "considering a requirement that independent labs validate the results of important preclinical studies as a condition of receiving grant funding."[32] That statement was made exactly ten years ago, but as far as one can tell, the "validation idea" has never been put into practice. As far as I know, it was never even piloted.

Another recurring theme is the idea of setting up some sort of central Research Integrity Organization. For the United States, C. K. Gunsalus (director of the National Center for Professional and Research Ethics at the University of Illinois at Urbana-Champaign), Marcia McNutt (president of the National Academy of Sciences), and several colleagues have been calling out for the establishment of a national research policy board that focuses on robustness and quality. The first call was made in 1992; the call was repeated periodically over the years, with a major and rather desperate-sounding manifesto published in 2019.[33] Yet no such board has ever been put in place.

Although these efforts and initiatives are well-intended and make sense, they call for relatively small, incremental advances. And none of these initiatives and suggestions go beyond what has been discussed ad nauseum for several decades. In addition, the most crucial aspect is rarely discussed: who, exactly, is supposed to put these changes in place, and what are the incentives and rewards for doing so? As long as the entire research framework and infrastructure and incentive structure remain the same, anyone who places a serious emphasis on these initiatives will be left behind when competing for extramural grants and awards. The current system rewards fast and flashy papers (even if the authors have been cutting corners and the stated findings will turn out to be irreproducible) over slower and more meticulous ones (even if they are the more reliable.)

Incremental changes are well intentioned, of course. But clearly, if they *worked*, they would have worked by now. If they *worked*, we would not be in the present situation.

Many scientists are waiting for some sort of savior to magically appear and "change the game." Some believe that artificial intelligence, machine learning, big data analysis, and complex network analysis will be this savior. In biomedical science, it has been proposed for some time that these new tools and approaches will be able to do independent hypothesis generation, experimental design, and interpretation of the findings at a level that is inconceivable for a human being.[34] Clearly, AI is better at processing massive amounts of data, and it would be great if new connections, new ideas, and new testable hypotheses could be produced by it. It may also be employed, at some point in the future, to improve the evaluation of grant applications and manuscripts. But with respect to the central question of this book—reproducibility—my big question is the following. If a massive portion of the published literature on which these systems are trained is bogus, fake, or at a minimum unreliable, how is any intelligence (human, non-human, cyborg, alien . . .) going to make *any* sense of it? Remember: "garbage in—garbage out."

Whatever happens, wherever science is heading, the current body of literature must be cleaned up retrospectively and must be replaced proactively with a more reliable and reproducible body of data. Then and only then can the new techniques and approaches be unleashed on them. Maybe AI could be used to help us point out some of the irreproducible science. This direction has already been taken with some of the image analysis programs that can find duplicate images by comparing new submissions with material on the internet. Jay Puara's research group at the University of Southern Carolina's Information Sciences Institute aims to take this a few steps further. A machine learning tool called SCORE (systematizing confidence in open research and evidence) hopes to assess the likely reproducibility of a paper by examining micro and macro information. Micro information, in this case, means things like P-values, sample sizes, and number of citations a paper receives. Macro information is assessed by evaluating citation relationships between papers, affiliated relationships between authors and organizations, and authorship. The system then produces what they call "Knowledge Graphs."[35] It sounds very promising, although—as far as I know—this system has not yet been tested in the "real world." Interestingly, some of the parameters used by this program seem to encompass some of the same warning signs that data detectives look for on a smaller scale, and signs and features the red-flag method also employs.

CALL FOR REAL REFORMS

What, then, should be done? Let's reiterate the main problems and try to come up with some solutions that address the root causes.[36] From a historical standpoint, the root of the problem is that all elements of today's scientific process—hypothesis-generation, experimentation, the academic institutions, the publication process, the way grants and scientific prizes are awarded—are rooted in the old concept of the *independent researcher*, which was established around the eighteenth century and worked reasonably well for a long time.[37] But many elements of this system are no longer compatible with today's environment. Scientists have changed; their motivations and incentives have changed; the methods of experimentation have changed; and everything has become more complex and highly specialized. Despite these changes, the foundations of how science is done is stuck in the old framework. If we want to improve the system, we must reform the system for all stakeholders—and we must do so simultaneously.

The *first stakeholder* is the scientist doing the research—or a group of scientists working together. As discussed at length, one root of the problem at the level of an individual scientist is hypercompetition and the ever-increasing pressure for publications and funding. The other principal root is that reproducibility is just not something most scientists care about, for a variety of reasons. It is unrealistic to expect that radically more funding will be allocated for science in the future, so the only workable solution I see is to (a) reduce the number of individuals working in life science research, and (b) reallocate some of the funding to studies focusing on validation and reproducibility.[38] The entire process would have to be reformed, from the way we select scientist trainees and how we train them to the way their career is managed.

First, we have to acknowledge that science as a career is not for everyone; in fact, it is only for a few. This means that the sheer number of life scientists working in academia should be reduced: quantity should be replaced with quality.[39] As pointed out by the renowned Oxford statistician Douglas G. Altman in 1994, we need "less research, better research and research done for the right reasons."[40]

Second, we must improve the selection process so individuals who should become scientists are incentivized to choose life science as a career rather than choosing more stable and more lucrative careers.

Third, individuals who are candidates for a career in science should be trained drastically differently, with core principles such as team science, collaboration, and data integrity and reproducibility.

Fourth, an entirely new scientific career track should be established—one that has statistics, experimental design, quality assurance, quality control, data integrity, and reproducibility at its core. Let's face it, the process of designing and interpreting experiments is becoming more and more complex—and it will be even more complex if reproducibility is added as a core parameter. It is getting to the point when specialized mathematical and biostatistical knowledge

6.2 On the general subject of scientific integrity ...
Source: Leonid Schneider / ForBetterScience.com, reproduced by permission.

will be required. This is exemplified, for instance, by the comparison of various experimental design models documented by Steinfath and colleagues in 2018.[41] An Average Joe/Jane scientist, with fundamental knowledge of probability and statistics but with no specialized training in the finest aspects of the process, may quickly become lost in all of this, and unintended errors and mistakes will be made. Why not train and recruit specialized individuals who are really good at this? Even better, let's train a specialized breed of laboratory personnel who are not only good at this but are also good in other related subjects, such as laboratory quality assurance and quality control, forensic analysis of scientific data, and proactive prevention of questionable practices or misconduct (figure 6.2).

A new profession of this type should be created at the PhD training level. After the undergraduate degrees are completed, individuals could choose between a track focused on scientific discovery and one focused on science integrity. This latter track would produce a degree in Scientific Integrity, and people who graduate with this degree would be employed as Science Quality Officers at universities, research centers, and in the industry. They could also be employed as professional referees for grant-giving bodies and scientific journals.

Of course, all of this would only be possible if the other stakeholders in the scientific enterprise are undergoing radical reforms too. In the transitional phase, however, with the downsizing of the research enterprise, many former scientists could be retrained and employed on this new Science Integrity Officer track.

The *second group of stakeholders* are the scientific institutions—universities, hospitals, private and governmental research centers, etc.—where the scientists work.

First, all of these institutions should realize that it is not the goal of an institution to expect everybody with a higher degree (MD or PhD) to be a scientist and have a publication record. Publications should not be held as a universal currency—a sentinel marker of quality or as an essential requirement for promotion. Other parameters (clinical excellence, teaching excellence, public service, etc.) should be established instead.

Second, the institutions must stop looking at research as a cash cow that must bring money into the institution. They must accept that they may just "break even" or may even—oh, shock of shocks!—*cost* institutional money. Still, research should be supported at top universities. Why? On one hand, because it increases the reputation and visibility of the institution, and on the other hand, it may produce patents and licenses and a stream of long-term income—but, of course, only if the findings described in the publications are real and reproducible.[42] Part of this change of thought can be forced upon the institutions externally by the grant-giving bodies (more on this later).

Third, institutions must reform the way scientists are employed and the way they conduct their research. Scientists should be employed on contracts or on variations of tenure arrangements—but their work and career should not be dependent on obtaining individual extramural grants (more on this later). A working group should consist of a principal investigator, many staff scientists, some postdocs, and students (but these should not make up the majority of the group). As new members of each team, science integrity officers will work closely with the principal investigator and all group members and actively participate in all aspects of research, from experimental design to data analysis. The focus of the research should be divided between discovery research and replication work—perhaps in a

70/30 arrangement. The topic of these replication projects should be coordinated with the grant-giving bodies (more on this later).

Fourth, the scientific institutions should drastically enhance their processes related to data collection and data integrity, from enhanced training in these areas all the way to data collection and storage/retrieval. Cameras in each laboratory and office, recording and keeping every piece of data collected in a laboratory and every keystroke hit, and every single computer file created by every member of the research group—these ideas may sound a bit too harsh for an academic environment, even though workplace surveillance laws usually allow companies to use video monitoring for legitimate business purposes. In any case, electronic notebooks should be set up as a matter of routine and should be archived daily.[43] Random spot-checks and data reviews should be set up. And a penalty system much more stringent than the current one—perhaps even one that elevates the most egregious cases to the level of criminalization—should be set up to deal with scientific misconduct.

Why would any institution—especially one that already has a steady support of grant funding—be willing to set up such a system? The incentive would have to come from the radical reform of the entire grant system, which is the *third stakeholder* in the necessary reform. In most countries, biomedical research is usually supported by a single state or federal granting agency. In the United States, it is mainly the NIH; in Switzerland, it is mainly the SNSF. Based on strong data that prove this, granting agencies should finally acknowledge that the current way of awarding grants is close to complete randomness. Even though the individual principal investigator awardee names are different, historically the same institutions receive most of the grant funding year after year. All of this can only bring us to one conclusion, which will raise some eyebrows, but here it is: Centers of Excellence.[44] No individual grant applications and evaluations are needed. Instead, I advocate for direct intrainstitutional arrangements that establish and support Centers of Excellence. Let's say the NIH designates Johns Hopkins University as an NIH Center of Excellence and agrees to support its research with a certain amount of money, perhaps $700 million per year for a period of five years.

But this money would come with certain stipulations. The university designated as a Center of Excellence would accept a lower overhead rate so more of the money goes to support actual research and would also agree that a certain percentage of research—say 20 percent of the total award—would be supplemented by intramural funds (e.g., from charity donations or from interest on its endowment). The university would also have to agree to reform their training schemes and the supervision of each research group with a clear emphasis on integrity and reproducibility—along the lines of the changes I proposed earlier. The funders may even make suggestions on what topics of research should be emphasized based on socioeconomic needs, for instance. And the funders would set up a system that tracks scientific productivity, which not only considers scientific novelty and excellence but also translatability, and, of course, *reproducibility*: future funding of the institution would depend on the institution's performance on these parameters.

I believe this system would be enthusiastically supported by scientists working at these institutions, who would enjoy less pressure and more job security and stability. Their true work productivity would increase by at least 30 percent because they no longer had to write individual grant applications.

The grant-giving bodies would also have to readjust their infrastructure and focus and assume the role of a collaborator. They would play an integral role in organizing and coordinating independent replication efforts. These would be conducted, in part, by other NIH-funded Centers of Excellence (i.e., one center may check the reproducibility of key findings produced at another center); other parts may be conducted by intramural NIH laboratories; and still other work could be done by independent contract laboratories. The outcome of these replication efforts would factor heavily in evaluation of the center's performance and determine its future funding.

Smaller granting agencies could join in. For instance, the American Heart Association may decide to supplement an NIH-funded Center of Excellence or to fund parts of a center that focuses on research on heart diseases.

This system could be implemented on a voluntary basis initially, or changes may happen at selected academic institutions. It could be started at the NIH or at similar federal research institutions where many essential components of the process are already in place: comprehensive employee training/retraining mechanisms, close management of expenditures, access to powerful information technology systems, and an existing office of research integrity.

Surely, a reform of this magnitude would be subjected to criticism. For example, critics may say that it is set up to "conserve the status quo" and the funding priorities of various institutions. To this criticism, I respond that the status quo has been and will be conserved anyway. If one compares the list of U.S. institutions that receive the most R01 funding now to that list a decade ago, the overlap is highly significant.[45]

I am convinced that the advantages of this new system would far outweigh the disadvantages. The institutions operating under this new funding system would not only attract the best scientific minds but could spend their time doing what they are best qualified for—i.e., *actual research*—as opposed to, for example, the endless writing of research proposals.

The *fourth stakeholder*, of course, is the publishing industry. The way scientific results are published should be subject to complete reform. Every stakeholder should realize that the current system—even with honest efforts of screening and policing the submissions—is unable to filter out suspicious and fraudulent material. Nor is it capable of (or, in most cases, even interested in) separating manuscripts that contain likely reproducible data from manuscripts that contain likely irreproducible findings. These days, data integrity guidelines and submission checklists that try to prevent misconduct are plentiful. All of the major publishers have them. And some organizations focus on this subject, with COPE (Committee on Publication Ethics, based in the UK) being one (publicationethics.org). All major scientific publishers are part of COPE. Among other things, COPE publishes flowcharts that lay out what to do when suspicion of misconduct or fraud emerges in a publication. The problem is *not* that we don't have enough guidance, but that the journals—even those published by COPE members—do not, or do not always, or not necessarily, or only with massive inertia follow these guidelines.

Another part of the problem is, of course, that a significant portion of the scientific community engages in borderline scientific behaviors, and many do not comply with COPE guidance. On top of all of this, there is a whole army of science criminals, such as paper millers, who brazenly bypass and subvert these guidelines (or any other guidelines).

Many incremental advances have been proposed to improve the scientific publication process (figure 6.3). For instance, the replacement of anonymous reviews with named ones has been suggested repeatedly—but has only rarely been implemented. Similarly, insisting on data sharing and full access to all raw data in a publication has been suggested for at least the last two decades. In January 2023 the NIH began requiring a data management plan to be included in their grant applications—and eventually they have plans to make all data that have been generated by NIH-funded research publicly available.[46] Other countries have been implementing similar policies for several years now. For instance, Switzerland's main granting agency, the SNSF, has been mandating data management plans for the last several years.

6.3 "Most scientists regarded the new streamlined peer review process as "quite an improvement."
Source: Nick Kim / scienceandink.com, reproduced by permission.

Let's hope that mandatory open data policies will make some difference (figure 6.4), even though it is possible that the availability or nonavailability of the data is *not* what matters the most from the standpoint of the reproducibility crisis. Of course, the premise makes sense: mandatory open data policies should, in principle, facilitate the identification and correction of various errors and would be expected to decrease the rate of subsequent retractions. The only problem is that it might not work in practice. Requiring authors to submit their raw data together with their manuscript may filter out fraudsters or bad actors from *some* journals, but these papers might (and probably will) find a place in a publication eventually.[47] Also, open data policies might not change the rate of retractions (figure 6.5). A recent study by Berberi and Roche compared the retraction rate before and after various journals have implemented mandatory data sharing policies. It made no difference; the retraction rate remained the same. Admittedly, this was a relatively small study (forty journals, 620 retractions), and it focused only on papers published in certain scientific fields (ecology and evolution).[48]

Some more interesting and increasingly radical ideas are being presented by the journal *eLife*. The Nobelist Randy Schekman, at the University of California,

6.4 On the subject of data sharing…

Source: Leonid Schneider / ForBetterScience.com, reproduced by permission.

6.5 Further on the subject of open access and data sharing…

Source: Leonid Schneider / ForBetterScience.com, reproduced by permission.

Berkeley, started the journal in 2012 after working at *PNAS* as editor in chief for a number of years. One of his new ideas in 2018 was to strip away the usual accept/reject role of the referees and simply ask them to identify the strong and weak points of a submission. This type of system—with some variations—is now in place at several other journals. In 2020, a more radical *eLife* idea was put in place, the "publish then review" concept, wherein submissions were only reviewed after they were made available for everyone to read at a preprint server. This means, in essence, that the material has a publication date when the reviewing process starts: this way the reviewing process does not delay any colleagues' access to the material.

From late 2022, *eLife* raised some eyebrows when its new editor in chief, Michael Eisen, decided to start the "no-reject publishing" model. In this model, the journal publishes all of the papers that are selected for peer review. These papers, called Reviewed Preprints, are refereed by *eLife* reviewers; in parallel, public comments can be added as well. When the review process is completed, the reviewers' comments are made public together with the article. This new format received a lot of praise, but also some sharp criticism—even from some of the journal's own senior editors who have threatened to resign if the new system is implemented. Several of them even demanded Eisen's firing. In the end, Eisen was forced out, although the stated reason did not have anything to do with *eLife*.[49] Nevertheless, the new publication system has survived, and the journal's website already features a significant number of Reviewed Preprints. One problem with the system is that indexing databases, such as PubMed, only include final versions of papers. Thus, if a paper is accepted by *eLife*, a final "version of record" will have to be generated, and only then will the paper be included in official searchable databases.

These initiatives are certainly well intended and are small steps in the right direction. But in the short to medium run, I think the way to deal with the various problems plaguing the publishing industry is for each major grant-giving body to select a group of quality-controlled journals in which grant awardees are allowed to publish. The selection should be made by the grant-giving bodies, perhaps via a consensus meeting of the major grant-giving organizations. To be included in this select list, the journal would have to agree to a stringent set of standards, including a strict data integrity assessment, a standardized evaluation process, and inclusion of *all* primary data with each published manuscript. A sub-subsection of PubMed could be set up going forward that only contains papers published in the journals that have been approved by the granting organizations.

218   THE WAY FORWARD

6.6 On the subject of retroactive "cleaning up" of the scientific literature . . .
Source: Leonid Schneider / ForBetterScience.com, reproduced by permission.

Going back, the entire body of published literature must be systematically cleaned up (figure 6.6). This should not be left to individual data detectives; this should be a concentrated effort spearheaded and funded by a consortium of the major publishers and conducted systemically, and very thoroughly, by professional data detectives. All papers that contain detectable "image irregularities" or other signs of misconduct should be flagged as well as *all* publications of those authors who regularly publish suspect material. Even today there are filters in PubMed allowing people to select only review articles or only clinical articles, or other types of articles. One could easily set up another filter that eliminates problematic and likely unreliable papers from a search.[50]

Those journals that make the cut would have to agree to certain reforms, one of them being a transition, over a number of years, to a system where—instead of the current practice of using unpaid, volunteer referees from academia—they use *paid, professional reviewers* (who, in the short run, would be retrained from scientists who leave the pool of individuals who are doing science, and would be part of a new, emerging class of scientific personnel—the previously proposed "science integrity officers" in the long-run). A professional organization of scientific reviewers could be set up, with training, certification, and continuing education processes in place. Journals in this preferred tier would also have to agree to transition to a postpublication review/comment system.

Ideally, a better future version of the publication process could (should) also *incorporate a replication component* as a matter of routine. If the scientific establishment would embrace the general idea of replication, and the 70/30 percent arrangement I outlined earlier, this could come about almost naturally. In practice, this would mean that a main manuscript would be accompanied at manuscript submission by a smaller, supplemental paper with a different set of authors who have performed—in a blinded and randomized fashion, which would be described in detail—a repeat of some key experiments of the main paper. One could call them the "Replication Supplement." The indexing databases would have to find a way to link these "little sister" papers to the main ones. Even in the short-term such replication supplements could easily be introduced; all it would require is some willingness from the journals' editorial boards. They would be refereed together with the main paper, and they should be viewed as a strong positive component of the manuscript during the evaluation process.

In addition, the selected journals would have to agree to set up and operate a *mandatory misconduct detection and reporting system*. For instance, a network could be set up—accessible to all journals involved in this select group—where suspected instances of misconduct would be entered, and all journals in the select group would be prewarned about the suspect material and their authors.[51] As part of this new system, journals would also have to agree to *mandatory reporting* to the offices of research integrity when they detect fraud or misconduct in a submitted manuscript. They would have to report any such suspicion to two places: to the scientific integrity office at the authors' own institution and to the scientific integrity office of the agency that funded the work. Similar to sports, a warning system could be put in place (verbal warning, yellow card, red card). Red-carded repeat offender authors could be put on a temporary halt of *all of their submissions to all of the journals* until the integrity office investigates their data and laboratory practices.

Once again, the criticism of this system is that it will conserve the status quo as far as the status and hierarchy of scientific journals go, and it would create an arbitrary new circle of select journals. It may also be criticized for not allowing new, exciting journals to be started. But with tens of thousands of scientific journals already in existence, I do not see this as a major problem: other than the scientific publishers, nobody needs more journals. What we need are better journals that publish material that is reliable.

Another criticism is that this reform would *not* eliminate the "second-tier" or "gray zone" journals, let alone predatory publishers. A large group of journals would continue to exist in which fraudulent and unreliable material would continue to litter the literature. This may be true, but these journals would quickly turn into depositories of marginal materials that nobody takes seriously.[52] The example of the failure of legal action with the OMICS group discussed earlier clearly shows that there is just no way to ban and shut down dishonest publishers. The only thing that can be done is to establish a completely separate new system in which high scientific integrity is maintained and into which the results of all legitimate research are concentrated.

In the long run, I am not even sure that scientific journals should continue to exist in their current form. I know this may sound unthinkable today. But if we get down to the fundamentals, what is the sole goal of the entire scientific publication process? It is to disclose new scientific information—i.e., data and their interpretation—and make that available to other scientists (and, more important, to the public, which supports the whole process through its tax money). That is all, nothing more. Everything else that has been "built on" this over time is irrelevant. Scientific journals were originally created because paper-based printing was the only way to disseminate information.[53]

This is no longer the case; the same goals could be achieved by cutting out the middle-man—i.e., the profit-oriented publishing industry. Data depositories, online information storage systems, and project boards could be set up, where various studies—including their raw datasets—are posted, updated, expanded, corrected, or modified (if necessary). In addition, replication attempts—linked to the original study—may be posted. Other researchers may comment on the results or even reanalyze them from a different perspective. Before the internet this was not a possibility, but now it can be easily implemented. The datasets can be assigned with author names and time stamps to keep track of who did what and when.[54] And, yes, PubMed and similar databases would have to find a way to keep track of this new system.

The way review articles should be written and used could also undergo a radical reform. (In fact, this could be easier and quicker to implement than reform of the original papers.) A new system of web-based reviews (for which I am recommending the term "Living Reviews") would serve the scientific community much better than the current system of review articles, which tend to become obsolete a few years after their publication. In the system I describe, experts in a field would cover a particular topic in an online Review Depository, which would be available for everyone to read and download. The authors would regularly update the page with the latest information on the topic. The prior versions would be archived with time stamps but still be available if a reader wants to track the changes and developments in the field. It would be a bit like Wikipedia—but only the original authors themselves would be able to update the material. The readers could comment on the article and perhaps engage in public discussions with the authors.[55]

In this new system, the best scientific journals could reinvent themselves as scientific news organizations, where scientifically trained journalists review the latest entries in the data depositories and prepare alerts and news articles to help scientists keep track of new trends and findings.[56]

The *fifth stakeholder*—often forgotten—is the public, who, by the way, pays for all of this through their taxes. There should be more—and better—participation by the public in the process of science. For all the money spent, they deserve something much better. With the enormous public costs involved, at least some of the research that is being supported should be directed toward the most pressing areas of biology and medicine, e.g., emerging diseases, diseases that affect the aging population, and diseases where the research is disproportionally less intensive than their socioeconomic imprint. All stakeholders should also be doing a better job of disseminating their findings to the public in a more effective way, i.e., with less hype and more focus on the—hopefully increasingly reproducible— findings the new system generates.

All of the changes and reforms proposed here seem like radical departures from today's practices. But it is plain to see that the current system is broken—or at least massively wasteful and astonishingly unreliable—and does not serve the interests of the public well.

Who should enact these changes? *Whoever holds the purse calls the shots.* As we have seen, one way or another most of the entire research infrastructure operates on government money. Government money pays for the work at the research institutions, government money pays the publication fees of the journals, government money pays the salary of the peer review scientists who evaluate the submissions for the journals, and government money pays for the libraries who subscribe to the journals. Thus governments and the politicians who are in charge must realize that *they* have the power—perhaps even more power than they themselves realize—to affect change.[57]

The change could be implemented in four steps. *First*, the public and the various governments—who are, ultimately, funding the largest grant-giving organizations in life sciences and should expect a transparent and effective system in return—would have to take an earnest look at the situation and accept that the current system is plagued with systematic problems. *Second*, they would have to accept the fact that the current system is full of self-perpetuating "perverse incentives" that left to their own devices will *not* self-correct. Hence coordinated efforts must be made to correct the system. *Third*, they would have to accept that minor, gradual adjustments won't work. These have been tried for decades, with no visible effect. And *fourth*, drastically different alternative systems—perhaps the ones I propose or possibly even more radical ones—must be put in place.

How should this new system be put in place? I don't have a clear answer. The change must come from the top and from the side with the money. If we look at the U.S. biomedical science system, the NIH is at the top of the pyramid. The new NIH director, Monica Bertagnolli, must be immensely familiar with many aspects of scientific misconduct in biology and medicine. Among others, she has spent much of her scientific career studying the effects of a class of anti-inflammatories called cyclooxygenase-2 inhibitors (COX-2 inhibitors) that prevent colon cancer. COX-2 inhibitors, in fact, have been plagued by a massive pharma scandal and subsequent drug recall.[58] (The adverse effect of these drugs has nothing to do with her scientific work in the field, but surely she is familiar with the whole sordid affair, as well as with all levels of scientific misconduct.) As discussed earlier in this book, her deputy, Lawrence Tabak, has been bringing attention to the reproducibility crisis for at least a decade. Thus there may be some interest from the top to take a renewed look at this issue. Perhaps, at some point, there may be a politician out there willing to ring the alarm bells about "cheating scientists, stonewalling universities, and billions of wasted tax dollars." Perhaps some cataclysmic event or the misconduct investigation of a prominent figure will be needed to trigger a change.[59] I hope that something, at some point, will put things in motion. In parallel, perhaps some new regulatory or harmonization entity may be set up with sufficient power to make decisions and catalyze a real change.[60] It would be mandated with a single goal: *global life science reform to establish reproducibility and reliability*. Perhaps this new entity could facilitate the cooperation of various stakeholders interested in reforming the system—such as reform-minded policymakers, granting agencies, members of the scientific community, and the general public.

AFTERWORD

In science, everything seems to gradually becoming different than what it is supposed to be. I have documented these changes here. Sadly, life science research—which is supposed to be a process to find new truths in biology and medicine—is becoming a career advancement tool for various stakeholders in academia.[1] Research grants are supposed to be sources of funds to support the search, but they have turned into money-making operations for universities. Scientific publishers are supposed to be vehicles for disseminating new findings, but they have become, at best, money-making operations, and at worst, disseminators of "paper mill products," i.e., an utter mockery of science.[2]

As a result of the processes describe in this book, the entire biomedical literature has become unreliable—not just by a little but by a lot. And here is the kicker. *The entire system exists on taxpayer money.* The people who bear the cost of the entire scientific ecosystem deserve better than this for their money. Much better.

What can we conclude at the end of this investigation? How much of the literature is unreliable? My own gut feeling is that about 50 percent of the literature I actively read and (try to) use in my own day-to-day work is not reproducible. The small survey I did with colleagues produced a similar number, just slightly above 50 percent—with a very large range: 10 percent to 90 percent. (This wide range may have something to do with different disciplines, but more likely it is due to differences in the temperament/level of frustration of my colleagues.)

This number—50 percent—is already scary. Based on the results of the reproducibility projects discussed in the book, the irreproducibility number is likely higher: about 70 percent—an even scarier number. However, the true number must be even higher for several reasons. First, reproducibility projects typically

226 AFTERWORD

MISCONDUCT-O-METER

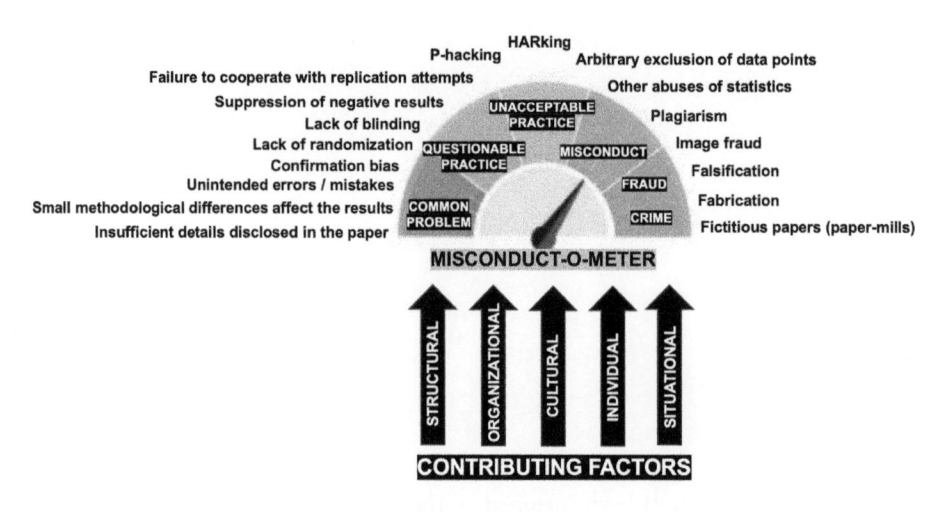

- HARking
- P-hacking
- Arbitrary exclusion of data points
- Failure to cooperate with replication attempts
- Other abuses of statistics
- Suppression of negative results — UNACCEPTABLE PRACTICE
- Lack of blinding — Plagiarism
- Lack of randomization — QUESTIONABLE PRACTICE — MISCONDUCT — Image fraud
- Confirmation bias — Falsification
- Unintended errors / mistakes — FRAUD
- Small methodological differences affect the results — COMMON PROBLEM — Fabrication
- Insufficient details disclosed in the paper — CRIME — Fictitious papers (paper-mills)

CONTRIBUTING FACTORS: STRUCTURAL, ORGANIZATIONAL, CULTURAL, INDIVIDUAL, SITUATIONAL

7.1 Key factors contributing to the unreliability of the literature.
Source: Author's own analysis.

select high-profile papers published in leading journals; these papers may be somewhat more reproducible than the average publication. Second, there are many low-quality journals and low-quality publications contained in them that are not even listed in proper scientific indexing databases, so nobody or almost nobody reads them, but they are also part of the published body of scientific literature. This bottom of the barrel material—a lot of it completely imaginary paper mill products or plagiarized copies of previously published material—has near-zero reliability. Taking all of this into account, my own estimation is that 80 to 90 percent of the life science/biomedical literature published *today* is irreproducible. This is similar to the unofficial estimation of John Ioannidis—a respected scientist who has probably studied this question more than anyone else on the planet.[3]

This irreproducibility comes from a variety of factors, many of which are baked into the process of conducting biomedical research (figure 7.1). Others arise from the various pressures and perverse incentives that are an integral part of the current scientific system. Another significant component is the unethical behavior of many scientists, which may be facilitated by poor training, enabled by lax controls, and stimulated by the combination of significant upsides (unfair advantages, as long as the misconduct goes undetected) and relatively mild downsides and penalties (if it is, in fact, detected).

TABLE 7.1 Potential factors contributing to irreproducibility in life science research

Potential contributor to irreproducible research	Always/often contributes (%)	Sometimes contributes (%)
Selective reporting	68	25
Pressure to publish	64	30
Low statistical power or poor analysis	57	33
Not replicated enough in original lab	54	34
Insufficient oversight / mentoring	52	39
Methods, code unavailable	45	37
Poor experiment design	45	42
Raw data not available from original lab	41	38
Fraud	41	28
Insufficient peer review	38	44

Source: Results of a 2016 survey of 1,576 anonymous responders. See Monya Baker, "1,500 Scientists Lift the Lid on Reproducibility," *Nature* 533, no. 7604 (May 2016): 452–54, https://doi.org/10.1038/533452a.

What, then, is the relative proportion of the various factors that contribute to the irreproducibility of the literature? *What proportion* of the irreproducible literature is made up of well-intentioned, well-designed, and honestly conducted studies that are irreproducible due to the inherent complexity of their underlying methods? *How much of this literature* is irreproducible due to biased and "sketchy" methods employed by the investigators? *How much of it* is intentionally grossly fraudulent or completely fictitious? In the absence of comprehensive data, this is anybody's guess; my guess—stated earlier in this book—is 25 percent–25 percent–50 percent. My own minisurvey of trusted colleagues produced the following proportions: 40 percent–40 percent–20 percent. A survey conducted in *Nature* produces similar numbers (see table 7.1), confirming that my colleagues are less pessimistic about the rate of deliberate fraud than I am. Of course, they are not familiar with all the relevant literature that I have had the (self-inflicted) misfortune of wading through while writing this book.

I have focused on basic science, i.e., laboratory-based studies typically done on cells or animals or model systems. The fact that these studies are irreproducible—in most cases—does not *directly* threaten human lives. Human lives are threatened indirectly, however, because of lost opportunities. Consider what would happen if this money was spent on something else, something better—for example, on *reproducible* studies that would be useful for improving health care.

The published body of *clinical* research is not looking better than the preclinical literature in terms of reproducibility. In this area, the problems are similar to the issues plaguing laboratory research: investigator bias, selective reporting (only those trials that show the expected drug effect), and statistical maneuvering, as well as outright fraud. But now patient lives directly come into play. One case in point is the massive worldwide lethal cluster of fraud and misconduct and regulatory problems that is euphemistically called the "opioid crisis."[4]

There is also the in-between zone of preclinical and clinical research, the area in which preclinical studies are directly linked to the development and approval of new drugs. This area, too, is often implicated in lapses of scientific integrity or potential misconduct. One field that has received much attention lately is the therapy for Alzheimer's disease. Long-accepted fundamentals in the field—such as the pathogenetic role of a modified amyloid protein, claimed in a 2006 *Nature* paper that has been cited more than 2,600 times—has been questioned based on image forensics. The paper was retracted in July 2024; the retraction note mentions "excessive manipulation, including splicing, duplication and the use of an eraser tool" and states that "the data cannot be verified from the records." As I was correcting the galley proofs of this book in October 2024, questions were raised in a lengthy *Science* article about more than one hundred papers published by another "giant" of the neurodegeneration and Alzheimer field, NIH's National Institute on Aging neuroscience chief Eliezer Masliah. The question is whether the very limited progress in the therapy of this condition may be due to the fact that the field has been misdirected for almost two decades.[5]

I tried to write this book in a way that a lay reader—i.e., someone with basic biology training and a general understanding of science—can follow it. At the same time, I tried to remain scientifically correct and avoided oversimplifications so I don't annoy readers with a deeper biomedical science background. Endnotes highlight selected references, but the referencing is not comprehensive; it is not at the level of a scientific paper. Also, I suggested many follow-up reading materials. And I tried to lighten the mood with a few cartoons and some comedy sections in the appendix. Most of it is dark humor, of course: this is the only kind of humor that fits the topic.

This book also contains fragments of a personal story. The journey of a starry-eyed young scientist entering the field of science, and continuing with the scientist working hard and hopefully contributing to the field over thirty years, but during all this time gradually realizing that the entire system suffers from major problems. And now that same scientist—not so young anymore, unfortunately—has written a book concluding that about 30 percent of the papers that come out every year are fake garbage and that 70 to 90 percent of the published scientific literature is not reproducible.

I tried not to write this as a "gotcha" type of story. The important point is not naming the individuals who engage in questionable practices or fraud. The important point is identifying the misaligned processes, misplaced priorities, perverse initiatives, and delineation of the forces and institutions who maintain them.

This is the only way improvements can be implemented. And we can only expect significant changes if multiple components are adjusted simultaneously. If we keep trying the same things that did not work in the past—i.e., small, gradual, superficial improvements—we should not expect a different outcome. According to Einstein, an approach of this type is the true definition of insanity.

And one more thing. The scientific approach remains *the only game in town*. The goal of science reform is *not* to create mistrust in the fundamentals of the scientific process in the eyes of the public. In his 2014 interview, Paul Brookes—one of the early science detectives who was quickly outed and silenced—makes an important point that remains valid to this day:

> Discussing possible science integrity problems in public is potentially damaging in the long term, because if the public gets the idea that a major portion of the science literature is just wrong, then maybe they will refuse to fund science. This is a catch-22, because if we do nothing, then the bad science will be perpetuated. If we speak out, then we risk the public will not believe in science anymore, so we need to be careful.[6]

I want to conclude with some encouraging notes.[7] The most positive point is that the problem, at last, is being acknowledged. Every fellow scientist I talked to privately during the preparation of this book agrees with me that the field is facing a severe crisis. Opinion papers in major scientific journals talk about various aspects of the crisis. And several books discuss various aspects of the problem, from laboratory studies to clinical research.[8] Dozens of bloggers and vloggers are raising the public's attention to the problem. Today, jocular excuses such as "even Newton, Pasteur, and Mendel 'cooked' some of their data" are no longer in fashion.[9] Some old mantras—"science is a self-correcting environment" and "this is only a few bad apples"—are not being repeated anymore. The problem is no longer being swept under the rug. The symptoms of the disease are abundantly clear, and many of its root causes have been identified. Several manifestos have been published that outline a better world of scientific processes—unfortunately, most of them fail to provide any clear ideas on how to get there or who should manage the change. This is where we are today.

And here are a few more things. None of the misaligned processes and none of the specific examples of individuals' actions discussed in this book were discovered by me. All of this information was *already in the open*; some of it for quite some time. Many of these examples are regularly used as training materials in seminars or appear in training documents related to scientific integrity. It is possible that the picture painted in this book is darker than many prior assessments. I agree. Some of my suggested solutions are radical and likely controversial because I don't have much hope that the traditional, gradual solutions will work. I do not consider myself a whistleblower, nor someone who gave away internal scientific secrets; nor was my aim to discredit biomedical science or the scientific community. I simply collected and presented the data *that are already out there* to document the current situation. As for the grant system discussion, I focused on the U.S. grants and the NIH because this is the system that is best documented, this is the one I am most familiar with, and—given the massive amounts of money it expends—improvements in its efficiency would make a significant difference. But many other problems covered in the book—such as the reasons contributing to the intrinsic unreliability of the publications and the problems of the publishing industry—are global problems.

I also want to stress that I wrote this book as a private person. None of the analysis, ideas, and suggestions contained here represent the positions or suggestions of my current employer, the University of Fribourg, in Switzerland.

So please, don't shoot the messenger. Instead, let's try to solve the problem.

APPENDIX

APPENDIX 1: A HANDY TOOL FOR MALIGNANT GRANT REVIEWERS

This is one of my feeble attempts at dark humor. This table should serve as a tool for grant reviewers in need of excuses to reject applications (no need to get down into the nitty-gritty of the actual science proposed!).

Characteristic of the application	What your comments should say	What you really mean by it
Early-stage scientist, with a meteoric rise in the field.	Ambitious applicant.	They made me wait for my turn; this applicant should wait too.
Brilliant new scientific ideas are presented.	Ambitious application. (or: The translational value and potential impact of the application is unclear.)	Maybe I can use these ideas later in my own research, although I may have signed a piece of paper that says I will never do that.
True discovery—a well justified screening project—is proposed.	The proposal is not hypothesis-driven.	We are not in the business of supporting "fishing expeditions."
Confirmation of a recently published concept or finding is proposed.	The proposal is not innovative.	We are not in the business of supporting derivative nonsense.
A repurposing effort is proposed.	The proposal is not innovative.	A drug company should pay for this.
Meticulously prepared application, supported by strong preliminary data. The applicant seeks to find answers to the next questions in the field.	This project will only produce an incremental advance in the field.	I almost fell asleep. Boring.
The applicant is down to his last grant, has been struggling to maintain his group. He published a few solid papers, but is no longer in his halcyon days.	The productivity of the applicant, in recent years, has been only moderate.	Time for early retirement.

(Continued)

Characteristic of the application	What your comments should say	What you really mean by it
The proposed project uses expensive methods and requests a higher than usual budget.	The budget is unrealistic relative to the estimated requirements for equipment, supplies, and personnel.	In my whole career I never got a grant with this high of a budget, why should anybody else be different?
Seems like a great grant, but in a field that I don't know much about and did not bother read up on either.	The application is convoluted and contains excessive repetitions.	Nobody can prove later that the application was written well and I wasn't doing my job.
There are a few typos and formatting issues.	The application is prepared with carelessness and with a lack of attention to detail. The same attitude might extend to the execution of the proposed study.	The formatting problems are sufficient to downgrade the application. This way I don't have to think about the actual science and can move on to the next grant in my pile.
The authors' hypothesis does not agree with the concept of the referee, or with a concept of a friend.	The applicant is unfamiliar with the background of their proposed research.	I can't let this applicant challenge my buddy's science, let alone my own track record.
In the "Introduction to a revised application" part, the applicant rebuts some points where the original reviewer was factually incorrect or unreasonable.	The applicant is argumentative and does not take the critiques seriously.	How dare you disagreeing with the reviewers?
The applicant moves into a new field and tries to apply existing skills to a new area.	The applicant does not have sufficient expertise in the proposed field of investigation.	How dare you moving into somebody else's field?
The applicant has decades of research experience and assumes that you can believe the applicant knows how to do statistics. Prior stats papers are cited.	It is a moderate weakness of the application that a power calculation and a full statistical analysis plan is missing.	I have finally found something I can kill this application with.
Most proposals either (a) use standard, validated methods or (b) propose new methodological approaches.	If (a): A weakness is that the methods are not innovative. If (b): A weakness is that the implementation of the proposed new methods is risky.	One way or another, I can kill this application on the methods aspects.
A great grant, with strong preliminary data, strong applicant, well written, well presented. You cannot even find a single typo.	It is a weakness of the proposal that the results of the proposed studies are not included as preliminary data. The application is rated as excellent.	I can always ding them for not yet completing what they are proposing. Also, "excellent" does not mean "outstanding" or "exceptional." It will probably not be funded.

APPENDIX 2: HOW TO USE AI TO CREATE ABSTRACTS OUT OF THIN AIR

Another feeble attempt at humor. I fed this prompt into ChatGPT. It did not have the slightest problem with the fact that I asked about a completely nonexistent entity called "nanocapsulated zumbroid." In a few seconds it confabulated a wonderful abstract on its anticancer effects out of thin air—with actual data, numbers, fake authors, and everything else.

PROMPT to ChatGPT:

Please write me a five-hundred-word abstract for a scientific meeting. The authors should be Frau D. Ster and four colleagues whose names should be selected from common names contained in cancer biology papers. The abstract should demonstrate that in cell-based and mouse models of cancer the nanocapsulated zumbroid shows significant efficacy. It should mention at least five different molecular and cellular methods used, and it should have some numbers with mean and standard error.

RESPONSE from ChatGPT:

Title: Enhanced Efficacy of Nanocapsulated Zumbroid Demonstrated in Cell-Based and Mouse Models of Cancer

Authors: Frau D. Ster, John A. Thompson, Sarah E. Davis, Michael J. Lee, Emily R. Morgan

The pursuit of innovative cancer therapies remains crucial in addressing the complex challenges posed by this debilitating disease. This study unveils the significant therapeutic potential of nanocapsulated zumbroid, a novel agent, in both cell-based and mouse models of cancer. The data presented here underscore the promise of zumbroid-loaded nanocapsules as a compelling strategy for targeted cancer treatment.

To comprehensively assess the therapeutic efficacy of nanocapsulated zumbroid, we harnessed a diverse range of molecular and cellular methodologies. Initially, we evaluated the impact of zumbroid on multiple cancer cell lines representing distinct tumor types. High-throughput screening assays revealed a substantial reduction in cell viability upon zumbroid treatment. Notably, the mean IC_{50} value across various cell lines was 12.5 ± 2.3 μM, highlighting the potent anti-proliferative effects of zumbroid.

Mechanistic investigations were conducted to elucidate the underlying pathways contributing to zumbroid's efficacy. Immunoblotting analyses demonstrated that zumbroid treatment led to the downregulation of critical pro-survival proteins, including Akt and ERK, by 45.8 percent ± 6.2 percent and 38.6 percent ± 4.7 percent, respectively. Conversely, pro-apoptotic markers such as Bax and caspase-3 were significantly upregulated by 2.3-fold ± 0.4-fold and 3.7-fold ± 0.6-fold, respectively, confirming zumbroid's multifaceted anti-tumor mechanism.

Furthermore, the efficacy of zumbroid was validated in vivo using mouse models of cancer. Intravenous administration of zumbroid-loaded nanocapsules led to a remarkable 62.4 percent ± 4.5 percent reduction in tumor volume compared to controls. Notably, zumbroid treatment extended median survival by 28.6 percent ± 3.2 percent in mice bearing xenograft tumors. Histopathological assessment of tumor sections revealed a 3.2-fold ± 0.7-fold decrease in Ki-67 staining, indicating reduced cellular proliferation following zumbroid treatment.

To assess the nanocapsules' targeted drug delivery efficiency, we employed advanced imaging techniques. Fluorescence microscopy demonstrated preferential accumulation of nanocapsules within tumor tissues, achieving a mean tumor-to-background ratio of 7.2 ± 0.9. Additionally, pharmacokinetic analysis revealed a prolonged half-life of zumbroid in the bloodstream, with a mean t½ of 6.7 ± 0.5 hours, thereby supporting sustained therapeutic efficacy.

In conclusion, this study showcases the significant therapeutic potential of nanocapsulated zumbroid in cell-based and mouse models of cancer. The mean IC_{50} value of 12.5 µM, coupled with its multifaceted mechanism involving Akt, ERK, Bax, and caspase-3, emphasizes its promising anti-tumor activity. The substantial reduction in tumor volume and extended median survival validate the in vivo efficacy of zumbroid-loaded nanocapsules. Our findings highlight the potential of nanocapsulated zumbroid as an effective targeted therapeutic strategy in the battle against cancer.

APPENDIX 3: A HANDY TOOL FOR SCAMFERENCE ORGANIZERS

Yet another of my feeble attempts at dark humor. With each letter, a triple goal must be achieved: (a) to annoy the target, (b) to make them feel bad about not responding to you, and (c) to make them question their own sanity, or in the very least their own command of the English language. With simple modifications, a similar letter can be used for predatory journal article invitations. (Everything in the left column was taken from actual emails I received in the last year.)

Component of the letter	Purpose of the component
Hi Csaba Szabo, Join International cast of speakers	Already at the email's topic line, it is best to throw in some bad English and incorrect capitalization. This way you will immediately gain the trust of the recipient (who are we kidding, let's call it the "target"). It is also essential to stress the "international" character of the scamference. Most attendees of scamferences are in it only for the travel component anyway.
Dear Szabo, (Alternative 1: Hello Dr. Szabo Csaba,) (Alternative 2: Dear Dr. Papapetropoulos,)	Make it absolutely clear already that the letter is a product of name/email fishing from the internet. Make sure that the email address and the target's name do not match; best to use a collaborator's name; this gives some degree of familiarity to the message. Don't bother to find out if your target is Dr, or Professor, or whatever. That would take actual work.
Greetings of the day! I hope you are doing well.	You will immediately win over the target if you start your letter with chummy nonsense.
We had recently reached out to you regarding your participation to the research topic but unfortunately, we have not heard back yet so I wanted to reach out to you personally to see if you may be interested. (Alternative: Hope all is good and healthy on your side. I apologize for the inconvenience if the letter disturbed you. I'm writing to follow-up my last invitation as below, would you please give me a tentative reply? Thank you.)	Let's put the pressure on the target early on by calling them rude and lazy for not responding earlier. (There may or may not have been a prior email—in any case, that is not the point).

We are happy to invite you to the upcoming 23nd Edition of XXX World Conference scheduled from September 4th to 6th, 2024.
 	
   It is an honour and privilege to have your participation in this Conference as Speaker/Delegate/Student.
	It always good if the target knows that the scammers are happy while they are trying to scam you into paying for a predatory conference.
	
	   "XXX" should be some organ that the organizers assume many people have heard about, like the brain or the heart.

(Continued)

Component of the letter	Purpose of the component
The program committee went above and beyond to put forth a meaningful and comprehensive program having keynote sessions, oral and poster presentations. As a member of the Scientific Committee, you will have the privilege of serving as the Chair/Co-chair for the session of your choice, as well as the chance to represent as a Scientific Adviser for the conference.	Put in a high number on the conference series because it may project some degree of track record (does not need to be a real number; nobody can check, because prior scamferences are quickly deleted from the internet). It is a good idea to mix British and American spelling in the same letter. This will broaden the appeal of the message. If all possible levels of scientist are listed, from student to professor, then the target can be sure that the scammer did not bother looking up anything about the target: the email is 100 percent robot-generated. It also helps to confuse matters a little: let's phrase the letter so that the target has no idea if writer is on the Scientific Committee or if the letter is written by the Scientific Committee (in any case, it does not really matter because in reality there is no scientific committee).
The conference will be conducted in San Francisco, USA. (Alternative: You will experience mythical nature and culture of this most developed country in Asia, such as Mount Fuji, Shrines and Temples, Hot Springs, Kimono, Traditional Drama, and Amazing Food.)	American cities are good places to organize a scamference, but Japanese cities are good too. If the latter, make sure you cram all possible stereotypes about that country into a single, badly constructed and incorrectly capitalized sentence.
The conference will be conducted in a hybrid format. (Alternative 1: This HYBRID EVENT allows you to participate In-person at Tokyo, Japan or Virtually from your home or workplace.) (Alternative 2: We are organizing this year's conference in hybrid format with both on-site and online versions and we're hoping for a tremendous turnout.)	Make sure the target understands that they don't actually need to be at the conference; all that matters is that they pay the registration fee. Also, the target will appreciate if you are kind enough to offer them several options to log into your scamference: either from home or from their workplace.
Conference theme: "Recent Advances in XXX & associated Disorders and Treatment."	Make sure the scamference has an extremely broad theme. Capitalize as many words as possible.
Conference website: ZZZ	Put in a website link to the conference, but make sure that no information is to be found in terms of any credible organizing committee or respectable invited speakers. Best to leave these parts blank.

Component of the letter	Purpose of the component
	Put in some broad information on the sessions to be held, which should include everything under the sun that the XXX organ can do. At the same time make sure the registration link, where the money is to be paid, is working well.
Your participation will be a great honor, promising valuable collaboration with experts in the field. The event addresses disorders, psychology, chemistry, and more, providing informative sessions, discussions, and presentations for a fruitful learning experience that significantly contributes to the conference's success.	Nothing works better than vacuous flattery: tell the target that their participation will be a great honor—of course, without offering any contribution to their travel or accommodation.
The meeting aims to explore frontier topics in disruptive technology. We will invite more than 300 leading scientists, together with students and postdoctoral fellows, to participate in a series of cutting-edge discussions. We hope that participants will be able to explore technological applications during the 3-day scientific program and post-conference workshops.	The listed topics should be ridiculously broad and as unrelated as possible. You must always include "and more." This way, no matter what the target's field is, there is always a possibility to submit something.
	It also is useful if it is explained what conferences can do in terms of "fruitful learning experiences." It is definitely a good move to treat the target like an infant who has no clue what conferences are.
	They will *invite* 300 people; this does not mean that anybody will actually attend. Make sure to throw in cool phrases like "cutting edge" and "technological applications." "Disruptive" is a good one too.
We want to ask you to submit an abstract confirming your participation. Your prompt response to this email would be greatly appreciated. (Alternative 1: As a positive acknowledgement from your side, please send us the title of your talk.) (Alternative 2: It will be a privilege to have you at the event. We hope that you will honor us with your presence. We are expecting your timely confirmation and talking if there are any queries or questions.)	Let's continue scattering some bad English in there, while simultaneously stressing the high urgency of the matter.
We are waiting for your favorable consideration and eagerly anticipate your potential involvement. (Alternative: If you have any questions or need clarification, please feel free to contact me anytime. Waiting for your swift and favourable response.)	It is best to wrap it up with the same flowery nonsense that worked so well in the beginning.

(Continued)

Component of the letter	Purpose of the component
Best regards, Karen Smith (alternative: Nova Iris) Email, Phone US address	The name of the scammer has to be an English common name. Female names are better. Any flower reference is always a winner. The scammer's address should be a U.S. address, in a large city that the target may have heard of. PO boxes are okay: nobody will check it, and nobody uses snail mail anyway.
PS- If you don't prefer to receive mails from us reply us back with subject line: Remove/Unsubscribe. We concern for your privacy.	This way they will know they have a "live" email address. They can sell the target's contact info, and their scamference invitations will multiply.

APPENDIX 4: A HANDY LIST OF TRIED AND TRUE RESPONSES

All of them are taken, verbatim, from the written responses of scientists whose papers have been "flagged" for various problems, mostly "image irregularities."

- "The data within the paper were rigorously and thoroughly reviewed by the Journal and the investigator's institution, and no merits for these concerns were identified."
- "The figures were original at all. The particles look similar, but the brightness, distance and contours were different, under the high-power field. Thanks again."
- "Part of the experiments in this paper, including western blot and flow cytometry test, is entrusted to a third party to complete and provide the experimental results to us."
- "We wrote an excellent article. It is unfortunate that we got scammed by [an outside] editing party."
- "We are delighted to reperform the experiments to further verify our results."
- "We will check our data files for errors and make corrections as needed."
- "We apologize for unable to offer the original films, caused they were spoiled in a damaged box."

- "We did a thorough search for the original data and could not find it as, at the time, the data on the microscope used were not archived."
- "Although it was determined by the investigating committee that these issues likely did not affect the overall conclusions of the study, the authors have requested that the paper be retracted."
- "All the journals that published our papers did not have any problems with our data. But the [internal] investigation committee members forced these journals to retract our papers."
- "We inadvertently inserted an incorrect image."
- "The top left panel was inadvertently mislabeled."
- "Some of the data has been inappropriately superimposed."
- "A panel of the figure was inadvertently duplicated during the review process."
- "... representative illustrations presented in Figure [XX] might be mistaken for original data. As this panel was used strictly for demonstrative purposes, it has been removed from the figure for clarity."
- "We are facing fraud accusations because of duplications in data tables that are assembled after a paper has been completed, not questions about the actual data."
- "We believe that the overall conclusions of the paper remain valid, but we are retracting the work due to these underlying concerns about the figure."
- "The conclusions of this manuscript have been supported by other peer reviewed publications."
- "The principal observations of this article were further confirmed in [further] publications."
- "I will only confirm that every conclusion offered in papers on which I am one of the authors is sound and can be relied upon."
- "Conclusions of this peer reviewed manuscript are not altered by the above."
- "The unreliable data is an aberration of the excellent patient care and academic research at [our institution]."
- "I have no comment and will allow the appropriate processes to address any concerns raised about my scientific research."
- "What is the world of science coming to?"
- "I see no need to defend my work against a group of anonymous detractors who negate a priori my integrity, yet are not even prepared to identify themselves."

- "Self-appointed "peers" are more interested in mud-slinging than in post-publication review and quality control of science."
- "PubPeer is now a platform for resentful, anonymous, petty, failed scientists to harass those who actually make discoveries and occasionally make mistakes because they are human."
- "Yup, we are being hounded."

APPENDIX 5: SHENANIGANOMETRY

Another feeble attempt at dark humor. Here are some types of shenanigans one encounters in the world of life sciences, handily organized alphabetically.

Type	Characteristics
The Boss-Man	Makes it clear that he is only interested in seeing data that confirm his preconceived hypothesis. Works with Yes-Men and Middle-Men.
The Buffoon	No longer capable/willing to do science—yet poses as the group leader.
The Circle-Former	Participates in a "club" where a group of referees accept each other's papers.
The Coauthor	Insists on authorship when it is not justified according to common rules.
The Copy-Paster	Reuses his own published material or lifts other authors' material.
The Crafty Discussant	Introduces the topic in the paper's Introduction as if it was completely novel; hides the relevant prior art in a remote part of the Discussion section.
The Disliker	Rejects other people's papers/grants due to jealousy.
The Dogfighter	Makes two postdocs compete on the exact same project: the one who finishes first gets to publish.
The Entrepreneur	Secretly uses his academic lab for industry-sponsored research; pockets the contract money. In the long term, this works best if there are several Picker-Choosers and Middle-Men in the group. (Later, the client company will wonder why others can't replicate the results.)
The Expert Referee	Evaluates—and accepts!—his own papers—by providing his/her own fake referees' email accounts to the journals during the submission.

Type	Characteristics
The Fashionist(a)	Jumps on the scientific bandwagon promising the most grant money and fame.
The Fierce Competitor	Rejects other people's papers/grants due to direct competition in the same field.
The Friend	Overlooks problems in papers/grants of friends, colleagues, and collaborators.
The Flipper	Copy/pastes, flips and reinserts various Western blot lanes, until a "perfect experiment" is created. Good friend of the Photoshop Artist.
The Forgetful	Agrees to collaborate on a topic, but in the end publishes on his own.
The Globetrotter	Travels the world and delivers the same talk over and over.
The HARKer	Forms his scientific hypothesis after the results are known, and presents it as if the hypothesis was formulated prior to the beginning of the project.
The Historian	Inappropriately reuses historic controls. These are often many years old, and often were not even conducted under the exact same experimental conditions.
The Hype-Man	Overinterprets and overhypes research findings.
The IF-Booster (Referee)	As manuscript referee, forces the authors to cite his/her own papers.
The IF-Booster (Editor)	As journal editor, pads the journal with reviews/guidelines to elevate its impact factor. In addition, forces the authors to cite other papers from the same journal—whether or not they fit the manuscript.
The Ignorer	Omits relevant literature in his/her paper to artificially boost the perceived importance of the findings.
The Keeper	Ignores or rejects colleagues' requests to share a reagent or research tool.
The Magician	"Produces" a complete study from a limited number of preliminary datapoints.
The Meeting-Buddy	Attends scientific meetings, hunts for interesting posters where the material has not yet been published. Extracts all relevant detail and information from the unsuspecting victim. Then, at home, repeats the experiments (or not . . .), and quickly publishes a paper.
The Method Man	Has a particular technique and runs various things through it indiscriminately. The resulting publications have a striking resemblance to each other. Good friend of the Salami-Man.

(Continued)

Type	Characteristics
The Middle-Man	Acts as a selective data filter of the work of Yes-Men and presents the "correct" datasets to the Boss-Man. Makes sure the Yes-Men stay "on message" and "in line" and never get a chance to interact with each other to discuss the data.
The Multiplier	Claims more n-numbers per experiment than what was performed.
The Omitter	Leaves out those experiments that "did not work" from the statistical analysis.
The Patent-Man	Uses the results from his university laboratory to apply for patents for his own company.
The P-Hacker	Keeps increasing the experimental replicate size or keeps running the stats software until something significant appears.
The Photoshop Artist	Manipulates histological pictures or Western blots to fit a certain hypothesis.
The Picker-Chooser	Picks and chooses data to include in papers/grants; leaves out bits that "don't fit."
The Politician	Spends most effort on networking, committees, and self-promoting.
The Recycler	Uses the same histological or immunohistochemical picture or Western blots in different papers, and labels them to depict completely different experiments.
The Salami-Man	Divides up the data from a project that belongs together and publishes several different papers.
The Self-Citer	Cites his own work even though the correct citation was someone else's paper.
The Sellout	Paid editor for predatory journals. Invited speaker at scamferences.
The Shopaholic	Purchases devices or reagents from a particular vendor for perks / kickbacks.
The Stealer-Rejecter	Rejects grants/papers but uses the ideas contained in them for his own research.
The Taskmaster	When refereeing others' papers, requests time-consuming or impossible additional experiments. Often moonlights as a Stealer-Rejecter.
The Yes-Man	Selectively picks and presents datasets that the Boss-Man wants to see.
The Wizard	Conjures entire datasets and whole papers out of thin air.

NOTES

PREFACE

1. John P. A. Ioannidis, "Why Most Published Research Findings Are False," *PLOS Medicine* 2, no. 8 (August 2005): e124, https://doi.org/10.1371/journal.pmed.0020124.
2. Florian Prinz, Thomas Schlange, and Khusru Asadullah, "Believe It or Not: How Much Can We Rely on Published Data on Potential Drug Targets?," *Nature Reviews Drug Discovery* 10, no. 9 (August 2011): 712, https://doi.org/10.1038/nrd3439-c1.
3. C. Glenn Begley and Lee M. Ellis, "Drug Development: Raise Standards for Preclinical Cancer Research," *Nature* 48, no. 7391 (March 2012): 531, https://doi.org/10.1038/483531a.
4. Timothy M. Errington et al., "An Open Investigation of the Reproducibility of Cancer Biology Research," *eLife* 3 (December 2014): e21627, https://doi.org/10.7554/eLife.04333.
5. Monya Baker, "1,500 Scientists Lift the Lid on Reproducibility," *Nature* 533, no. 7604 (May 2016): 452–54, https://doi.org/10.1038/533452a.
6. The selection of responders may not have been fully representative of the entire scientific community: it is possible that people who are more concerned or frustrated with this matter responded to the survey.
7. So, please, don't get me started on the topic of the pseudoscience called "alternative medicine."

1. CAREERS AND CAREER PRESSURES IN BIOMEDICAL SCIENCE

1. As I realized later, this taught me useful skills such as using a healthy dose of skepticism and the ability to "read between the lines"—excellent tools for reading the scientific literature.
2. The Rubik's cube was invented in 1979 by the Hungarian architect Ernő Rubik.
3. The Western approach to medical genetics was not exactly a favorite scientific topic in the Soviet Union at the time, but in Hungary it was supported and popularized.
4. This is, in fact, only a Hungarian "urban myth." If we count Nobels per capita, at least ten countries do better than Hungary, including Switzerland, Austria, and the United Kingdom. However, Hungary is ranked pretty high on this list, above Germany, the United States, and France.

5. I highly recommend this excellent book on Szent-Györgyi's life: Ralph W. Moss, *Free Radical: Albert Szent-Györgyi and the Battle Over Vitamin C* (New York: Paragon, 1988). It is out of print and hard to find, but it is worth seeking out.
6. Mihály Beck, *Tudomány-Áltudomány* (Budapest: Akadémiai Kiadó, 1977).
7. Most scientists who use this term don't even know where it comes from.
8. Paul M. Vanhoutte, "Could the Absence or Malfunction of Vascular Endothelium Precipitate the Occurrence of Vasospasm?," *Journal of Molecular and Cellular Cardiology* 18, no. 7 (July 1986): 679–89, https://doi.org/10.1016/s0022-2828(86)80940-3.
9. Csaba Szabó et al., "Noradrenaline Induces Rhythmic Contractions of Feline Middle Cerebral Artery at Low Extracellular Magnesium Concentration," *Blood Vessels* 27, no. 6 (1990): 373–77.
10. Richard M. J. Palmer, A. G. Ferrige, and Salvador Moncada, "Nitric Oxide Release Accounts for the Biological Activity of Endothelium-Derived Relaxing Factor," *Nature* 327, no. 6122 (June 1987): 524–26, https://www.nature.com/articles/327524a0.
11. This is a famous Nobel omission with relevance for how scientific meetings are held and how and why scientists have become extremely guarded about disclosing anything unpublished at any meeting. This topic is covered in more detail later.
12. Just like nitric oxide, initially these molecules were considered highly toxic and were not considered relevant for the regulation of biological processes in human cells. Later on we and others discovered that in small concentrations they do play interesting roles as biological regulators in human cells and tissues.
13. At that time, the Candidate of Science degree was typically awarded to midcareer scientists rather than to recent medical school graduates. Submission of my thesis at the age of twenty-five raised a few eyebrows at this powerful institution. It was put on hold for several years before I was finally allowed to defend it. From those times to the present day, let's just say that I am not the most favorite person in the eyes of this hallowed academy. It also may have had something to do with the fact that in 2006, when I was already a scientist of international stature, I called for a full reform of the whole, rather obsolete, Soviet-style academic system in Hungary. See Csaba Szabo, "Hungary: Academy Needs More Than Internal Reform," *Nature* 442, no. 7101 (July 2006): 353, https://doi.org/10.1038/442353b.
14. Attributed to Szent-Györgyi by Robert Cohen, in *The Development of Spatial Cognition* (New York: Taylor and Francis, 1985)
15. Siddhartha Roy and Marc A. Edwards, "NSF Fellows' Perceptions About Incentives, Research Misconduct, and Scientific Integrity in STEM Academia," *Scientific Reports* 13, no. 1 (April 2023): 5701, https://doi.org/10.1038/s41598-023-32445-3.
16. Sometimes universities still award PhD degrees in fields that are generally considered extinct or nearly extinct—or at the very least embarrassingly unfashionable—such as biochemistry, physiology, anatomy, or my own field of pharmacology. This field, just like departments with names such as Department of Pharmacology, is now typically folded into more fashionable entities labeled "Molecular Medicine" or "Molecular Pharmacology."
17. In some countries a PhD can be obtained in a shorter time period. For example, in the UK or in Switzerland a PhD can be completed in under three years.

18. Nick Riddiford, "A Survey of Working Conditions Within Biomedical Research in the United Kingdom," *F1000Research* 6 (March 2017): 229, https://doi.org/10.12688/f1000research.11029.2.
19. Chris Woolston, "PhDs: The Tortuous Truth," *Nature* 575, no. 7782 (November 2019): 403–6, https://doi.org/10.1038/d41586-019-03459-7.
20. Katie Langin, "U.S. Labs Face Severe Postdoc Shortage," *Science* 376, no. 6600 (2022): 1369–70. There are many reasons for this shortage. It may have something to do with the availability of jobs outside of academia—such as in the pharma or biotech industry, other private institutions, or perhaps in the academic publishing industry where salaries are much higher and a better work/life balance is promised. It may also have to do with the fact that postdocs or prospective postdocs are getting wiser, and they realize the poor odds for fully making it in academia and choose a different career path. The COVID-19 pandemic did not help either, but the postdoc shortage problem began in pre-COVID years.
21. Maximiliaan Schillebeeckx, Brett Maricque, and Cory Lewis, "The Missing Piece to Changing the University Culture," *Nature Biotechnology* 31, no. 10 (October 2013): 938–41, https://doi.org/10.1038/nbt.2706.
22. The Royal Society, "The Scientific Century: Securing Our Future Prosperity," *The Royal Society Policy Document 02/10*, March 2010, DES1768.
23. Richard C. Larson, Navid Ghaffarzadegan, and Yi Xue, "Too Many PhD Graduates or Too Few Academic Job Openings: The Basic Reproductive Number R0 in Academia," *Systems Research and Behavioral Science* 31, no. 6 (November-December 2014): 745–50, https://doi.org/10.1002/sres.2210; and Shulamit Kahn and Donna K. Ginther, "The Impact of Postdoctoral Training on Early Careers in Biomedicine," *Nature Biotechnology* 35, no. 1 (January 2017): 90–94, https://doi.org/10.1038/nbt.3766.
24. For example, in Hungary, the three consecutive levels are labeled teacher's assistant, adjunktus, docens and professor. In the French part of Switzerland, where I work now, they are labeled senior scientist, maitre-assistant, stabilized maitre-assistant, and two different levels of professor.
25. Yes, there is some degree of correlation between the quality of science performed by a group, the quality of the journals in which the work is published, and the number of grants the group receives. But this is not a tight correlation; many groups that seem to do good science have difficulty bringing in grant money. The recent Nobel laureate Katalin Karikó—whose book *Breaking Through: My Life in Science* (New York: Crown, 2023), gives a vivid example of this matter—is only one case in point. There are also highly polished "grant-writer-type" individuals who consistently bring in vast sums of extramural funds even though their scientific output may not be all that impressive.
26. I cover overheads and the entire "grants game" in chapter 2.
27. In the NIH system in the United States, a grant score is still called a "pink slip"—even though it has been sent as an electronic file since the mid-nineties. If faculty members end up moving, they take their grant money with them. This is not viewed favorably by the investigator's current institution, but by law the institution has to release the money.
28. For unknown reasons, a lot of scientists in the United States call themselves or each other "principle" investigators rather than "principal" investigators.

29. Lay audiences are looked down upon and scorned by many scientist, forgetting that it is their taxes, after all, that pay for the vast majority of their work.
30. A former classmate of mine runs a clinical laboratory in a county hospital in a small town in the United States. One of his duties is to facilitate the "generation" of publications of residents, research fellows, and everybody else who needs it. The hospital does not have any meaningful research infrastructure, so the kinds of publications produced in this process are retrospective clinical analyses, case reports, and review articles on various topics in which the authors don't have any direct hands-on expertise. They publish in lower-tier journals. I am sure all of this activity can advance the career of multiple individuals, but I am also pretty sure it does not contribute much to the advancement of biomedical science.
31. Wei Quan, Bikun Chen, and Fei Shu, "Publish or Impoverish: An Investigation of the Monetary Reward System of Science in China (1999–2016)," *Aslib Journal of Information Management* 69, no. 5 (2017): 486–502, https://doi.org/10.1108/AJIM-01-2017-0014.
32. Smriti Mallapaty, "China Bans Cash Rewards for Publishing Papers," *Nature* 579, no. 7797 (March 2020): 18, https://doi.org/10.1038/d41586-020-00574-8.
33. Dahui You, "Work-Life Balance: Can You Actually Make That Happen?," *Frontiers in Pediatrics* 6, no. 3 (January 2016): 117, https://doi.org/10.3389/fped.2015.00117.
34. Woolston, "PhDs: The Tortuous Truth," *Nature*, November 13, 2019 (https://www.nature.com/articles/d41586-019-03459-7).
35. Richard N. Pitt, Yasemin Taskin Alp, and Imani A. Shell, "The Mental Health Consequences of Work-Life and Life-Work Conflicts for STEM Postdoctoral Trainees," *Frontiers in Psychology* 12 (November 2021): 750490, https://doi.org/10.3389/fpsyg.2021.750490.
36. Jack Grove, "Nobelist: Scientific Success 'No Barrier' to Work-Life Balance," *Times Higher Education*, June 29, 2023, https://www.timeshighereducation.com/news/nobelist-scientific-success-no-barrier-work-life-balance.
37. From time to time throughout the book I include humorous elements such as irony and sarcasm. This is one of many such examples. I won't bookmark every instance.
38. "Science and everyday life cannot and should not be separated." (Rosalind Franklin). "*Futbol Is life.*" (Dani Rojas, from the television series *Ted Lasso*, season 1, episode 6, "Two Aces," first aired September 4, 2020.)
39. Hungarian-born immigrants working in key positions on the Manhattan Project included Theodore von Kármán, John von Neumann, Leo Szilard, Edward Teller, and Eugene Wigner. As a Hungarian-born scientist, I cannot leave out this anecdote: During a top-secret meeting Oppenheimer had to step out to the gents' room, and Leo Szilard then said: "Gentlemen, perhaps we should now continue in Hungarian!"
40. Peter B. Medawar, *Advice to a Young Scientist* (New York: Basic Books, 1981).
41. Thankfully, after I completed my research period at the William Harvey Research Institute in London, I was recruited by the University of Cincinnati for a proper faculty position, so I began my U.S. career on a higher-level visa, the H-1. But this was an exception; the general rule is to recruit earlier-stage investigators on J-1s.
42. Press Trust of India, "US Announces Overhaul of Its Annual Lottery for H-1B Visas to Reduce Fraud," January 31, 2024, https://www.business-standard.com/world-news/us-announces-overhaul-of-its-annual-lottery-for-h-1b-visas-to-reduce-fraud-124013100175_1.html.

43. Like most of my colleagues, I regularly receive requests from prospective green card applicants to write support letters for them. Even though most of them are completely unknown to me, and by browsing through their CVs many don't seem particularly "exceptional," many requests come with obviously lawyer-prepared letters that "only need my signature." (I don't respond to these requests unless I truly believe that the person in question is exceptional.)
44. Adam Marcus, "Braggadacio, Information Control, and Fear: Life Inside a Brigham Stem Cell Lab Under Investigation," Retraction Watch, May 30, 2014, https://retractionwatch.com/2014/05/30/braggadacio-information-control-and-fear-life-inside-a-brigham-stem-cell-lab-under-investigation/.
45. Aaron Mobley et al., "A Survey on Data Reproducibility in Cancer Research Provides Insights Into Our Limited Ability to Translate Findings from the Laboratory to the Clinic," PLOS One 8, no. 5 (May 2013): e63221, https://journals.plos.org/plosone/article?id=10.1371/journal.pone.0063221.
46. An example, focusing on the NCI alone (i.e., just one of the many NIH research institutes), is Natalie B. Aviles, *An Ungovernable Foe: Science and Policy Innovation in the U.S. National Cancer Institute* (New York: Columbia University Press, 2023).
47. Before returning to Europe, my last academic title at the University of Texas was "Tenured Full Professor & Laura B. McDaniel Chair, Departments of Anesthesiology and Pharmacology." It is way cooler-sounding than, say, "Section Head, NIH," isn't it?
48. John Carreyrou, *Bad Blood: Secrets and Lies in a Silicon Valley Startup* (New York: Knopf, 2018); and *The Dropout*, Hulu movie miniseries, 2022, https://www.imdb.com/title/tt10166622/.

2. HYPERCOMPETITION FOR RESEARCH GRANTS

1. Call me paranoid, but knowing how things worked in Hungary, this was not by accident. It was done on purpose to prevent outsiders—such as myself—from applying.
2. For reasons like this and several others, I am no fan of the Hungarian Academy of Sciences.
3. Maybe not countless, but in my thirty years in science, I guestimate that I must have written at least 150 full grant applications. I must have also worked on another 150 as a collaborator, mentor, or consultant advisor. When I look around at my colleagues, I don't think these numbers are exceptionally high.
4. Sometimes institutions can apply for separate equipment or infrastructural grants that enable the successful applicant (typically a whole institution) to get funds to buy a new piece of expensive equipment. In Switzerland, the SNSF paid 50 percent of the cost of a particular mechanism, and the other 50 percent had to be covered by the recipient institution. (This grant mechanism was closed down in 2023, and now all scientists will have to try to squeeze the money from their own institution if a large piece of equipment is needed.)
5. With NIH grants, applicants must prove that their proposed research is linked to human health or disease. With the NSF, there is no such requirement. Many bizarre studies have been funded by the NSF over the years, from studies on whether body size affects how fast animals urinate (it does not), to studies on how social status affects the mental health of adolescents in Nepal (it does), to the ethical and social impact of robot kitchens (we don't know yet, this study is currently ongoing).

6. Many investigators call these grants "RO1"—with the capital letter "O" in the middle. Often these are the same folks who call themselves "principle" rather than "principal" investigators.
7. Often, when one applies for $250,000 per year, one only receives $220,000, for example. Funding institutes may decide to cut everybody by a certain percent to "squeeze in" a few more awards.
8. Jay Greene and John Schoof, "Indirect Costs: How Taxpayers Subsidize University Nonsense," *Heritage Foundation, Center for Education Policy*, no. 3681, January 18, 2022.
9. Beneficiaries include organizations created by Mark Zuckerberg (Chan Zuckerberg Initiative), Jeff Bezos (Bezos Family Foundation), George Soros (Open Society Foundation), and Bill Gates (Bill and Melinda Gates Foundation).
10. Brian Resnick, "Trump Wants to Cut Billions from the NIH. This Is What We'll Miss Out On if He Does," *Vox*, March 11, 2019.
11. National Institutes of Health, "The NIH Data Book," https://report.nih.gov/nihdatabook/.
12. For stats connoisseurs, the formula for calculating the chance of receiving at least one grant when a scientist submits six grants and each grant has a 30 percent chance of funding (i.e. a 70 percent chance of rejection) is $(1-0.70^6) \times 100$.
13. This is an often heard adage in academia: "The best time to start writing your next research grant application is the day you receive one."
14. John Allen Paulos, *Innumeracy* (New York: Penguin, 2000), and James C. Zimring, *Partial Truths* (New York: Columbia University Press, 2022), are excellent starting points on this topic.
15. Once again these pesky Hungarian-Americans are everywhere. For an excellent, recent book about John von Neumann's life, see Ananyo Bhattacharya, *The Man from the Future* (New York: Norton, 2022).
16. Ken Stone, "Scripps Research Institute to Pay 10M for Improper Use of NIH Grants," *Times of San Diego*, September 11, 2020, https://timesofsandiego.com/tech/2020/09/11/scripps-research-institute-to-pay-10m-over-improper-use-of-nih-grants/.
17. In my own anonymous mini-survey, I asked fifteen scientist colleagues of mine who work around the world about a few things. All of them are full professors, are successful in their careers, and are exemplary scientists. One of my questions was this: "What percent of your time is spent on grant-writing and editing?" The average was 30 percent.
18. But investigators must tread carefully. Each field of biology and medicine has its own "popes" and "mini-popes." In addition, some grant reviewers may be seriously offended or feel threatened if an applicant tries to move into a new field.
19. While at the University of Texas, I held a joint affiliation with the Galveston Shriners Institute and received funding for several projects focusing on burn injury (lung disease, wound healing). I even had the rare pleasure of meeting a Grand Potentate, a high-ranking leader of the organization—donning a fez and ornate chest-decorations and all the works. This Freemason organization seems to have some unusual rituals and processes; nevertheless, they have devoted a lot of money to support biomedical research over the years.
20. In my own case, this new country is Switzerland. My current university covers my salary as well as the salary of a small group of research and administrative personnel. At the age of fifty-one, for the *first time* in my career my own salary did not need to be covered—partially or fully—by external grant sources.

21. In addition to the regular R01 evaluations, some investigators are invited to evaluate for other granting mechanisms, for example, to evaluate RFAs, SBIRs, or other special grants of the NIH grant system.
22. Everyone's understanding (or hope) is that their chances of funding at Special Study Sections are higher than average. I could not find statistical data for these sessions alone, but I feel that there may indeed be a higher chance of funding through them. However, there is also a higher degree of randomness than what occurs in the regular system, which, as you will see later, is already wildly random.
23. Imagine the sort of discussions that take place when a group of twenty scientists, most of whom have not even properly read the application, try to come to an agreement on whether an application is "strong with essentially no weakness" or is only "strong with negligible weaknesses."
24. I have no doubt that this will be a contentious point. Let's say that an experienced investigator's proposal is supported in part based on the consistent track record of the investigator. Early-career investigators may consider that this "reputational bias" works against their own competing proposal, whereas later-career investigator recipients may consider it to be a "well-deserved recognition of leading experience in the field." Let's be honest. Whether a project environment is sufficient for a project is not a simple "yes or no" question. One environment may be sufficient, but it may require significant amounts of troubleshooting to start up the experiments, whereas another investigator's environment may be ready to go, with minimal ramping up. One environment may be sufficient but utilizes somewhat outdated equipment that will slow the rate of progress, whereas another environment may have state-of-the art equipment in place that generates results faster and at less cost.
25. "Impact" can be difficult to define, but in the NIH grant jargon it means "a potential for making a real difference in the real world, or at least in the particular field of study." In other words, an impactful project is supposed to do more than merely advancing basic scientific knowledge.
26. This type of situation can happen to rather extreme degrees. One of my colleagues applied for a standard scientific grant in a country in Europe that shall not be named but has an excellent overall reputation for scientific excellence, integrity, and adherence to the rule of law. He happens to be a world expert in his field, and his proposal was novel, contained solid preliminary data, and superbly applied all the usual grantsmanship as well. The first version of his application was evaluated by three reviewers. Each had minor criticisms but were generally supportive. The application was rejected.

    The revised application was sent to five reviewers, including the three original ones. Four of them were supportive, and one brought up some new, relatively minor issues. Rejection again.

    My colleague revised the application again, responded to all points, and resubmitted it. This second revision was sent out to seven reviewers—including the five prior reviewers who had supported the previous application and were puzzled why this was coming back to them after they had already recommended it for funding. Out of the seven reviewers, six supported the application, and one (once again, a newly recruited one) brought up a series of brand new issues that had not been mentioned in any of the previous evaluation cycles. The committee, in turn, excluded the comments of the five original reviewers due to a "lack of scientific substance" and essentially ended up using only two reviews (one positive, one negative) to achieve what they called a "fair and balanced evaluation." The result of this fair and balanced evaluation was, of course, rejection of the application.

At this point my colleague gave up, even though, he said, it would have been interesting to find out if next time the granting agency would go to the extreme of engaging nine reviewers.

27. Sometimes these so-called new projects bear a striking resemblance to an earlier one, perhaps with a different title and a slightly different abstract.
28. If you take your task seriously, you will have spent about fifty to one hundred hours of your own time reviewing the applications prior to the meeting. Next you will sit in a room and engage in intensive discussions for another twelve to sixteen hours. Your travel, hotel, and meals are covered, and you will receive an honorarium of a few hundred dollars. It computes to an hourly rate of $5 to $8: your efforts are generously compensated . . . well below the minimum hourly wage level. This is infinitesimally better, however, than peer-reviewing other investigators' manuscript submissions to scientific journals, which scientists are doing completely for free.
29. Mike Lauer, "Outcomes of Amended (A1) Applications," Extramural Nexus (blog), March 23, 2017, https://nexus.od.nih.gov/all/2017/03/23/outcomes-of-amended-a1-applications/.
30. Thank God for the "find and replace" function in MS Word.
31. A collection of hundreds of thousands of compounds is considered an average library, and a million-compound library is a great one. The relevant NIH center, NCATS, has a library of about 500,000 compounds.
32. RAFs specifically requesting proposals centering on screening efforts are rarely announced. One of the NIH institutes, NCATS, maintains an intramural screening center, and external applicants may be able to access it. But, overall, the interest of the NIH (and other grant-giving bodies worldwide) in investing in screening-based efforts is rather lukewarm.
33. Andreas Papapetropoulos and Csaba Szabo, "Inventing New Therapies Without Reinventing the Wheel: The Power of Drug Repurposing," *British Journal of Pharmacology* 175, no. 2 (January 2018): 165–67, https://doi.org/10.1111/bph.14081.
34. It is done, almost exclusively, by pharma companies today.
35. Nicola J. Curtin and Csaba Szabo, "Poly(ADP-Ribose) Polymerase Inhibition: Past, Present and Future," *Nature Reviews Drug Discovery* 19, no. 10 (October 2020): 711–36, https://doi.org/10.1038/s41573-020-0076-6.
36. Nathan A. Berger et al., "Opportunities for the Repurposing of PARP Inhibitors for the Therapy of Non-Oncological Diseases," *British Journal of Pharmacology* 175, no. 2 (January 2018): 192–222, https://doi.org/10.1111/bph.13748.
37. In light of common public knowledge about the "replication crisis," this position is simply no longer tenable.
38. This is exemplified by the 1926 Nobel Prize of James Fibiger, the 1946 Nobel Prize of Wendell Meredith Stanley, and the 1949 Nobel Prize of Egas Moniz. Each story deserves a separate chapter in some other book, not this one.

39. Some focused scientific efforts and some dedicated sources of funds have finally begun to emerge that support reproducibility and replication projects. More on that later.
40. As mentioned previously, this book contains humorous elements of irony and sarcasm. This is another one of those examples, and there are many more. I won't bookmark every instance.
41. Michael Price, "A Shortcut to Better Grantsmanship," *Science*, June 4, 2013, https://www.science.org/content/article/shortcut-better-grantsmanship.
42. To take a break from the serious aspects of this book, have a look at Appendix 1 for some dark humor related to grant reviewers' comments. They lampoon actual examples I have come across during my career.
43. The NIH grant system is supposed to give some "breaks" to early-stage investigators, but (as shown later) this does not really happen.
44. If you zoom in to the bottom right part of figure B, it looks like a few grants were funded with priority scores of fifty or even higher. This is rather amazing because above fifty is generally considered the "triage zone" where grants don't even get discussed at the Study Sections.
45. A brief sentence, most commonly attributed to Johnny Depp, comes to mind: "You may strategically place your wonderful lips upon my posterior and kiss it repeatedly."
46. This is the same reason scientific meetings are getting duller and duller; few scientists dare to show any truly novel, unpublished information anymore—as you will see later in this book.

47. Not a real country.

48. Likewise, in many other grant systems, the expert reviewers can be overruled by (named or unnamed) members of a subsequent, additional committee. I have seen some puzzling examples in which overwhelmingly positive external reviewers' comments were ignored by the next level committee and funding was denied.
49. Here is an interesting fact. Over a period of time, even having coffee or tea available for the Study Section was not allowed, which surely saved a tremendous amount of money for the NIH.
50. Maybe it is clear by now that among my many hobbies I am also a science history buff. And everybody should stop reading this book immediately and research the life and work of the Nobel laureate Peter Mitchell, one of the pioneers of cellular bioenergetics. A good place to start is with this book by John Prebble and Bruce Weber, *Wandering in the Gardens of the Mind: Peter Mitchell and the Making of Glynn* (Oxford: Oxford University Press, 2003).
51. David J. Galton, "Did Mendel Falsify His Data?," *QJM: An International Journal of Medicine* 105, no. 2 (February 2012): 215–16, https://doi.org/10.1093/qjmed/hcr195.
52. "Case Summaries" by the Office of Research Integrity at the National Institutes of Health can be accessed at https://ori.hhs.gov/content/case_summary. For example, Potts Kant's case is disclosed in Federal Register Notice 84, no. 216, 2019; Spirli's case is disclosed in Federal Register Notice 88, no. 71, 2023; and Armstead's case is disclosed in Federal Register Notice 88, no. 129, 2023.
53. More about them later.
54. This last "penalty" is almost meaningless for two reasons: (1) most of the people listed on the ORI website do not have the scientific stature to be NIH grant reviewers, and (2) now that the sordid affair is public no self-respecting granting agency will ever invite them.

55. Ferric C. Fang and Arturo Casadevall, "Research Funding: The Case for a Modified Lottery," *mBio* 7, no. 2 (April 2016): e00422-16, https://doi.org/10.1128/mBio.00422-16.
56. Elizabeth L. Pier et al., "Low Agreement Among Reviewers Evaluating the Same NIH Grant Applications," *Proceedings of the National Academy of Sciences of the United States of America* 115, no. 12 (March 2018): 2952–57, https://doi.org/10.1073/pnas.1714379115.
57. Elena A. Erosheva, Patricia Martinková, and Carole J. Lee, "When Zero May Not Be Zero: A Cautionary Note on the Use of Inter-Rater Reliability in Evaluating Grant Peer Review," *Journal of the Royal Statistical Society—Series A* 184, no. 3 (July 2021): 904–19, https://doi.org/10.1111/rssa.12681.
58. However, human factors come into play, and bitter and jealous reviewers *do* exist.
59. However, many granting agencies prohibit the same investigator from submitting multiple applications in the same cycle. Even if this is not prohibited, it is often frowned upon by the reviewers.
60. Kelsey Piper, "Science Funding Is a Mess. Could Grant Lotteries Make It Better? We Might Be Better Off Funding Scientific Research by Choosing Projects at Random. Here's Why," *Vox*, January 18, 2019.
61. Joshua M. Nicholson and John P. A. Ioannidis, "Research Grants: Conform and Be Funded," *Nature* 492, no. 7427 (December 2012): 34–36, https://doi.org/10.1038/492034a.
62. C. Lee, "Slump in NIH Funding Is Taking Toll on Research," *Washington Post*, May 28, 2007.
63. Over time, the rich get richer and the poor get poorer. A 2022 NIH analysis shows that the top 10 percent of institutions that got NIH grant funding receive more than 60 percent of the overall funding for research. The bottom 50 percent receives less than 5 percent. See Max Kozlov, "NIH Plans Grant Review Overhaul to Reduce Bias," *Nature* 612, no.7941 (December 2022): 602–3, https://doi.org/10.1038/d41586-022-04385-x.
64. More about her unholy deeds later.
65. Charlie Mom and Peter van den Besselaar, "Do Interests Affect Grant Application Success? The Role of Organizational Proximity," *arXiv* 2206 (2022): 03255, https://doi.org/10.48550/arXiv.2206.03255.
66. Depending on complicated political negotiations, the Swiss and the British enjoy variable levels of EU grant access, which seem to have gone back and forth in recent years with a disturbingly high frequency.

3. "DOING SCIENCE": FROM HYPOTHESIS TO PUBLICATION

1. Over the years, for unclear reasons, all three of these disciplines—physiology, pathophysiology, and pharmacology—have become less and less "fashionable."
2. To those who decide to go against my advice and will skip over these passages, let me say goodbye with an adage attributed to the English statistician George E. P. Box: "All experimental models are wrong, but some are useful."
3. Robert F. Furchgott, and John V. Zawadzki, "The Obligatory Role of Endothelial Cells in the Relaxation of Arterial Smooth Muscle by Acetylcholine," *Nature* 288, no. 5789 (November 1980): 373–76, https://doi.org/10.1038/288373a0.
4. I was unable to locate the original source of this well-known Szent-Györgyi quote. It is cited and discussed in *The Scientist Speculates: An Anthology of Partly-Baked Ideas*, ed. I. J. Good (New York: Basic Books, 1962).

5. John R. Vane, "Adventures and Excursions in Bioassay: The Steppingstones to Prostacyclin," Nobel lecture, December 8, 1982, https://www.nobelprize.org/uploads/2018/06/vane-lecture.pdf.
6. Singular: medium, plural: media (incorrectly used by so many investigators). By the way, it is not called "medium" because it has a crystal ball and sees the future. The one who is expected to see the future is the investigator—especially if the investigator wishes to apply for multiyear grants, as explained in previous chapters.
7. By changing key constituents of the medium, one can generate special, resistant versions of cells, for example, stem cell–like cancer cell clones. There are also protocols that call for the temporary omission and replacement of FBS, either for fundamental scientific or practical reasons (some FBS components interfere with some methods).
8. No, not always.
9. No, not really.
10. The word "omics" is currently being hijacked and discredited by the predatory publishing/conference group OMICS, in much the same way that the word "theranostic" was hijacked and discredited by a now defunct company called Theranos.
11. Bruce Beutler, "How Mammals Sense Infection: From Endotoxin to the Toll-Like Receptors," The Nobel Prize, December 7, 2011, https://www.nobelprize.org/uploads/2018/06/beutler-lecture.pdf.
12. This was demonstrated in animal models. Later, in human trials, antibodies that neutralize TNFα did not induce a statistically detectable protective effect in patients with sepsis. However, in other diseases, such as arthritis, the same antibodies have been shown to be highly effective. This story highlights the well-known fact that the findings observed in cell models do not always translate to animal models, and the findings in animal models do not always translate to humans.
13. Haichao Wang et al., "HMGB-1 as a Late Mediator of Endotoxin Lethality in Mice," *Science* 285, no. 5425 (July 1999): 248–51, https://doi.org/10.1126/science.285.5425.248; and Cassie Shu Zhu et al., "Identification of Procathepsin L-Neutralizing Monoclonal Antibodies to Treat Potentially Lethal Sepsis," *Science Advances* 9, no. 5 (February 2023): eadf4313, https://doi.org/10.1126/sciadv.adf4313.
14. Steve C. Linn et al., "Transcriptional Regulation of Human Inducible Nitric Oxide Synthase Gene in an Intestinal Epithelial Cell Line," *American Journal of Physiology* 272, no. 6, part 1 (June 1997): G1499-508, https://doi.org/10.1152/ajpgi.1997.272.6.G1499.
15. Csaba Szabó et al., "DNA Strand Breakage, Activation of Poly (ADP-Ribose) Synthetase, and Cellular Energy Depletion Are Involved in the Cytotoxicity of Macrophages and Smooth Muscle Cells Exposed to Peroxynitrite," *Proceedings of the National Academy of Sciences of the United States of America* 93, no. 5 (March 1996): 1753–58, https://doi.org/10.1073/pnas.93.5.1753.
16. Francisco Garcia Soriano et al., "Diabetic Endothelial Dysfunction: The Role of Poly(ADP-Ribose) Polymerase Activation," *Nature Medicine* 7, no. 1 (January 2001): 108–13, https://doi.org/10.1038/83241.
17. Csaba Szabo et al., "Tumor-Derived Hydrogen Sulfide, Produced by Cystathionine-β-Synthase, Stimulates Bioenergetics, Cell Proliferation, and Angiogenesis in Colon Cancer," *Proceedings of the National Academy of Sciences of the United States of America* 110, no. 30 (July 2013): 12474–79, https://doi.org/10.1073/pnas.1306241110.

254   3. "DOING SCIENCE"

18. Theodora Panagaki et al., "Overproduction of H_2S, Generated by CBS, Inhibits Mitochondrial Complex IV and Suppresses Oxidative Phosphorylation in Down Syndrome," *Proceedings of the National Academy of Sciences of the United States of America* 116, no. 38 (September 2019): 18769–71, https://doi.org/10.1073/pnas.1911895116.
19. *In vivo* vs. *in vitro* is a relative matter. In the eyes of many biochemists who spend most of their lives working with isolated proteins or enzymes, a cell-based experiment is often considered *in vivo* (or, let's say, "*in vivo*" enough).
20. Inbred colonies sometimes undergo "genetic drift": one supplier's Black 6 may, in fact, not be exactly the same as another supplier's Black 6.
21. Here is a little known fact: the cells were obtained from the mammary gland of a sheep, and the reason for the name "Dolly" is Sir Ian Wilmut's reference to the American country singer Dolly Parton, who happened to be a very curvaceous person. A poor taste sexist joke like this would not fly today. I also don't have space to get into the fact that later it also turned out that Sir Ian had a much smaller role in the creation of Dolly than initially announced, and other investigators deserve most of the credit. Finally, I also cannot get into the fact that even clones are not 100 percent the same as the parental cell due to dedifferentiation and epigenetic factors.
22. Many scientists and philosophers have questioned whether a hypothesis is even possible in the most fundamental sense. Famously, Newton wrote "*Hypotheses non fingo*," and Karl Popper talks about "*hypotheto-deductive interpretation*" rather than *de novo* hypothesis generation. Anyway, let's assume there *is* such a thing as a hypothesis because millions of papers in PubMed use this word.
23. Although I am all for replication studies, I don't recommend repeating this one.
24. An instructive interview with Marshall appeared in *Discover* magazine. See Pamela Weintraub, "The Doctor Who Drank Infectious Broth, Gave Himself an Ulcer, and Solved a Medical Mystery," *Discover*, April 8, 2010. Every year I bring dozens of photocopies of this interview to one of my medical student lectures and ask the students to take one home and read it. Some of them take a copy, many don't.
25. It turns out that even the noncoding regions of the DNA also have biological functions; this entire field of research remains mostly unexplored.
26. Paula Dobosz and Tomasz Dzieciątkowski, "The Intriguing History of Cancer Immunotherapy," *Frontiers in Immunology* 10 (December 2019): 2965, https://doi.org/10.3389/fimmu.2019.02965.
27. This happens quite often, even if the biological phenomenon detected by the brand new method is not particularly novel.
28. Jeromie M. Ballreich et al., "Allocation of National Institutes of Health Funding by Disease Category in 2008 and 2019," *JAMA Network Open* 4, no. 1 (January 2021): e2034890, https://doi.org/10.1001/jamanetworkopen.2020.34890.
29. Clearly, in reality, this is nonsense. By definition, there have to be infinitely more uninteresting topics in the world than interesting ones. More on the topic of hype later.
30. There are dozens of different types of biases in biomedical research. Many of them—but not all—are discussed in various parts of this book. For a fairly comprehensive discussion on biases, see Justin Morgenstern, "Bias in Medical Research," First 10 EM, July 2, 2021, https://first10em.com/bias/.

31. "Experimental day" is not the same as the "calendar day." It is how many times the investigator repeats the whole experiment, using new reagents, cells, etc. With some planning, one can squeeze several "experimental days" into a calendar day.
32. The technical terms for these are "independent" and "dependent" variables, respectively.
33. In hundreds of publications, the annoying and nonsensical term "very unique" is used.
34. I first saw this clearly explained in the Nobelist Peter Medawar's 1964 BBC talk: see Peter Medawar, "Is the Scientific Paper a Fraud?," in *Experiment: A Series of Scientific Case Histories First Broadcast in the BBC Third Programme*, ed. D. Edge, 7–13, London: BBC, 1964.
35. "Statistical significance" and real "significance" are not at all the same thing. More on stats later.
36. "Data," originally, was a plural word. One should say "data show" or "data tell you," rather than "data shows" or "data tells you." Over the years—largely due to the destructive workmanship of Americans—the singular version has begun to take over.
37. I feel that this basic need is deeply human. Perhaps it comes from the same source that makes a person see various objects in a Rorschach test, see faces in cloud patterns, or imagine monsters in the forest when the leaves are rattling.
38. This expectation is real no matter how much scientists try to fight it. Biostatisticians force scientists to formulate their idea as a "null hypothesis," in which the expectations are reversed and statements are made, such as "My null hypothesis is that the addition of Hormone Y will *not* change the expression of Gene Z." But semantics won't change what is actually in the scientists' head about the whole experiment.
39. Peter B. Medawar, *Advice to a Young Scientist* (New York: Basic Books, 1981).
40. You see already here that a sense of involvement is expressed. Scientists talk about "*their* data" and not "*the* data."
41. Daniel Kahneman, recipient of the Nobel Memorial Prize in Economic Sciences, talks about theory-induced blindness. "Once you have accepted a theory and used it as a tool in your thinking, it is extraordinarily difficult to notice its flaws. If you come upon an observation that does not seem to fit the model, you assume that there must be a perfectly good explanation that you are somehow missing." Daniel Kahnemann, *Thinking, Fast and Slow* (New York: Farrar, Straus and Giroux, 2013).
42. In today's general approach to biomedical science, most investigators don't have the interest, time, or money to explore such "side-roads." This is also part of the problem.
43. For a humbling introduction to this topic, see Robert Winston, *Human Instinct: How Our Primeval Impulse Shapes Our Modern Lives* (London: Bantam, 2003).
44. In the abstract sense, every conclusion in science is an approximation or a model anyway. Think of physics, with the Newtonian system followed by the Einsteinian system of relativity. Think of geometry, with the Euclidean system and then the non-Euclidian. Even quantum physicists will readily admit that their "standard model"—however closely it models the observed quantum events—is still (only) a model. Some important experimental proofs only occur infrequently. A few months ago, I visited the Large Hadron Collider in Geneva. My physicist guide explained that Higgs bosons are only formed once in every one billion collisions.
45. Jo Marchant, "Powerful Antibiotics Discovered Using AI. Machine Learning Spots Molecules That Work Even Against 'Untreatable' Strains of Bacteria," *Nature* (February 20, 2020), https://doi.org/10.1038/d41586-020-00018-3.

46. The topic of hyping science to the general public goes beyond the scope of this book. It is certainly not a new topic. See Andrew Moore, "Bad Science in the Headlines. Who Takes Responsibility When Science Is Distorted in the Mass Media?," *EMBO Reports* 7, no. 12 (December 2006): 1193–96, https://doi.org/10.1038/sj.embor.7400862). A recent analysis of the problem can be found in Fiona Fox, *Beyond the Hype: The Inside Story of Science's Biggest Media Controversies* (London: Elliott & Thompson, 2022).
47. Leonard P. Freedman, Iain M. Cockburn, and Timothy S. Simcoe, "The Economics of Reproducibility in Preclinical Research," *PLOS Biology* 13, no. 6 (June 2015): e1002165, https://doi.org/10.1371/journal.pbio.1002165.
48. Patrick D. Lyden et al., "A Multi-Laboratory Preclinical Trial in Rodents to Assess Treatment Candidates for Acute Ischemic Stroke," *Science Translational Medicine* 15, no. 714 (September 20, 2023): eadg8656, https://doi.org/10.1126/scitranslmed.adg8656.
49. John P. A. Ioannidis, "Why Most Published Research Findings Are False," *PLOS Medicine* 2, no. 8 (August 2005): e124, https://doi.org/10.1371/journal.pmed.0020124.
50. Some estimations of the lost money start out from the retracted publications. An analysis of retracted NIH-funded publications estimates that only 1 percent of the research funding is "lost." See Andrew M. Stern et al., "Financial Costs and Personal Consequences of Research Misconduct Resulting in Retracted Publications," *Elife* 3 (August 2014): e02956, https://doi.org/10.7554/eLife.02956. However, such analysis does not take into account that a large percentage of the literature is not retracted or corrected; it is just not reliable and not reproducible—as proven by the low rate of replication in the direct replication attempts discussed in this chapter.
51. Derek Lowe, "The Sirtris Compounds: Worthless? Really?" *Science*, In the Pipeline (blog), January 12, 2010.
52. When I was working as CSO of a U.S. biotech company, the company relied on a series of patents from an American university for a clinical development program. The program was subsequently stopped in the clinical trial stage. Later on, it turned out that the group whose work supported—at least in part—the program, and from which the patents were licensed, had included "doctored" figures in several key publications. A series of papers were retracted, but these retractions happened many years after the clinical program's termination.
53. Many years ago, a former colleague of mine was hired by a series of investors to manage such a replication attempt. The investors saw some promising publications and a patent from a well-respected university, but first they wanted to do an independent confirmation of the effects. It turned out that experiments, when conducted in a blinded manner by an independent contract research organization, could not reproduce the published effect. Naturally, the potential investors lost their interest in the program (and my former colleague lost a lucrative consulting job).
54. Lev Osherovich, "Hedging Against Academic Risk," *Science-Business eXchange* 4 (April 2011): 416, https://doi.org/10.1038/scibx.2011.416.
55. "Enhancing Reproducibility Through Rigor and Transparency," Policy & Compliance, NIH Grants & Funding, last modified January 2023, https://grants.nih.gov/policy/reproducibility/index.htm.
56. Brian A. Nosek and Timothy M. Errington, "The Best Time to Argue About What a Replication Means? Before You Do It," *Nature* 583, no. 7817 (July 2020): 518–20, https://doi.org/10.1038/d41586-020-02142-6.

57. In the same study, for negative findings (i.e., reported no effects), the replication attempts produced a replication rate of about 50 percent. For negative findings, this is a disturbingly low rate of reproducibility.
58. Timothy M. Errington et al., "Investigating the Replicability of Preclinical Cancer Biology," *eLife* 10 (December 2021): e71601, https://doi.org/10.7554/eLife.71601.
59. Asher Mullard, "Half of Cancer Studies Fail High-Profile Replication Test," *Nature* 600, no. 7889 (December 2021): 368–69, https://doi.org/10.1038/d41586-021-03691-0.
60. Open Science Collaboration, "Estimating the Reproducibility of Psychological Science," *Nature* 536 (2015), 371–86, https://doi.org/10.1038/s41586-015-0694-7; Nosek and Errington, "The Best Time to Argue About What a Replication Means?" *Nature*, July 21, 2020. https://www.nature.com/articles/d41586-020-02142-6
61. Pyrogens are biological substances that are known to cause fever (at extremely low doses) when given to animals. At higher doses, they can cause systemic inflammation and circulatory shock as well.
62. Inadvertently, the system has become a model of the vasculature during septic shock rather than a model of a healthy blood vessel.
63. Thus, inadvertently, the system is no longer a model of a healthy blood vessel but is an accidental model of aging or diabetic blood vessel complications.
64. "Laboratory Water. Its Importance and Application," Division of Technical Resources, NIH, March 2013, https://orf.od.nih.gov/TechnicalResources/Documents/DTR%20White%20Papers/Laboratory%20Water-Its%20Importance%20and%20Application-March-2013_508.pdf.
65. One common trick is to buy a large batch of FBS, then freeze it and use it for several years. Eventually it runs out and one needs to get a new batch. Some companies market artificial sera as an alternative. Most of my colleagues are wary of this because switching over to a new system would mean repeating a lot of earlier experiments to achieve a new baseline. Also, once a group of investigators is hooked on an artificial serum from a commercial company, what happens if the company changes the recipe or shuts down and the material is no longer available?
66. There are various sophisticated—and expensive—LPS removal systems, which are sometimes used if the topic of the investigation is directly related to LPS or LPS contamination. However, few laboratories can afford to use them routinely.
67. Tonyia Eaves-Pyles et al., "Flagellin, a Novel Mediator of Salmonella-Induced Epithelial Activation and Systemic Inflammation: I Kappa B Alpha Degradation, Induction of Nitric Oxide Synthase, Induction of Proinflammatory Mediators, and Cardiovascular Dysfunction," *Journal of Immunology* 166, no. 2 (January 2001): 1248–60, https://doi.org/10.4049/jimmunol.166.2.1248.
68. Csaba Szabó, "Hydrogen Sulphide and Its Therapeutic Potential," *Nature Reviews Drug Discovery* 6, no. 11 (November 2007): 917–35, https://doi.org/10.1038/nrd2425.
69. Giuseppe Cirino, Csaba Szabo, and Andreas Papapetropoulos, "Physiological Roles of Hydrogen Sulfide in Mammalian Cells, Tissues, and Organs," *Physiological Reviews* 103, no. 1 (January 2023): 31–276, https://doi.org/10.1152/physrev.00028.2021.
70. A few years ago we conducted a screening campaign using one of these commercial compound libraries. We had identified a very interesting, highly effective inhibitor of inflammatory mediator production. The effect was reproducible as long as we repeated the experiment using material from the same well of the same library. When we purchased

the same compound from the supplier again, the new compound turned out to be completely inactive. Maybe the compound in the library was mislabeled, or maybe the original compound degraded over time and a degradation product was responsible for the observed effect. We never got to the bottom of this enigma.

71. This is a quantitatively different thing than producing—through chemical synthesis—a new batch of a chemical compound once the previous stock is sold out. Chemical synthesis is fairly reliable even though it can also have different contaminant and impurity profiles, from batch to batch. Also, an animal's immune response can be different in different experiments. The clone produced to make a monoclonal antibody can also be different each time, and the antibody generated may be different. It may bind to the target protein weaker or stronger, possibly even at a different area ("molecular site") than the previous one.

72. *Literally* by the book. Because immunohistochemistry and Western blotting are also used in routine clinical diagnosis, there are dozens of books focusing on the pitfalls and methods related to these techniques. When a patient's life or choosing the correct therapy is at stake, quality control becomes paramount. But academic research laboratories don't tend to use the same level of stringency as that of a clinical pathology laboratory.

73. Leonard P. Freedman et al., "Reproducibility: Changing the Policies and Culture of Cell Line Authentication," *Nature Methods* 12, no. 6 (June 2015): 493–97, https://doi.org/10.1038/nmeth.3403.

74. Sara Reardon, "A Mouse's House May Ruin Experiments," *Nature* 530, no. 7590 (February 2016): 264, https://doi.org/10.1038/nature.2016.19335.

75. Cornelia Kiank et al., "Seasonal Variations in Inflammatory Responses to Sepsis and Stress in Mice," *Critical Care Medicine* 35, no. 10 (October 2007): 2352–58, https://doi.org/10.1097/01.ccm.0000282078.80187.7f.

76. Okko Alitalo et al., "A Wake-Up Call: Sleep Physiology and Related Translational Discrepancies in Studies of Rapid-Acting Antidepressants," *Progress in Neurobiology* 206 (November 2021): 102140, https://doi.org/10.1016/j.pneurobio.2021.102140.

77. Maria-Luisa Alegre, "Mouse Microbiomes: Overlooked Culprits of Experimental Variability," *Genome Biology* 20, no. 1 (May 2019):108, https://doi.org/10.1186/s13059-019-1723-2.

78. This would possibly make it a useful drug for treating patients with inflammatory diseases. There are many attempts along these lines, but no "IL-10 drug" has been approved yet. However, many anti-inflammatory drugs on the market do work, at least in part, by boosting the body's own IL-10 production. Even the age-old class of anti-inflammatories such as "steroids" (glucocorticoids) has IL-10 as a component of its mode of action.

79. Zequan Yang, Basilia Zingarelli, and Csaba Szabo, "Crucial Role of Endogenous Interleukin-10 Production in Myocardial Ischemia/Reperfusion Injury," *Circulation* 101, no. 9 (March 2000): 1019–26, https://doi.org/10.1161/01.cir.101.9.1019.

80. The idea of doing all experiments in germ-free mice to get rid of all microorganisms inside and outside the animal is also not realistic. It will create other problems, which I don't have the space to explain. Not to mention that this would not be a model of real life; we all have our own intestinal microbiome, and we all live in an environment that contains myriad microorganisms.

81. Yet another Hungarian. It comes as no surprise that, once again, he did his work outside his home country, in his case in Canada. One of his main books is Hans Selye, *The Stress of Life* (New York: McGraw-Hill, 1956).
82. Bonnie L. Hylander, Elizabeth A. Repasky, and Sandra Sexton, "Using Mice to Model Human Disease: Understanding the Roles of Baseline Housing-Induced and Experimentally Imposed Stresses in Animal Welfare and Experimental Reproducibility," *Animals (Basel)* 12, no. 3 (February 2022): 371, https://doi.org/10.3390/ani12030371.
83. Robert E. Sorge et al., "Olfactory Exposure to Males, Including Men, Causes Stress and Related Analgesia in Rodents," *Nature Methods* 11, no. 6 (June 2014): 629–32, https://doi.org/10.1038/nmeth.2935.
84. It turns out that even this is not a perfect solution. Not to get into too much scatological detail, but mice can eat their own feces right as it comes out. In this case, wire floors do not help. See Said R. Bogatyrev, Justin C. Rolando, and Rustem F. Ismagilov, "Self-Reinoculation with Fecal Flora Changes Microbiota Density and Composition Leading to an Altered Bile-Acid Profile in the Mouse Small Intestine," *Microbiome* 8, no. 1 (February 2020): 19, https://doi.org/10.1186/s40168-020-0785-4.
85. Irving Zucker and Annaliese K. Beery, "Males Still Dominate Animal Studies," *Nature* 465, no. 7299 (June 2010): 690, https://doi.org/10.1038/465690a.
86. Fernanda Ferreira Cruz, Patricia Rieken Macedo Rocco, and Paolo Pelosi, "Anti-Inflammatory Properties of Anesthetic Agents," *Critical Care* 21, no. 1 (March 2017): 67, https://doi.org/10.1186/s13054-017-1645-x.
87. Aliah M. Alhayyan et al., "The Effect of Anesthesia on the Magnitude of the Postoperative Systemic Inflammatory Response in Patients Undergoing Elective Surgery for Colorectal Cancer in the Context of an Enhanced Recovery Pathway: A Prospective Cohort Study," *Medicine (Baltimore)* 100, no. 2 (January 2021): e23997, https://doi.org/10.1097/MD.0000000000023997.
88. This also applies to human anesthesia to some degree, by the way.
89. Viola Volpato and Caleb Webber, "Addressing Variability in iPSC-Derived Models of Human Disease: Guidelines to Promote Reproducibility," *Disease Models & Mechanisms* 13, no. 1 (January 2020): dmm042317, https://doi.org/10.1242/dmm.042317.
90. This is because CBS is encoded on chromosome 21, which Down syndrome people have an extra copy of.
91. Theodora Panagaki et al., "Role of the Cystathionine β-Synthase / H_2S Pathway in the Development of Cellular Metabolic Dysfunction and Pseudohypoxia in Down Syndrome," *Redox Biology* 55 (July 2022): 102416, https://doi.org/10.1016/j.redox.2022.102416.
92. Sara G. Susco et al., "Molecular Convergence Between Down Syndrome and Fragile X Syndrome Identified Using Human Pluripotent Stem Cell Models," *Cell Reports* 40, no. 10 (September 2022): 111312, https://doi.org/10.1016/j.celrep.2022.111312.
93. Morva Mansoury et al., "The Edge Effect: A Global Problem. The Trouble with Culturing Cells in 96-Well Plates," *Biochemistry and Biophysics Reports* 26 (March 2021): 100987, https://doi.org/10.1016/j.bbrep.2021.100987.
94. In my laboratory, we don't use the outside rows. We put culture medium there but do not use them for experimental purposes. This way we lose some of the plate for experiments

(now we have only sixty useful wells per 96-well plate), but the experiments are more reliable.

95. In scientific terms, μM means 1 100,000th molar, while mM means 1 1000th molar. In the MS Word program, "μ" is created by typing the letter 'm' and changing the font to the Greek character. When text is copy-pasted into another text, which uses another font type, "μM" can revert to "μM." If this creeps into an actual experiment protocol, this would mean applying 1,000 times more reagent than the correct amount. (If one uses "uM" with a letter u, although it is not technically correct, at least it stays the same after transfer.)

96. There are even worse situations. The following horror story, which took place in a German laboratory, falls into the category of sabotage. When one of the lab members was dismissed from the group because of his inappropriate attitude, he muttered the following under his moustache on his way out, "You know, all white powder looks the same..." All the chemicals in the lab had to be disposed of and replaced with new ones.

97. Elisabeth Davenas et al., "Human Basophil Degranulation Triggered by Very Dilute Antiserum Against IgE," *Nature* 333, no. 6176 (June 1988): 816–18, https://doi.org/10.1038/333816a0.

98. Don't get me started on homeopathy—at least not in this book. Even though it has been disproven time and again, even though hundreds of scientific societies have called out against it, in most countries it is still a thriving business. You can find homeopathic "medicines" everywhere. Homeopathic effects are placebo effects, which are not the same as "no effects," by the way. Mind-blowingly, homeopathy is still considered a valid discipline in many Swiss universities. Together with other "complementary and alternative" approaches, it is considered a regular part of the Swiss health care system—after a national referendum mandated it. In 2016, two-thirds of the citizens voted for this.

99. Stuart J. Hirst et al., "Human Basophil Degranulation Is Not Triggered by Very Dilute Antiserum Against Human IgE," *Nature* 366, no. 6455 (December 1993): 525–27, https://doi.org/10.1038/366525a0.

100. Junhee Seok et al., "Genomic Responses in Mouse Models Poorly Mimic Human Inflammatory Diseases," *Proceedings of the National Academy of Sciences of the United States of America* 110, no. 9 (February 2013): 3507–12, https://doi.org/10.1073/pnas.1222878110.

101. Keizo Takao and Tsuyoshi Miyakawa, "Genomic Responses in Mouse Models Greatly Mimic Human Inflammatory Diseases," *Proceedings of the National Academy of Sciences of the United States of America* 112, no. 4 (August 2014): 1167–72, https://doi.org/10.1073/pnas.1401965111.

102. Keizo Takao, and Tsuyoshi Miyakawa, "Correction for: Genomic Responses in Mouse Models Greatly Mimic Human Inflammatory Diseases," *Proceedings of the National Academy of Sciences of the United States of America* 112, no. 10 (February 2015): E1163-7, https://doi.org/10.1073/pnas.1502188112.

103. Two questions were to be addressed: (1) To what extent is the growth of nestling blue tits influenced by competition with siblings? and (2) How does grass cover influence *Eucalyptus* seedling recruitment?

104. Anil Oza, "Reproducibility Trial: 246 Biologists Get Different Results from Same Data Sets. Wide Distribution of Findings Shows How Analytical Choices Drive Conclusions," *Nature*, October 2023, https://doi.org/10.1038/d41586-023-03177-1.

105. Michael J. Curtis et al., "Experimental Design and Analysis and Their Reporting II: Updated and Simplified Guidance for Authors and Peer Reviewers," *British Journal of Pharmacology* 175, no. 7 (April 2018): 987–93, https://doi.org/10.1111/bph.14153.
106. Harvey J. Motulsky and Martin C. Michel, "Commentary on the BJP's New Statistical Reporting Guidelines," *British Journal of Pharmacology* 175, no. 18 (September 2018): 3636–37, https://doi.org/10.1111/bph.14441.
107. David Robert Grimes and James Heathers, "The New Normal? Redaction Bias in Biomedical Science," *Royal Society Open Science* 8, no. 12 (December 2021): 211308, https://doi.org/10.1098/rsos.211308.
108. In this particular example, the treatment was not a very good one. It is fairly easy to find drugs that effectively suppress tumor growth in mouse models. Mouse cancer has been cured hundreds of times. It is a whole different challenge to translate it into an effective human drug.
109. Of course, in a well-designed study, this can be prevented. Each mouse can be permanently labeled (e.g., by ear punches) and should remain fully identifiable throughout the experiment.
110. One "belief system" says that it is okay to exclude an outlier that is outside two standard errors, whether there is a scientific reason for it or not. Another one says that outliers are outside the "1.5 times the interquartile range (IQR)" rule. According to this "rule," data points falling below $Q_1 - 1.5 * IQR$ or above $Q_3 + 1.5 * IQR$ are considered outliers and may be excluded.
111. When a clinical trial turns out to be negative, the investigators often start "fishing" for signs of efficacy—even though the overall hypothesis has failed. The goal is to try to find a sub-group of patients within the trial subjects, for whom it seems that the drug may have been effective. This form of data massaging, called post hoc subgroup analysis, belongs to clinical study-related shenaniganry and won't be covered in this book.
112. Domokos Gerö et al., "Identification of Pharmacological Modulators of HMGB1-Induced Inflammatory Response by Cell-Based Screening," *PLOS One* 8, no. 6 (June 2013): e65994, https://doi.org/10.1371/journal.pone.0065994.
113. E. J. Masicampo and Daniel R. Lalande, "A Peculiar Prevalence of p Values Just Below .05," *Quarterly Journal of Experimental Psychology (Hove)* 65, no. 11 (2012): 2271–79, https://doi.org/10.1080/17470218.2012.711335; and Megan L. Head et al., "The Extent and Consequences of p-Hacking in Science," *PLOS Biology* 13, no. 3 (March 2015): e1002106, https://doi.org/10.1371/journal.pbio.1002106.
114. Regina Nuzzo, "How Scientists Fool Themselves—and How They Can Stop," *Nature* 526, no. 7572 (October 2015): 182–85, https://doi.org/10.1038/526182a.
115. Brian Wansink, "The Grad Student Who Never Said No," web.archive (blog), November 21, 2016, https://web.archive.org/web/20170312041524/http:/www.brianwansink.com/phd-advice/the-grad-student-who-never-said-no.
116. Brett Dahlberg, "Cornell Finds Laboratory Head Falsified Data," WRVO, November 6, 2018, www.wrvo.org.
117. After a few years of hiatus, Wansink began publishing in various journals again, based on material that apparently had been collected during his tenure at Cornell. In one of the new papers, his affiliation is listed as "Applied Economics and Management, Cornell

262   3. "DOING SCIENCE"

University," and in the other his affiliation is listed as "Behavioral Psychology, Cornell University."

118. Angelica M. Stefan and Felix D. Schönbrodt, "Big Little Lies: A Compendium and Simulation of p-Hacking Strategies," *Royal Society Open Science* 10, no. 2 (February 2023): 220346, https://doi.org/10.1098/rsos.220346.

119. Optional stopping is another form of flawed practice that is often associated with clinical trials. Clinical trial related misconduct is not discussed in this book.

4. SCIENTIFIC FRAUD—AND THE FRAUDULENT FRAUDSTERS

1. There are cultural differences in practically every aspect of life. Science is no exception. In some cultures, it is more difficult to say no to a boss than in others. Real or perceived pressure to *please the boss* is considered to be one of many root causes of scientific fraud.

2. This is true—unless the fake paper claims to be confirming previously published real results. In this case, the conclusions of the fake paper may be correct, but the paper still does not have anything to do with reality. Also, too much of this type of fake confirmation would distort the balance of the scientific literature. But as discussed earlier, confirmatory papers do not generate much excitement in any field and are difficult to get published. Thus, I believe, confirmatory fake papers must be relatively rare.

3. The Summerlin affair was published almost fifty years ago. See Joseph Hixson, *The Patchwork Mouse* (New York: Anchor & Doubleday, 1976). Some of these cases have even become teaching materials. For example, the Hwang, Potti, and Obokata cases were described in detail in appendix D ("Detailed Case Histories") of a consensus report of the National Academies of Science. See National Academies of Sciences, Engineering, and Medicine, *Fostering Integrity in Research* (Washington, DC: National Academies Press, 2017), https://doi.org/10.17226/21896. The University of Stanford maintains a "Best Practices in Science" page where summary information and hundreds of weblinks are presented on the Hwang, Hauser, and Das cases (as well as many others). See "Instances of Scientific Misconduct," https://bps.stanford.edu/?page_id=7155. Jeffrey Flier describes the Darsee and Anversa cases in detail in his 2021 article. See Jeffrey S. Flier, "Misconduct in Bioscience Research: A 40-Year Perspective," *Perspectives in Biology and Medicine* 64, no. 4 (Autumn 2021): 437–56, https://doi.org/10.1353/pbm.2021.0035).

   Further information on the Darsee case can be found in the following articles: Richard Knox, "The Harvard Fraud Case: Where Does the Problem Lie?," *Journal of the American Medical Association* 249, no. 14 (April 1983): 1797–99, 1802–7; Barbara J. Culliton, "Coping with Fraud: The Darsee Case," *Science* 220, no. 4592 (April 1983): 31–35, https://doi.org/10.1126/science.6828878; Eugene Braunwald, "On Analysing Scientific Fraud," *Nature* 325, no. 6101 (January 1987): 215–16, https://doi.org/10.1038/325215a0; and Charles Gross, "Scientific Misconduct," *Annual Review of Psychology* 67, no. 1 (2016): 693–711, https://doi.org/10.1146/annurev-psych-122414-033437.

   The Penkowa case is covered in the following articles: Ewen Callaway, "Danish Neuroscientist Challenges Fraud Findings," *Nature*, 2012, https://doi.org/10.1038/nature.2012.11146; Ewen Callaway, "Danish Court Quashes Ruling Against Physiologist," *Nature*, 2015, https://doi.org/10.1038/nature.2015.16960; Alison McCook, "Copenhagen Revokes

Degree of Controversial Neuroscientist Milena Penkowa," Retraction Watch, September 12, 2017, https://retractionwatch.com/2017/09/12/copenhagen-revokes-degree-controversial-neuroscientist-milena-penkowa/; and Adam Marcus, "Years Later, Researcher at Center of Highly Publicized Case Has Another Paper Retracted," Retraction Watch, September 24, 2020, https://retractionwatch.com/2020/09/24/years-later-researcher-at-center-of-highly-publicized-case-has-another-paper-retracted/.

The Anversa affair was widely covered in the news. See Kelly Servick, "$10 Million Settlement Over Alleged Misconduct in Boston Heart Stem Cell Lab. Brigham and Women's Hospital to Pay Government After Probe of Manipulated and Fabricated Data," *Science*, April 2017, https://www.science.org/content/article/10-million-settlement-over-alleged-misconduct-boston-heart-stem-cell-lab; Gina Kolata, "He Promised to Restore Damaged Hearts. Harvard Says His Lab Fabricated Research," *New York Times*, October 29, 2018, https://www.nytimes.com/2018/10/29/health/dr-piero-anversa-harvard-retraction.html; Carolyn Y. Johnson, "Harvard Investigation Finds Fraudulent Data in Papers by Heart Researcher," *Washington Post*, October 15, 2018, https://www.washingtonpost.com/science/2018/10/15/harvard-investigation-finds-fraudulent-data-papers-by-heart-researcher/; and Marisa Taylor and Brad Heath, "Years After Brigham-Harvard Scandal, U.S. Pours Millions Into Tainted Stem-Cell Field," *Reuters Investigation*, June 21, 2022, https://www.reuters.com/investigates/special-report/health-hearts-stem-cells/.

The Hwang case was covered, in detail, in the following articles: "Timeline of a Controversy. A Chronology of Woo Suk Hwang's Stem-Cell Research," *Nature*, December 19, 2005, https://doi.org/10.1038/news051219-3; "Hwang Indicted for Fraud, Embezzlement. Korean Prosecutors Also Charge Five Other Members of Stem Cell Team," *Science*, May 12, 2006, https://www.science.org/content/article/hwang-indicted-fraud-embezzlement; David Cyranoski, "How a Scientific Fraud Reinvented Himself," *New York Times*, June 21, 2023, https://www.nytimes.com/2023/06/21/opinion/cloning-science-fraud.html; and Kang Hyun-kyung, "Whistleblower Recalls Price Paid for Revealing Hwang Woo-suk's Scientific Misconduct," *Korea Times*, July 27, 2023, https://www.koreatimes.co.kr/www/nation/2024/04/113_355797.html.

The Hauser case is covered in the following articles: David Dobbs, "Marc Hauser, Monkey Business, and the Sine Waves of Science," *Wired*, August 11, 2010, https://www.wired.com/2010/08/marc-hauser-monkey-business-and-the-sine-waves-of-science/; Carolyn Y. Johnson, "Former Harvard Professor Marc Hauser Fabricated, Manipulated Data, US Says," Boston.com, September 5, 2012, https://www.boston.com/uncategorized/noprimarytagmatch/2012/09/05/former-harvard-professor-marc-hauser-fabricated-manipulated-data-us-says/; and Carolyn Y. Johnson, "Harvard Report Shines Light on Ex-Researcher's Misconduct," *Boston Globe*, May 29, 2014, https://www.bostonglobe.com/metro/2014/05/29/internal-harvard-report-shines-light-misconduct-star-psychology-researcher-marc-hauser/maSUowPqL4clXrOgj44aKP/story.html.

The Potti case has been covered in many articles including these: Michael McCarthy, "Former Duke University Oncologist Is Guilty of Research Misconduct, US Officials Find," *British Medical Journal* 351 (November 2015), https://doi.org/10.1136/bmj.h6058; "Duke Officials Silenced Med Student Who Reported Trouble in Anil Potti's Lab," *Cancer Letter*, January 09, 2015, https://cancerletter.com/the-cancer-letter/20150109_1/;

Kerry Grens, "Oncologist Found Guilty of Misconduct. A Government Investigation Concludes That Anil Potti Faked Data on Multiple Grants and Papers," *The Scientist*, November 9, 2015, https://www.the-scientist.com/oncologist-found-guilty-of-misconduct-34523; and Jocelyn Kaiser, "Potti Found Guilty of Research Misconduct. Federal Report Closes Long Cancer Research Saga," *Science*, November 9, 2015, https://www.science.org/content/article/potti-found-guilty-research-misconduct.

Further details regarding the Das case can be found in the following articles: Tony Isaacs, "A Tragic and Stunning Case of Scientific Fraud in Studies on Red Wine and Resveratrol," *Natural News*, March 22, 2012, https://www.naturalnews.com/035315_red_wine_resveratrol_scientific_fraud.html; Ryan Jaslow, "Red Wine Researcher Dr. Dipak K. Das Published Fake Data," *CBS News*, January 12, 2012, https://www.cbsnews.com/news/red-wine-researcher-dr-dipak-k-das-published-fake-data-uconn/; Ewen Callaway, "In Vino Non Veritas? Red Wine Researcher Implicated in Misconduct Case," *Nature*, January 12, 2012, https://blogs.nature.com/news/2012/01/in-vino-non-veritas-red-wine-researcher-implicated-in-misconduct-case.html; and Scott Hensley, "UConn Claims Resveratrol Researcher Falsified Work," *NPR News*, January 12, 2012, https://www.npr.org/sections/health-shots/2012/01/12/145117068/uconn-claims-resveratrol-researcher-falsified-work.

The Potts-Kant's affair has been covered in the following articles: Alison McCook, "Whistleblower Sues Duke, Claims Doctored Data Helped Win $200 Million in Grants. Alleging That Faked Pulmonary Research Led to Grants from NIH and Other Federal Agencies, Scientist Could Win Huge Payout from False-Claims Suit and Set a Precedent," *Science*, September 16, 2016, https://www.science.org/content/article/whistleblower-sues-duke-claims-doctored-data-helped-win-200-million-grants; and Edward Martin, "Deceit at Duke: How Fraud at a University Research Lab Prompted a $112M Fine," *Business North Carolina*, August 2019, https://businessnc.com/deceit-at-duke-how-fraud-at-a-university-research-lab-prompted-a-112m-fine/.

Han's case has been covered in several articles, see Rachel Bernstein, "HIV Researcher Found Guilty of Research Misconduct Sentenced to Prison," *Science*, July 6, 2015, https://www.science.org/content/article/hiv-researcher-found-guilty-research-misconduct-sentenced-prison; Sara Reardon, "US Vaccine Researcher Sentenced to Prison for Fraud," *Nature* 523 (2015): 138–39, https://doi.org/10.1038/nature.2015.17660; and Tony Leys, "Ex-Scientist Sentenced to Prison for Academic Fraud," *USA Today*, July 1, 2015, https://eu.usatoday.com/story/news/nation/2015/07/01/ex-scientist-sentenced-prison-academic-fraud/29596271/. The case is also listed on the U.S. Attorney's Office website, https://www.justice.gov/usao-sdia/pr/former-iowa-state-researcher-sentenced-making-false-statements.

The Obokata case has been widely discussed, both in the scientific and the lay literature; selected examples follow: Ian Sample, "Stem Cell Scientist Haruko Obokata Found Guilty of Misconduct," *The Guardian*, April 1, 2014, https://www.theguardian.com/science/2014/apr/01/stem-cell-scientist-haruko-obokata-guilty-misconduct-committee; Kirk Spitzer, "Science Scandal Triggers Suicide, Soul-Searching in Japan," *Time Magazine*, August 8, 2014, https://time.com/3091584/japan-yoshiki-sasai-stem-cells-suicide-haruko-obokata/; David Cyranoski, "Accusations Pile Up Amid Japan's Stem-Cell Controversy," *Nature*, 2014, https://doi.org/10.1038/nature.2014.15163; John Rasko

4. SCIENTIFIC FRAUD—AND THE FRAUDULENT FRAUDSTERS

and Carl Power, "What Pushes Scientists to Lie? The Disturbing but Familiar Story of Haruko Obokata," *The Guardian*, February 18, 2015, https://www.theguardian.com/science/2015/feb/18/haruko-obokata-stap-cells-controversy-scientists-lie; and Dana Goodyear, "The Stress Test," *New Yorker*, February 29, 2016, https://www.newyorker.com/magazine/2016/02/29/the-stem-cell-scandal.

The Office of Research Integrity website of the NIH (https://ori.hhs.gov) also has active or archived entries on the majority of the above cases. Internet searches (where the person in question is entered together with the word "misconduct") will produce plenty of news articles on these cases. Additional information on these cases—as well as on many more recent ones—can also be found on websites that focus on scientific misconduct, such as Retractionwatch.com and ForBetterScience.com.

4. R. Knox, "The Harvard Fraud Case: Where Does the Problem Lie?" *JAMA* 249, no. 14 (April 1983): 1797–807.
5. See also the recent case of William Armstead, which was discussed earlier in the ORI grant cases.
6. Most likely, this massive outlier of a penalty was related to the fact that Han's fraud, if undetected, could have led to an ineffective human vaccine that could have endangered human lives. In fraud cases involving clinical research and patients, additional prison sentences were issued, which are not discussed in this book. Here I am focusing on preclinical and bench research only.
7. Marisa Taylor and Brad Heath, "Years After Brigham-Harvard Scandal, U.S. Pours Millions Into Tainted Stem-Cell Field," *Reuters Investigation*, June 21, 2022, https://www.reuters.com/investigates/special-report/health-hearts-stem-cells/. Later in this book I introduce a "swamp creatures" and "swamp ecosystems" analogy on this matter.
8. You see, plausible deniability is a powerful weapon.
9. Marc D. Hauser and Justin N. Wood, "Replication of 'Rhesus Monkeys Correctly Read the Goal-Relevant Gestures of a Human Agent,'" *Proceedings of the Royal Society B: Biological Sciences* 278, no. 1702 (January 2011): 158–59, https://doi.org/10.1098/rspb.2010.1441.
10. In this case, the story is probably more complicated than the *Reuters* investigation presents it. Just because injected stem cells don't turn into functional heart cells does not (necessarily) mean that injected stem cells are inactive. For example, it is possible, at least theoretically, that stem cells can release various factors that may protect the heart or help with the generation of new blood vessels. Many of the follow-up directions of stem cell research focus on directions that do not directly depend on Anversa's discredited concept.
11. Schekman was already mentioned in chapter 2 of the book with his helpful advice of discouraging postdocs from posting cat pictures during laboratory experimentation. I also discuss his innovative initiatives on publication reform later, in chapter 5.
12. PubPeer defines itself as an online "Journal club," which was set up for anonymous "contributors" to discuss published scientific work. In reality, it has become a site where "science sleuths" flag and expose questionable papers.
13. Theo Baker, "Internal Review Found 'Falsified Data' in Stanford President's Alzheimer's Research, Colleagues Allege," *Stanford Daily*, February 17, 2023, https://stanforddaily.com/2023/02/17/internal-review-found-falsified-data-in-stanford-presidents-alzheimers-research-colleagues-allege/.

14. Theo Baker, "Genentech Discloses Unreported Research Misconduct in Stanford President's Lab, Says It Did Not Find 'Intentional Wrongdoing' in 2009," *Stanford Daily*, April 6, 2023, https://stanforddaily.com/2023/04/06/stanford-president-research-misconduct-concerns-genentech/.
15. Another comment: "I am a well-known neurobiologist who has, in the past, published papers 'juxta' the Alzheimer's field. I can tell you that, in general, and unfortunately, this field has some of the sketchiest research in all of neuroscience."
16. Gunjan Sinha, "Critics Challenge Data Showing Key Lipids Can Curb Inflammation," *Science* 376, no. 6593 (May 2022): 565–66, https://doi.org/10.1126/science.abq8439.
17. Holly Else, "Dozens of Papers Co-Authored by Nobel Laureate Raise Concerns," *Nature* 611, no. 7934 (November 2022): 19–20, https://doi.org/10.1038/d41586-022-03032-9.
18. A. Roberts, "Nobel Laureate, Hopkins Researcher Retracts Additional Articles, Bringing Total to Six in Two Years," *Baltimore Sun*, August 1, 2023. At the time of this writing, Semenza had not publicly commented or responded to any of the above issues. The language of the retractions follows the standard retraction language, e.g., "there were signs of background modification," "a lane was potentially artificially removed," "there were signs of potential splicing," the panel "presented artificially deleted bands," "the panel at the bottom presented signs of inappropriate manipulation," "lanes 1 and 4 were reused," etc. Most of the retractions conclude with the following sentence: "The original data provided did not resolve these issues." One of the retractions concludes with the following: "The authors apologize to the scientific community and deeply regret any inconveniences or challenges resulting from the publication and subsequent retraction of this article."
19. Michele Catanzaro, "Possible Misconduct in Papers from Italian Health Minister," *Science* 381, no. 6664 (September 2023): 1267–68, https://doi.org/10.1126/science.adk9441.
20. The original—often forgotten—meaning of the "bad apples" metaphor ("one bad apple can spoil the barrel") would be more applicable here, of course. William Broad and Nicholas Wade, *Betrayers of the Truth* (New York: Simon & Schuster, 1983), is an interesting time capsule for these early "times of denial." Among others, it details the 1981 testimony of Philip Handler, president of the National Academy of Sciences, in front of the U.S. Congress, who insisted that "the problem is grossly exaggerated."
21. This is exactly what happened in one of Tessier-Lavigne's now-retracted papers.
22. Actually, what most people do is just to ignore PubPeer altogether, or reply with "I will look into this" and leave it at that.
23. All of this leads us to a broader question related to the role of the supervisor and the overall climate of a research laboratory, which is based on some degree of trust in each other. No supervisor looks at group members with the constant suspicion of intentional fraud. This topic is revisited later.
24. While I was writing this book, I learned that Antonio—who has subsequently returned to his native country and climbed to significant heights of the academic ladder—had to suddenly resign from some of his university's administrative positions. Not for the many image problems contained in his papers but because of accusations of channeling public research money into a family-owned corporation.
25. As you will see later, this does not mean *all* journals—far from it.

26. The blog of Paul Brookes contains an illustrative example of this type of fakery. See Paul S. Brookes, "Going Beyond Faked Data . . . Faking the Raw Data!" November 10, 2020, https://psblab.org/?p=651. In the past, Brookes ran an extensive, anonymous blog, trying to expose fraud in science, which he was forced to shut down in response to various legal threats received from the lawyers of investigators whom he exposed. I return to his story in chapter 6 of this book.
27. Elisabeth M. Bik, Arturo Casadevall, and Ferric C. Fang, "The Prevalence of Inappropriate Image Duplication in Biomedical Research Publications," *mBio* 7, no. 3 (June 2016): e00809-16, https://doi.org/10.1128/mBio.00809-16.
28. Sholto David, "A Quantitative Study of Inappropriate Image Duplication in the Journal *Toxicology Reports*," bioRxiv, September 10, 2023, https://www.biorxiv.org/content/10.1101/2023.09.03.556099v2.full.
29. Uri Simonsohn, "Just Post It: The Lesson from Two Cases of Fabricated Data Detected by Statistics Alone," *Psychological Science* 24, no. 10 (October 2013): 1875–88, https://doi.org/10.1177/0956797613480366; James E. Mosimann, Claire V. Wiseman, and Ruth E. Edelman, "Data Fabrication: Can People Generate Random Digits?," *Accountability in Research* 4, no. 1 (1995): 31–55, https://doi.org/10.1080/08989629508573866; and Chris H. J. Hartgerink et al., "Detection of Data Fabrication Using Statistical Tools," PsyArXiv Preprints, August 19, 2019, https://osf.io/preprints/psyarxiv/jkws4/.
30. The exact same figure cannot mean two (or more) completely different things. In such cases, the only question left to determine is whether the underlying cause is intentional fraud or accidental error. The more duplications, copies, rotations, and rescalings there are in the same publication, and the more such papers are published by the same authors, the smaller is the possibility of "accidental error."
31. Dalmeet Singh Chawla, "8 Percent of Researchers in Dutch Survey Have Falsified or Fabricated Data," *Nature*, July 2021, https://doi.org/10.1038/d41586-021-02035-2.
32. Daniele Fanelli, "How Many Scientists Fabricate and Falsify Research? A Systematic Review and Meta-Analysis of Survey Data," *PLOS One* 4, no. 5 (May 2009): e5738, https://doi.org/10.1371/journal.pone.0005738.
33. Susan Eastwood et al., "Ethical Issues in Biomedical Research: Perceptions and Practices of Postdoctoral Research Fellows Responding to a Survey," *Science and Engineering Ethics* 2, no. 1 (January 1996): 89–114, https://doi.org/10.1007/BF02639320.
34. Michael W. Kalichman and Paul J. Friedman, "A Pilot Study of Biomedical Trainees' Perceptions Concerning Research Ethics," *Academic Medicine* 67, no. 11 (November 1992): 769–75, https://doi.org/10.1097/00001888-199211000-00015.
35. Yu Xie, Kai Wang, and Yan Kong, "Prevalence of Research Misconduct and Questionable Research Practices: A Systematic Review and Meta-Analysis," *Science and Engineering Ethics* 27, no. 4 (June 2021): 41, https://doi.org/10.1007/s11948-021-00314-9.
36. Melissa S. Anderson, Brian C. Martinson, and Raymond De Vries, "Normative Dissonance in Science: Results from a National Survey of US Scientists," *Journal of Empirical Research on Human Research Ethics* 2, no. 4 (December 2007): 3–14, https://doi.org/10.1525/jer.2007.2.4.3.
37. I am using the word "productive" here in the sense of turning out more publications per year, not in the sense of more actual information being contained in them.

38. Richard Smith, "Time to Assume That Health Research Is Fraudulent Until Proven Otherwise?," *BMJ* (blog), July 5, 2021, https://blogs.bmj.com/bmj/2021/07/05/time-to-assume-that-health-research-is-fraudulent-until-proved-otherwise/.
39. A delicious detail is that his book seems to contain plagiarized sections from the classic authors Raymond Carver and James Joyce but does not use quotes for them. The sources are not completely ignored, however; they are acknowledged separately in the appendix.
40. C. K. Gunsalus and Aaron D. Robinson, "Nine Pitfalls of Research Misconduct," *Nature* 557, no. 7705 (May 2018): 297–99, https://doi.org/10.1038/d41586-018-05145-6.
41. Ferric C. Fang and Arturo Casadevall, "Why We Cheat," *Scientific American Mind* 24, no. 2 (May 2013): 30–7, 2013; and Joan O'C Hamilton, "Why We Cheat," *Stanford Magazine*, September 2015.
42. Samuel J. Leistedt and Paul Linkowski, "Fraud, Individuals, and Networks: A Biopsychosocial Model of Scientific Frauds," *Science & Justice* 56, no. 2 (March 2016): 109–12, https://doi.org/10.1016/j.scijus.2016.01.002.

5. A BROKEN SCIENTIFIC PUBLISHING SYSTEM

1. The Nobel laureate John Vane, in whose institute I worked in the early 1990s in London, gave different advice. He opined: "If you think your paper is good, and the referees don't like it, the reason may be that they did not understand it. So, try to re-submit it to a better journal." (This approach worked for me sometimes over the years, but many other times it did not.)
2. Peter Medawar, "Is the Scientific Paper a Fraud?," in *Experiment: A Series of Scientific Case Histories First Broadcast in the BBC Third Programme*, ed. D. Edge, 7–13 (London: BBC, 1964).
3. Randy Schekman, "How Journals Like Nature, Cell and Science Are Damaging Science," *The Guardian*, December 9, 2013.
4. Björn Brembs, "Prestigious Science Journals Struggle to Reach Even Average Reliability," *Frontiers in Human Neuroscience* 12 (February 2018): 37, https://doi.org/10.3389/fnhum.2018.00037.
5. Rejection of an important paper by a top journal may not only hurt the feelings of the authors but may also extend the time until the work realizes its overall impact. It would be nice to believe that truly groundbreaking discoveries, even if originally published in a smaller journal, eventually were recognized.
6. Stories like this rarely get much publicity. The exception is Michael Dansinger of Tufts University in Boston, whose manuscript was stolen by a reviewer for the *Annals of Internal Medicine* and published under that person's own name. Dansinger did not let it go and fought on; the plagiarized work was ultimately retracted, and the journal apologized. See Michael Dansinger, "Dear Plagiarist: A Letter to a Peer Reviewer Who Stole and Published Our Manuscript as His Own," *Annals of Internal Medicine* 166, no 2 (January 2017): 143, https://doi.org/10.7326/M16-2551. It would be nice to think that Dansinger's case was an exceptional and rare case. But many similar cases have appeared on Retraction Watch. And, once again, this is just the tip of the iceberg. Unfortunately, many of my colleagues have had similar things happen to them.

7. Balazs Aczel, Barnabas Szaszi, and Alex O. Holcombe, "A Billion-Dollar Donation: Estimating the Cost of Researchers' Time Spent on Peer Review," *Research Integrity and Peer Review* 6, no. 1 (November 2021): 14, https://doi.org/10.1186/s41073-021-00118-2.
8. More on Brookes' science sleuthing activities later.
9. Lisa Parker, Stephanie Boughton, Rosa Lawrence, and Lisa Bero, "Experts Identified Warning Signs of Fraudulent Research: A Qualitative Study to Inform a Screening Tool," *Journal of Clinical Epidemiology* 151 (November 2022): 1–17, https://doi.org/10.1016/j.jclinepi.2022.07.006.
10. The actual calculation considers two preceding years. If a journal has a calculated impact factor of fifty for 2023 (released typically in Q1 of 2024), then on average each paper published in that journal during 2021 and 2022 received fifty literature citations in 2023.
11. Ewen Callaway, "Beat It, Impact Factor! Publishing Elite Turns Against Controversial Metric," *Nature* 535, no. 7611 (July 2016): 210–11, https://doi.org/10.1038/nature.2016.20224.
12. Impact factor is not the only indicator of a journal's influence in a field. A second factor, called Eigenfactor, follows a chain of citations that start out from a paper published in a given journal. The basic premise is that if a paper is influential it will be cited in other papers, and this will be cited in further papers, and so on. On the top of the journal scale, Eigenfactor and impact factor both identify the same journals: *Nature*, *Science*, and *PNAS*. *Cell* ranks lower on the Eigenfactor scale than on the impact factor scale. See Alan Fersht, "The Most Influential Journals: Impact Factor and Eigenfactor," *Proceedings of the National Academy of Sciences of the United States of America* 106, no. 17 (April 2009): 6883–84, https://doi.org/10.1073/pnas.0903307106.
13. Dalmeet Singh Chawla, "What's Wrong with the Journal Impact Factor in 5 Graphs. Scholars Love to Hate the Journal Impact Factor, but How Flawed Is It?" *Nature*, April 3 2018.
14. Ling Xin, "Young China Science Journal 'The Innovation' Joins International Rankings," *South China Morning Post*, July 3, 2023.
15. Michael Fire and Carlos Guestrin, "Over-Optimization of Academic Publishing Metrics: Observing Goodhart's Law in Action," *GigaScience* 8, no. 6 (June 2019): 1–20, https://doi.org/10.1093/gigascience/giz053.
16. Asghar Ghasemi, Parvin Mirmiran, Khosrow Kashfi, and Zahra Bahadoran, "Scientific Publishing in Biomedicine: A Brief History of Scientific Journals," *International Journal of Endocrinology and Metabolism* 21, no. 1 (December 2022): e131812, https://doi.org/10.5812/ijem-131812.
17. Mark A. Hanson, Pablo Gómez Barreiro, Paolo Crosetto, and Dan Brockington, "The Strain on Scientific Publishing," arXiv, September 2023, https://arxiv.org/ftp/arxiv/papers/2309/2309.15884.pdf.
18. Michael Park, Erin Leahey, and Russell J. Funk, "Papers and Patents Are Becoming Less Disruptive Over Time," *Nature* 613, no. 7942 (January 2023): 138–44, https://doi.org/10.1038/s41586-022-05543-x.
19. Max Kozlov, "'Disruptive' Science Has Declined—Even as Papers Proliferate," *Nature* 613, no. 7943 (January 2023): 225, https://doi.org/10.1038/d41586-022-04577-5.
20. As a rough estimate, perhaps we should consider the product of the number of articles published in a certain year and the CD index of the papers published in the same year

270   5. A BROKEN SCIENTIFIC PUBLISHING SYSTEM

as a potential measure. But sadly, even this derivative number is declining (it was about 3,500 in 1945; it declined to 1,750 by 2000, and further declined to a measly 300 by 2010).
21. Dimitrije Curcic, "Academic Publishers Statistics," Words Rated, June 21, 2023, https://wordsrated.com/academic-publishers-statistics/.
22. In most cases, these fees are paid by the academic centers that employ the authors or by the granting agencies that support the authors' research, not by the authors themselves.
23. Kyle Siler, "Demarcating Spectrums of Predatory Publishing: Economic and Institutional Sources of Academic Legitimacy," *Journal of the Association for Information Science and Technology* 71 (February 2020): 1386–401, https://doi.org/10.1002/asi.24339.
24. Andrew Silver, "Controversial Website That Lists 'Predatory' Publishers Shuts Down," *Nature*, January 2017, https://doi.org/10.1038/nature.2017.21328.
25. Andrea Manca et al., "Predatory Journals Enter Biomedical Databases Through Public Funding," *British Medical Journal* 371 (December 2020): m4265, https://doi.org/10.1136/bmj.m4265.
26. Adam Marcus and Ivan Oransky, "Retraction Watch Hijacked Journal Checker," Retraction Watch (blog), https://retractionwatch.com/the-retraction-watch-hijacked-journal-checker/.
27. Dalmeet Singh Chawla, "Hundreds of 'Predatory' Journals Indexed on Leading Scholarly Database," *Nature*, February 2021, https://doi.org/10.1038/d41586-021-00239-0.
28. David Mazières and Eddie Kohler, "Get Me Off Your Fucking Mailing List," Stanford Secure Computer Systems Group, February 4, 2005, https://www.scs.stanford.edu/~dm/home/papers/remove.pdf.
29. John Bohannon, "Who's Afraid of Peer Review?," *Science* 342, no. 6154 (October 2013): 60–65, https://doi.org/10.1126/science.2013.342.6154.342_60.
30. So far these manuscripts are not very good. But with the advancement of AI technology, surely new heights will soon be reached in the rarified art of generating scientific papers out of thin air. And sooner or later, AI-based image generators will be trained by some scientific crime syndicate or paper millers to produce all kinds of realistic-looking histological graphs and Western blots. From then on, these figures will not be copies of any previously published figures, and computer programs will not be able to detect them as frauds. I estimate that in two to three years it will be virtually impossible to distinguish between an immunohistochemical graph produced by AI and one that is the result of real experimental work.
31. Nicole Shu Ling Yeo-Teh and Bor Luen Tang, "An Active Aigiarism Declaration for Manuscript Submission," *Journal of Accounting Research* 8 (March 2023): 1–2, https://doi.org/10.1080/08989621.2023.2185776.
32. Gina Kolata, "The Price for 'Predatory' Publishing? $50 Million," *New York Times*, April 3, 2019.
33. Jennifer A. Byrne et al., "Protection of the Human Gene Research Literature from Contract Cheating Organizations Known as Research Paper Mills," *Nucleic Acids Research* 50, no. 21 (November 2022): 12058–70, https://doi.org/10.1093/nar/gkac1139.
34. This seems like a childish trick that could easily be prevented by checking the proposed referee's identity, or better yet, never use referees proposed by the submitter of a paper. It is mind-blowing that fraudsters can get away with this.

35. There are other reasons too. For example, suggested referees—even if real—can be friends, former colleagues, or supporters of the authors of the submission. I have heard rumors of so-called peer review circles, in which a group of academics get together to form a cartel and agree to provide uncritical positive refereeing of other cartel members' submissions. As long as the cartel members don't have prior joint papers or overlapping affiliations with the authors of the submitted paper—something editors can check—a cartel of this type is nearly impossible to spot.
36. Many actual scientists are becoming peer review fatigued, and it is getting harder and harder to find a good referee. A good scientist may smell a fake paper from miles away and will decline to referee it rather than taking up the arduous and unthankful task of exposing fraudsters.
37. Holly Else, "Scammers Impersonate Guest Editors to Get Sham Papers Published," *Nature* 599, no. 7885 (November 2021): 361, https://doi.org/10.1038/d41586-021-03035-y.
38. Founded in 1873, this journal used to be one of the premium journals in pharmacology. Its impact factor these days is only around three; even so, in pharmacology it was considered a respectable journal until the latest paper mill fiasco, which is discussed in this chapter. The journal is published by Springer, a legitimate scientific publisher, one of the largest in the world.
39. Roland Seifert, "How Naunyn-Schmiedeberg's Archives of Pharmacology Deals with Fraudulent Papers from Paper Mills," *Naunyn Schmiedebergs Archives of Pharmacology* 394, no. 3 (March 2021): 431–36, https://doi.org/10.1007/s00210-021-02056-8.
40. Dalmeet Singh Chawla, "How a Site Peddles Author Slots in Reputable Publishers' Journal," *Science* 376, no. 6590 (April 2022): 231–32, https://doi.org/10.1126/science.abq4276.
41. Hindawi was acquired in 2021 by the American scientific publishing company Wiley. Since its acquisition, at least eight thousand (!!!) Hindawi publications have been retracted, and more retractions are in the works. After millions of dollars of lost revenue that stemmed from this acquisition, in late 2023 Wiley decided to do away with the brandished name "Hindawi" altogether.
42. Scientists are not only commonly rated and ranked by the journals in which they publish but also by the number of literature citations their publications receive. Scientists who receive the largest number of citations end up on various prestigious highly cited lists. The number of citations is considered a crucial indicator of the scientists' work in their field and is often used by institutional promotional committees to rank applicants for promotion. Thus a market has sprung up to "assist" fraudulent investigators artificially boost their citation numbers.
43. Jeffrey Brainard, "New Tools Show Promise for Tackling Paper Mills," *Science* 380, no. 6645 (May 2023): 568–69, https://doi.org/10.1126/science.adi6523.
44. Bernhard A. Sabel, Emely Knaack, Gerd Gigerenzer, and Mirela Bilc, "Fake Publications in Biomedical Science: Red-Flagging Method Indicates Mass Production," MedRxiv, October 18, 2023, https://doi.org/10.1101/2023.05.06.23289563. (Disclaimer: please note that manuscripts posted on MedRxiv have not yet undergone peer review.)
45. As previously mentioned, some paper mills also offer citation services for their fake papers. This can be arranged by citing the fake paper in other fake papers, often in completely unrelated subjects. Such "citation farms" are also being uncovered from time

to time. For example, a recent investigation published in Forbetterscience.com revealed that a 2017 paper in *Current Biology* on moths releasing pheromones had suspiciously been cited more than seven hundred times, mostly by papers that have nothing to do with moths, pheromones, or even biology.

46. C. Szabó et al., "Suppression of Macrophage Inflammatory Protein (MIP)-1Alpha Production and Collagen-Induced Arthritis by Adenosine Receptor Agonists," *British Journal of Pharmacology* 125, no. 2 (September 1998): 379–87, https://doi.org/10.1038/sj.bjp.0702040.
47. Serge P. J. M. Horbach and Willem Halffman, "The Extent and Causes of Academic Text Recycling or 'Self-Plagiarism,'" *Research Policy* 48, no. 2 (March 2019): 492–502, https://doi.org/10.1016/j.respol.2017.09.004.
48. Jim Giles, "Taking on the Cheats," *Nature* 435 (2005): 258–59, https://doi.org/10.1038/435258a.
49. Vanja Pupovac and Daniele Fanelli, "Scientists Admitting to Plagiarism: A Meta-Analysis of Surveys," *Science and Engineering Ethics* 21, no. 5 (October 2015): 1331–52, https://doi.org/10.1007/s11948-014-9600-6.
50. In these cases, the authors will be incorrect, but at least the information contained in the article would not be altered when compared to the original publication.
51. Alan S. Brown and Dana R. Murphy, "Cryptomnesia: Delineating Inadvertent Plagiarism," *Journal of Experimental Psychology: Learning, Memory, and Cognition* 15, no. 3 (1989): 432–42, https://doi.org/10.1037/0278-7393.15.3.432.
52. This type of misconduct is so common that today this is not the exception but almost a rule. It is also a testament to the low quality of the whole manuscript evaluation system. For a while I tried to contact some of the authors and journals when this happened to my own work. Sometimes I could get a correction in the journal, other times a begrudging email apology from the authors. Here is one example: "I am sorry to have given you the wrong impression in our manuscript. I will try to do better and correct this in our next manuscripts." I can print it and frame it and stick it to the wall, but after a while, I gave up. I don't want to spend my time with this.
53. Dalmeet Singh Chawla, "Two-Thirds of Researchers Report 'Pressure to Cite' in Nature Poll," *Nature*, October 2019, https://doi.org/10.1038/d41586-019-02922-9.
54. In theory, a good editor could easily identify when the referees are pushing for the papers to cite their own work. But this would mean even more work for the editors. (Remember, most editors—just like the referees—work on a voluntary basis and are severely overwhelmed.)
55. Guilleamue Cabanac and Cyrille Labbé, "Prevalence of Nonsensical Algorithmically Generated Papers in the Scientific Literature," *Journal of the Association for Information Science and Technology* 72, no. 2 (May 2021) 1461–76, https://doi.org/10.1002/asi.24495.
56. David Bimler, "Better Living Through Coordination Chemistry: A Descriptive Study of a Prolific Papermill That Combines Crystallography and Medicine," Research Square, April 15, 2022, https://doi.org/10.21203/rs.3.rs-1537438/v1.
57. David M. Markowitz and Jeffrey T. Hancock, "Linguistic Obfuscation in Fraudulent Science," *Journal of Language and Social Psychology* 35, no. 4 (2016): 435–45, https://doi.org/10.1177/0261927X15614605.

58. Moncada's *Nature* paper has been cited more than fifteen thousand times. Furchgott's paper, eventually published in 1998 as a chapter in a book titled *Vasodilatation, Vascular Smooth Muscle, Peptides and Endothelium*, published by Raven Press, is almost never cited.
59. In the small informal survey I conducted with fifteen of my colleagues, the average number of suspect/predatory meeting invitations they receive by email is twelve *per day*.
60. Ruairi J. Mackenzie, "Inside a 'Fake' Conference: A Journey Into Predatory Science," Technology Networks, July 11, 2019, https://www.technologynetworks.com/tn/articles/inside-a-fake-conference-a-journey-into-predatory-science-321619.
61. It will not come as a surprise that the fraudsters who have invaded science with predatory publications and predatory conferences have now started passing out scam awards for paying customers. In this case, the scammers sift through the published literature and offer a publication award for the "outstanding publication"—of course, for a reasonable fee. Similar to other predatory industries, they don't bother to match the name of the award to the discipline of the victim's paper. For instance, I was recently offered a publication award in physics for one of my groups' papers in cancer pharmacology.
62. Ivan Oransky, "Retractions Are Increasing, but Not Enough," *Nature* 608, no. 7921 (August 2022): 9, https://doi.org/10.1038/d41586-022-02071-6.
63. Katia Audisio et al., "A Survey of Retractions in the Cardiovascular Literature," *International Journal of Cardiology* 349 (February 2022): 109–14, https://doi.org/10.1016/j.ijcard.2021.12.021.
64. The recent series of corrections and retractions of the 2019 Nobelist Gregg Semenza have already been covered. Other Nobelists with retracted papers include the 2004 Nobelist Linda Buck, the 2009 Nobelist Jack Szostak, the 2011 Nobelist Bruce Beutler, the 2013 Nobelist Thomas Südhof, and the 2018 Nobelist Frances Arnold. PubPeer lists numerous additional flagged papers published by the laboratories of various other Nobel laureates.
65. Andrew M. Stern, Arturo Casadevall, R. Grant Steen, and Ferric C. Fang, "Financial Costs and Personal Consequences of Research Misconduct Resulting in Retracted Publications," *Elife* 3 (August 2014): e02956, https://doi.org/10.7554/eLife.02956.
66. Janet D. Stemwedel, "Life After Misconduct: Promoting Rehabilitation While Minimizing Damage," *Journal of Microbiology & Biology Education* 15, no. 2 (December 2014): 177–80, https://doi.org/10.1128/jmbe.v15i2.827; and David Cyranoski, "China Introduces 'Social' Punishments for Scientific Misconduct," *Nature* 564, no. 7736 (December 2018): 312, https://doi.org/10.1038/d41586-018-07740-z.
67. Direct quotes from a text message of the authors of the article to another during preparation of the review article: "Although I hate when politics is injected into science—but it's impossible not to, especially given the circumstances." See U.S. House Committee on Oversight and Accountability, Interim Staff Report, "The Proximal Origin of a Cover-Up: Did the 'Bethesda Boys' Downplay a Lab Leak?," July 12, 2023, https://oversight.house.gov/report/interim-staff-report-the-proximal-origin-of-a-cover-up-did-the-bethesda-boys-downplay-a-lab-leak/.
68. Carl Elliott, *White Coat, Black Hat: Adventures on the Dark Side of Medicine* (Boston: Beacon Press, 2011).

6. THE WAY FORWARD

1. Elisabeth Pain, "Paul Brookes: Surviving as an Outed Whistleblower," *Science*, March 10, 2014, https://www.science.org/content/article/paul-brookes-surviving-outed-whistleblower, https://doi.org/10.1126/science.caredit.a1400061.
2. Thankfully, there is no such thing as truly deleting something from the internet. Using various time machines (e.g., the Wayback Machine), the blog can still be accessed.
3. Paul S. Brookes, "Internet Publicity of Data Problems in the Bioscience Literature Correlates with Enhanced Corrective Action," *PeerJ* 2 (April 2014): e313, https://doi.org/10.7717/peerj.313.
4. "Paper Mills and Research Misconduct: Facing the Challenges of Scientific Publishing," Hearing of the U.S. House of Representatives, Committee on Science, Space and Technology, Subcommittee on Investigations and Oversight, July 20, 2022, https://www.congress.gov/117/meeting/house/115022/documents/HHRG-117-SY21-20220720-SD002.pdf.
5. His resignation was connected to "violations of multiple Whitehead policies, including its sexual harassment policy." He denies the sexual harassment allegations and has filed a countersuit against the Whitehead Institute. In 2023, he began a new career in the Czech Republic, at the Organic Chemistry and Biochemistry Prague, a part of the Czech Academy of Sciences.

    See Shawna Williams, "Cell Biologist David Sabatini Fired for Sexual Harassment," *The Scientist*, August 22, 2021, https://www.the-scientist.com/news-opinion/cell-biologist-david-sabatini-fired-for-sexual-harassment-69117; and Meredith Wadman, "Despite Sexual Harassment Shadow, Biologist David Sabatini Lands Job at Top Czech Institute," *Science*, November 2023, https://www.science.org/content/article/despite-sexual-harassment-shadow-biologist-david-sabatini-lands-job-top-czech-institute.
6. This, indeed, may be the case in some instances. It is difficult to determine—just by looking at a photoshopped series of blots—what the original data might have been prior to photoshopping. The range could be from removing some "smudges" that had nothing to do with the results or conclusions to completely fictitious sets of copy-pasted blots (sometimes called Frankenblots) that have no relationship to any actual laboratory experiments.
7. His dark humor can be appreciated in many of the cartoons contained in this book.
8. Ulrich Herb, "Das Wissenschaftliche Publikations—und Reputationssystem Ist Gehackt," Telepolis, April 15, 2020, https://www.telepolis.de/features/Das-wissenschaftliche-Publikations-und-Reputationssystem-ist-gehackt-4701388.html.
9. As often happens, the first allegations were published in PubPeer and Schneider's blog, then the story was picked up by mainstream papers. But in this case the mainstream papers actually acknowledged Schneider's blog as the original source. Let's hope this trend continues in the future. Another interesting observation is that the timelines seem to accelerate. In the Fulda case, Schneider's blog came out on January 22, 2024, the German media reported on it a week later, and Fulda resigned on February 10, 2024. Another problematic cluster of papers, published by Harvard's Dana Farber Institute (the "farberifications" story) was first covered by Schneider's blog and then picked up by the *Boston Globe* and other mainstream journals. The director of the institute, Laurie Glimcher, retired in September 2024. (The official news release on this matter did not mention any misconduct issues or problematic publications.)

10. Elisabeth Bik, "Hindawi's Mass Retraction of 'Special Issues' Papers," *Science Integrity Digest*, August 10, 2023, https://scienceintegritydigest.com.
11. Nick Wise, nhwise.bsky.social X account, started April 2012, https://twitter.com/nickwizzo.
12. David Bimler, "Better Living Through Coordination Chemistry: A Descriptive Study of a Prolific Papermill That Combines Crystallography and Medicine," Research Square, April 15, 2022, https://europepmc.org/article/ppr/ppr482891#impact, https://doi.org/10.21203/rs.3.rs-1537438/v1.
13. Holly Else, "What Makes an Undercover Science Sleuth Tick?," *Nature* 608, no. 7923 (August 2022): 463, https://doi.org/10.1038/d41586-022-02099-8.
14. Dorothy Bishop, "Ramblings on Academic-Related Matters," *BishopBlog*, http://deevybee.blogspot.com.
15. Adam Mastroianni, "Experimental History," ongoing newsletter since 2022, https://www.experimental-history.com/.
16. Elie Dolgin, "PubMed Commons Closes Its Doors to Comments. The US National Institutes of Health Shutters Its Journal-Commenting Platform," *Nature*, February 2018, https://doi.org/10.1038/d41586-018-01591-4.
17. Malhar N. Kumar, "Dealing with Misconduct in Biomedical Research: A Review of the Problems and the Proposed Methods for Improvement," *Journal of Accounting Research* 16, no. 6 (November 2009): 307–30, https://doi.org/10.1080/08989620903328576.
18. Susan M. Kuzma, "Criminal Liability for Misconduct in Scientific Research," *University of Michigan Journal of Law Reform* 25, no 2 (1992): 357–421.
19. Zulfiqar A. Bhutta and Julian Crane, "Should Research Fraud Be a Crime?," *British Medical Journal* 349 (July 2014): g4532, https://doi.org/10.1136/bmj.g4532.
20. Justin T. Pickett and Sean Patrick Roche, "Questionable, Objectionable or Criminal? Public Opinion on Data Fraud and Selective Reporting in Science," *Science and Engineering Ethics* 24, no. 1 (February 2018): 151–71, https://doi.org/10.1007/s11948-017-9886-2.
21. See William Bülow and Gert Helgesson, "Criminalization of Scientific Misconduct," *Medicine, Health Care and Philosophy* 22, no. 2 (June 2019): 245–52, https://doi.org/10.1007/s11019-018-9865-7.
22. This type of attitude is well documented, See William Broad and Nicholas Wade, *Betrayers of the Truth* (New York: Simon & Schuster, 1983).
23. Gengyan Tang and Jingyu Peng, "Are the Lists of Questionable Journals Reasonable: A Case Study of Early Warning Journal Lists," *Journal of Accounting Research* 22 (September 2023): 1–24, https://doi.org/10.1080/08989621.2023.2261846.
24. Nicole Shu Ling Yeo-Teh and Bor Luen Tang, "Research Data Mismanagement—from Questionable Research Practice to Research Misconduct," *Journal of Accounting Research* 14 (January 2023): 1–8, https://doi.org/10.1080/08989621.2022.2157268.
25. Minal M. Caron, Sarah B. Dohan, Mark Barnes, and Barbara E. Bierer, "Defining 'Recklessness' in Research Misconduct Proceedings," *Journal of Accounting Research* 11 (September 2023): 1–23, https://doi.org/10.1080/08989621.2023.2256650.
26. Loreta Tauginienė and Inga Gaižauskaitė, "Jumping with a Parachute—Is Promoting Research Integrity Meaningful?," *Journal of Accounting Research* 30, no. 8 (December 2023): 548–73, https://doi.org/10.1080/08989621.2022.2044318.

27. Lex Bouter, "Research Misconduct and Questionable Research Practices Form a Continuum," *Journal of Accounting Research* 3 (March 2023): 1–5, https://doi.org/10.1080/08989621.2023.2185141.
28. Tommy Shih, "Recalibrated Responses Needed to a Global Research Landscape in Flux," *Journal of Accounting Research* 2 (August 2022): 1–7, https://doi.org/10.1080/08989621.2022.2103410.
29. Francis S. Collins and Lawrence A. Tabak, "Policy: NIH Plans to Enhance Reproducibility," *Nature* 505, no. 7485 (January 2014): 612–13, https://doi.org/10.1038/505612a.
30. This is another contentious topic that has been discussed for decades, but no clear consensus has yet emerged. Certainly, there would be some positive aspects, as it would eliminate some bias, perhaps increase the reviewers' focus on merit, may encourage diversity, and could reduce the reviewers' potential unconscious bias and conflict of interest. Critics of the approach argue, however, that the evaluation of many applications can only occur if the referees have contextual understanding (such as the investigator's background and expertise). Moreover, implementing a fully blinded review process can be logistically challenging. If preliminary data are included, reviewers who are true experts in the given field can easily identify the applicant group. Several granting agencies, including the NIH, the Canadian Institutes of Health Research, the European Research Council, the UK's Medical Research Council, and the Swiss SNSF have all piloted anonymous grant reviews; many of them appear to go back-and-forth between the idea of named and anonymous processes.
31. Leonard P. Freedman, Gautham Venugopalan, and Rosann Wisman, "Reproducibility2020: Progress and Priorities [version 1; peer review: 2 approved]," *F1000Research* 6 (May 2017): 604, https://doi.org/10.12688/f1000research.11334.1.
32. Meredith Wadman, "NIH Mulls Rules for Validating Key Results," *Nature* 500, no. 7460 (August 2013): 14–16, https://doi.org/10.1038/500014a.
33. C. K. Gunsalus et al., "Overdue: A US Advisory Board for Research Integrity," *Nature* 566, no. 7743 (February 2019): 173–75, https://doi.org/10.1038/d41586-019-00519-w.
34. Daniil A. Boiko, Robert MacKnight, Ben Kline, and Gabe Gomes, "Autonomous Chemical Research with Large Language Models," *Nature* 624 (2023): 570, https://doi.org/10.1038/s41586-023-06792-0.
35. Kexuan Sun et al., "Assessing Scientific Research Papers with Knowledge Graphs," *SIGIR '22: Proceedings of the 45th International ACM SIGIR Conference on Research and Development in Information Retrieval* (July 2022): 2467–72, https://doi.org/10.1145/3477495.3531879.
36. The following material mainly applies to the scientific ecosystem in the United States and Europe—ones with which I have direct experience.
37. The *myth of the solitary genius scientist* may have been somewhat true for the eighteenth and nineteenth centuries, when genius scientists such as Einstein, Maxwell, and Leibniz mainly worked alone. But this is no longer the case, even though movies (*Back to the Future*, etc.) and many popular science articles tend to perpetuate this false image. In reality, in modern biomedical science, most scientists work in multidisciplinary teams; in addition, it is rather common for several independent working groups to come up with the same discovery at around the same time.

38. Anyway, *why* would anyone want to invest more funding in an inefficient system that produces more unreliable results than reliable ones?
39. A military analogy comes to mind: instead of battalions of ground soldiers, we need smaller, but well-equipped and highly trained units: products of an extremely stringent selection process, followed by top-level training—kind of like navy seals.
40. Douglas G. Altman, "The Scandal of Poor Medical Research," *British Medical Journal* 308, no. 6924 (January 1994): 283–84, https://doi.org/10.1136/bmj.308.6924.283.
41. Matthias Steinfath et al., "Simple Changes of Individual Studies Can Improve the Reproducibility of the Biomedical Scientific Process as a Whole," *PLOS One* 13, no. 9 (September 2018): e0202762, https://doi.org/10.1371/journal.pone.0202762.
42. Research also produces new knowledge and societal benefits, of course, and will improve our civilization. But I have not seen many university administrators who had this as their core concern, so here I focused on more tangible products such as reputation and money.
43. Such processes—as well as the routine application of standard operating procedures and good laboratory practice guidelines for preclinical research—were first introduced in the laboratories of large pharmaceutical companies, but it would definitely be worth considering implementation of some of these more regulated approaches in academic laboratories as well.
44. I am sure that in academic circles it would soon be renamed the "Just give them the money approach," but the concept is much more than that. The "free" money would come with serious strings attached, which is explained later.
45. The list of the most NIH-funded U.S. institutions in 2013 and 2023 is almost identical. It includes Johns Hopkins University, University of Washington, University of Pennsylvania, University of Michigan, University of Pittsburgh, Stanford University, Duke University, Yale University, Harvard University, and several members of the University of California system.
46. Max Kozlov, "NIH Issues a Seismic Mandate: Share Data Publicly," *Nature* 602, no. 7898 (February 2022): 558–59, https://doi.org/10.1038/d41586-022-00402-1.
47. Tsuyoshi Miyakawa (Fujita Health University, Japan), as editor-in-chief of *Molecular Brain*, handled 180 manuscripts between 2017 and 2020. Over this period of time he selected forty-one submissions and—prior to sending them to peer review—requested that the authors provide raw data. Half of these submissions were withdrawn as a response. Miyakava, in turn, had to reject nineteen of the remaining twenty manuscripts because the raw data supplied were insufficient. See Tsuyoshi Miyakawa, "No Raw Data, No Science: Another Possible Source of the Reproducibility Crisis," *Molecular Brain* 13, no. 1 (February 2020): 24, https://doi.org/10.1186/s13041-020-0552-2.
48. Ilias Berberi and Dominique G. Roche, "No Evidence That Mandatory Open Data Policies Increase Error Correction," *Nature Ecology and Evolution* 6 (September 15, 2022): 1630–33, https://doi.org/10.1038/s41559-022-01879-9.
49. Alison Abott, "Strife at eLife: Inside a Journal's Quest to Upend Science Publishing," *Nature* 615, no. 7954 (March 2023): 780–81, https://doi.org/10.1038/d41586-023-00831-6. While I was writing this book, I sent an email to Eisen with a couple of questions. I wondered what the selection process was to choose submissions sent for peer review (and, simultaneously, published) and what their percentage of immediate "desk rejections"

was. I received no response, but a week later I read in *Science* that he had been let go as the editor-in-chief of the journal. —*not* for the publication reforms he initiated but for sharing a satirical tweet from *Onion News* on X (formerly Twitter) on the topic of the Hamas-Israel conflict.

50. It would be easy to set up a system in which even the extent of the "elimination stringency" could be selected.
51. As mentioned earlier, in the current system, if one journal rejects a likely fraudulent paper, the authors will keep trying their luck with the second, third, fourth, and umpteenth journal, until they finally get their paper published. Meanwhile, reviewers working for those journals that rejected the manuscript may provide unwitting help to the fraudsters in hiding their fake material better in the manuscript.
52. I know the paying authorship scheme and the paper mill industry will continue to flourish. There is no way to stop them. They operate in countries with opaque jurisdiction, and they can avoid detection by switching servers or countries. There will also be many individuals who may continue to pay for fake publications or buy authorships. But the proposal I present would at least completely marginalize them and set up a new, "clean" subset of literature that would serve the interest of all serious scientists.
53. In the old days (meaning before the internet), the only way to get scientific papers was through a university library. Everything was paper-based. A weekly booklet called *Current Contents* was published every week that listed the titles of papers published in various scientific journals. One would have to look through it in the library, find papers that looked interesting based on their title, and then get the magazine and make a photocopy of the article. If the publication in which the desired paper was published was not carried by the library, a small postcard could be sent to the authors, who could mail a copy of the reprint to you if they had extra copies (or not . . . it was a hit or miss situation). Everything was paper-based until the early 1990s.

    Later on *Current Contents* became a CD-ROM-based system, and the titles and abstracts of the articles could be searched on a computer. Every publication year was on a separate CD-ROM and had to be searched separately. All of this is now in the distant past. Every paper on the internet is now in the form of a pdf and is available at the push of a button. (What I don't understand is why PhD students and postdocs are still no better informed about the literature than were the young scientists of the past.) In any case, the scientific publishing industry is keeping some of the old formats, such as volumes, page numbers, galley proofs, even page charges, in the present day. But, of course, all of this represents shadows and remnants of the old paper-based system, which could rapidly be disrupted in the same way the music industry based on physical formats such as LPs and CDs was rapidly disrupted with the emergence of mp3s, iTunes, and Spotify.
54. Adam Kane and Bawan Amin, "Amending the Literature Through Version Control," *Biology Letters* 19, no. 1 (January 2023): 20220463, https://doi.org/10.1098/rsbl.2022.0463.
55. I cannot be the first person on the planet who has thought of this. I am also certain that the only reason this is not in place already is commercial: if this system came into existence, scientific journals would not be able to publish review articles on the same subject year after year, making profit from them and boosting their journal's impact factor.

56. Many journals already have sections of this type in which they discuss interesting new papers even if they were published in another journal.

57. Yes, I am familiar with the Ronald Reagan quote: "The nine most terrifying words in the English language are 'I'm from the Government, and I'm here to help.'" But, in this case, there is simply no alternative. If profit-oriented enterprises would be interested in scientific reform, it would have happened already.

58. Misconduct related to clinical trials and approved drugs and associated drug recalls are beyond the scope of this book. But here I will make an exception and briefly summarize this case. Selective inhibition of COX-2 seemed like a fantastic idea from the standpoint of basic science. COX-2 is the protein responsible for the inflammation-associated production of prostaglandins, and a closely related protein called COX-1 plays important roles in maintaining the health of the normal body, including protection of the stomach. Older inhibitors of this pathway (nonsteroidal anti-inflammatory drugs [NSAIDs] like aspirin) inhibit both proteins, and the inhibition of COX-1 is responsible for their stomach-irritating side effects. COX-2 inhibitors were developed with the intention of providing pain relief and anti-inflammatory effects without the gastrointestinal side effects associated with traditional NSAIDs.

Several pharmaceutical companies developed COX-2 inhibitors and marketed them aggressively. This included a drug named Vioxx by Merck, Celebrex by Pfizer, and Bextra from Pharmacia. The scandal emerged when concerns were raised about the cardiovascular safety of COX-2 inhibitors, particularly Vioxx. It turned out that COX-2 inhibitors increase the risk of heart attacks and strokes. The safety concerns surrounding COX-2 inhibitors, particularly Vioxx, were initially brought to light by Eric Topol, a cardiologist at the Cleveland Clinic, who published a letter in the *New England Journal of Medicine* in March 2001 on this topic. Several years later, in 2004, David Graham, a scientist at the U.S. Food and Drug Administration, presented several sets of data to an FDA advisory committee, revealing that there was a significant increase in the risk of heart attacks and strokes among patients taking Vioxx. It caused public outrage when later it emerged that the pharmaceutical companies marketing these drugs had been aware of potential safety issues well before Topol's and Graham's revelations. Internal documents from Merck, the maker of Vioxx, suggested that the company's own studies had already revealed increased cardiovascular risks associated with the drug, but the company did not fully disclose this information to the public or to the FDA. In September 2004, Merck withdrew Vioxx from the market, and in 2005 Bextra followed suit.

In 2005 the *Journal of the American Medical Association* estimated that Vioxx may have contributed to tens of thousands of excess cases of heart attacks and deaths in the United States. Legal consequences ensued for Merck when thousands of lawsuits were filed by individuals who claimed to have suffered harm as a result of taking Vioxx. In 2007, Merck agreed to a $4.85 billion settlement to resolve the majority of these lawsuits. The withdrawal of Bextra was also accompanied by legal actions and settlements.

In 2009, Pfizer (who by then had acquired Pharmacia, Bextra's original developer) agreed to pay $2.3 billion to settle allegations that it had illegally promoted Bextra and some other drugs for off-label uses. (Celebrex remains on the market but with a Black Box Warning that raises the safety issues. It is still prescribed and remains on the market

for conditions such as arthritis and acute pain, but physicians who use it today exercise greater caution and pay attention to cardiovascular risk factors.)
59. For a brief period, I was hoping that Tessier-Lavigne's recent misconduct investigation case would be such a triggering event. But apparently it was not. After Stanford's investigation, things have calmed down—once again.
60. "United Nations of Scientific Reproducibility?" I am afraid I cannot think of a good name for it, but I am sure this is the least of the problems to solve.

AFTERWORD

1. The 2023 Nobelist Katalin Karikó often present a witty English-language caricature in her lectures to illustrate this situation. On the left side of the cartoon, a large 19th century scientist with a white beard is musing as follows: "I must find an explanation for this phenomenon in order to understand Nature." On the right side of the figure, a younger, white-robed 21st century researcher mumbles, "I need to find results that are consistent with my concept so that we can publish the article in Nature."
2. Of course, present company is always an exception. I would like to sincerely thank Columbia University Press for expressing interest in the topic of this book and making possible its publication in English. I also thank Corvina Press (Budapest) for doing the same in my "old country," Hungary.
3. See Ronald Bailey, "Most Scientific Findings Are Wrong or Useless. Science Isn't Self-Correcting, It's Self-Destructing," Reason, October 26, 2016, https://reason.com/2016/08/26/most-scientific-results-are-wrong-or-use/.
4. This is a clinical topic that belongs in a future book. This crisis also had significant foundations in an unreliable/misleading publication. In this case, the publication was a letter in the New England Journal of Medicine in 1980 purporting that the addictiveness of opioid painkillers is much lower than generally believed. This misconception paved the way for widespread opioid use, contributing to a public health emergency that originated with the overprescription of opioid painkillers, notably driven by Purdue Pharma's aggressive marketing of OxyContin. Purdue Pharma, the maker of OxyContin, played a pivotal role by downplaying the drug's addictive nature and fueling its overprescription. Despite evidence of the growing epidemic, the FDA failed to exercise sufficient oversight, allowing Purdue Pharma to perpetuate misinformation. The consequences have been devastating, with more than 450,000 opioid-related deaths in the United States since the late 1990s. Purdue Pharma faced legal repercussions: it filed for bankruptcy in 2019 and reached settlements addressing its role in the crisis. However, the damage was done, and the crisis took a new turn when many of the prescription-opioid-addicted individuals later switched over to street drugs such as heroin and later on to synthetic opioids such as fentanyl, which are cheaper but also more dangerous and now cause more deaths each month than the prescription-opioids had in previous years.
5. Charles Piller, "Blots on a Field? A Neuroscience Image Sleuth Finds Signs of Fabrication in Scores of Alzheimer's Articles, Threatening a Reigning Theory of the Disease," Science 377, no. 6604 (July 2022): 358–63, https://doi.org/10.1126/science.add9993; Piller, "Picture Imperfect," Science 385, no. 6716 (September 2024): 1406–1412, https://doi.org/10.1126/science.adt3535.

6. Elizabeth Pain, "Paul Brookes, Surviving as an Outed Whistleblower," *Science*, March 10, 2014, https://www.science.org/content/article/paul-brookes-surviving-outed-whistleblower, https://doi.org/10.1126/science.caredit.a1400061.
7. The age-old joke in Hungarian goes something like this. The doctor presents a depressing diagnosis to Aunt Maria, a sweet old lady. "You have terminal cancer, and I am afraid you only have a few days to live." Aunt Maria turns pale and responds, "I understand, Doc, but can you tell me something encouraging?" So, the doctor pulls out a couple of cheerleader poms and starts to dance and chant: "LET'S GO, Aunt Maria, LET'S GO!!!"
8. See Ben Goldacre, *Bad Pharma: How Drug Companies Mislead Doctors and Harm Patients: How Medicine Is Broken, and How We Can Fix It* (New York: Harper Collins, 2013); Richard Harris, *Rigor Mortis: How Sloppy Science Creates Worthless Cures, Crushes Hope, and Wastes Billions* (New York: Basic Books, 2017); and Stuart Ritchie, *Science Fictions* (New York: Metropolitan, 2021).
9. These old stories make fascinating reading. See D. Emerick Szilagyi, "The Elusive Target: Truth in Scientific Reporting. Comments on Error, Self-Delusion, Deceit, and Fraud," *Journal of Vascular Surgery* 1, no. 2 (March 1984): 243–53, https://doi.org/10.1016/0741-5214(84)90055-7).

BIBLIOGRAPHY

Abott, Alison. "Strife at eLife: Inside a Journal's Quest to Upend Science Publishing." *Nature* 615, no. 7954 (March 2023): 780–81. https://doi.org/10.1038/d41586-023-00831-6.
Aczel, Balazs, Barnabas Szaszi, and Alex O. Holcombe. "A Billion-Dollar Donation: Estimating the Cost of Researchers' Time Spent on Peer Review." *Research Integrity and Peer Review* 6, no. 1 (November 2021): 14. https://doi.org/10.1186/s41073-021-00118-2.
Alegre, Maria-Luisa. "Mouse Microbiomes: Overlooked Culprits of Experimental Variability." *Genome Biology* 20, no. 1 (May 2019): 108. https://doi.org/10.1186/s13059-019-1723-2.
Alhayyan, Aliah M., Stephen T. McSorley, Rachel J. Kearns, Paul G. Horgan, Campbell S. D. Roxburgh, and Donald C. McMillan. "The Effect of Anesthesia on the Magnitude of the Postoperative Systemic Inflammatory Response in Patients Undergoing Elective Surgery for Colorectal Cancer in the Context of an Enhanced Recovery Pathway: A Prospective Cohort Study." *Medicine (Baltimore)* 100, no. 2 (January 2021): e23997. https://doi.org/10.1097/MD.0000000000023997.
Alitalo, Okko, Roosa Saarreharju, Ioline D. Henter, Carlos A. Zarate Jr., Samuel Kohtala, and Tomi Rantamäki. "A Wake-Up Call: Sleep Physiology and Related Translational Discrepancies in Studies of Rapid-Acting Antidepressants." *Progress in Neurobiology* 206 (November 2021): 102140. https://doi.org/10.1016/j.pneurobio.2021.102140.
Altman, D. G. "The Scandal of Poor Medical Research." *British Medical Journal* 308, no. 6924 (January 1994): 283–84. https://doi.org/10.1136/bmj.308.6924.283.
Anderson, Melissa S., Brian C. Martinson, and Raymond De Vries. "Normative Dissonance in Science: Results from a National Survey of US Scientists." *Journal of Empirical Research on Human Research Ethics* 2, no. 4 (December 2007): 3–14. https://doi.org/10.1525/jer.2007.2.4.3.
Audisio, Katia, N. Bryce Robinson, Giovanni J. Soletti, Gianmarco Cancelli, Arnaldo Dimagli, Cristiano Spadaccio, Roberto Perezgrovas Olaria et al. "A Survey of Retractions in the Cardiovascular Literature." *International Journal of Cardiology* 349 (February 2022): 109–14. https://doi.org/10.1016/j.ijcard.2021.12.021.
Aviles, Natalie B. *An Ungovernable Foe: Science and Policy Innovation in the U.S. National Cancer Institute*. New York: Columbia University Press, 2023.
Bailey, Ronald. "Most Scientific Findings Are Wrong or Useless. Science Isn't Self-Correcting, It's Self-Destructing." Reason, October 26, 2016. https://reason.com/2016/08/26/most-scientific-results-are-wrong-or-use/.

Baker, Monya. "1,500 Scientists Lift the Lid on Reproducibility." *Nature* 533, no. 7604 (May 2016): 452–54. https://doi.org/10.1038/533452a.

Baker, Theo. "Genentech Discloses Unreported Research Misconduct in Stanford President's Lab, Says It Did Not Find 'Intentional Wrongdoing' in 2009." *Stanford Daily*, April 6, 2023. https://stanforddaily.com/2023/04/06/stanford-president-research-misconduct-concerns-genentech/.

———. "Internal Review Found 'Falsified Data' in Stanford President's Alzheimer's Research, Colleagues Allege." *Stanford Daily*, February 17, 2023. https://stanforddaily.com/2023/02/17/internal-review-found-falsified-data-in-stanford-presidents-alzheimers-research-colleagues-allege/.

Ballreich, Jeromie M., Cary P. Gross, Neil R. Powe, and Gerard F. Anderson. "Allocation of National Institutes of Health Funding by Disease Category in 2008 and 2019." *JAMA Network Open* 4, no. 1 (January 2021): e2034890. https://doi.org/10.1001/jamanetworkopen.2020.34890.

Beck, Mihály. *Tudomány-áltudomány*. Budapest: Akadémiai Kiadó, 1977.

Begley, C. Glenn, and Lee M. Ellis. "Drug Development: Raise Standards for Preclinical Cancer Research." *Nature* 48, no. 7391 (March 2012): 531. https://doi.org/10.1038/483531a.

Berger, Nathan A., Valerie C. Besson, A. Hamid Boulares, Alexander Bürkle, Alberto Chiarugi, Robert S. Clark, Nicola J. Curtin et al. "Opportunities for the Repurposing of PARP Inhibitors for the Therapy of Non-Oncological Diseases." *British Journal of Pharmacology* 175, no. 2 (January 2018): 192–222. https://doi.org/10.1111/bph.13748.

Bernstein, Rachel. "HIV Researcher Found Guilty of Research Misconduct Sentenced to Prison." *Science*, July 6, 2015. https://www.science.org/content/article/hiv-researcher-found-guilty-research-misconduct-sentenced-prison.

Beutler, Bruce. "How Mammals Sense Infection: From Endotoxin to the Toll-Like Receptors." The Nobel Prize, December 7, 2011. https://www.nobelprize.org/uploads/2018/06/beutler-lecture.pdf.

Bhattacharya, Ananyo. *The Man from the Future*. New York: Norton, 2022.

Bhutta, Zulfiqar A., and Julian Crane. "Should Research Fraud Be a Crime?" *British Medical Journal* 349 (July 2014): g4532. https://doi.org/10.1136/bmj.g4532.

Bik, Elisabeth. "Hindawi's Mass Retraction of 'Special Issues' Papers." *Science Integrity Digest*. August 10, 2023. https://scienceintegritydigest.com.

Bik, Elisabeth M., Arturo Casadevall, and Ferric C. Fang. "The Prevalence of Inappropriate Image Duplication in Biomedical Research Publications." *mBio* 7, no. 3 (June 2016): e00809-16. https://doi.org/10.1128/mBio.00809-16.

Bimler, David. "Better Living Through Coordination Chemistry: A Descriptive Study of a Prolific Papermill That Combines Crystallography and Medicine." Research Square. April 15, 2022. https://doi.org/10.21203/rs.3.rs-1537438/v1.

Bishop, Dorothy. "Ramblings on Academic-Related Matters." *BishopBlog*. http://deevybee.blogspot.com.

Bogatyrev, Said R., Justin C. Rolando, and Rustem F. Ismagilov. "Self-Reinoculation with Fecal Flora Changes Microbiota Density and Composition Leading to an Altered Bile-Acid Profile in the Mouse Small Intestine." *Microbiome* 8, no. 1 (February 2020): 19. https://doi.org/10.1186/s40168-020-0785-4.

Bohannon, John. "Who's Afraid of Peer Review?" *Science* 342, no. 6154 (October 2013): 60–65. https://doi.org/10.1126/science.2013.342.6154.342_60.

Boiko, Daniil A., Robert MacKnight, Ben Kline, and Gabe Gomes. "Autonomous Chemical Research with Large Language Models." *Nature* 624 (2023): 570. https://doi.org/10.1038/s41586-023-06792-0.

Bouter, Lex. "Research Misconduct and Questionable Research Practices Form a Continuum." *Journal of Accounting Research* 3 (March 2023): 1–5. https://doi.org/10.1080/08989621.2023.2185141.

Brainard, Jeffrey. "New Tools Show Promise for Tackling Paper Mills." *Science* 380, no. 6645 (May 2023): 568–69. https://doi.org/10.1126/science.adi6523.

Braunwald, Eugene. "On Analysing Scientific Fraud." *Nature* 325, no. 6101 (January 1987): 215–16. https://doi.org/10.1038/325215a0.

Brembs, Björn. "Prestigious Science Journals Struggle to Reach Even Average Reliability." *Frontiers in Human Neuroscience* 12 (February 2018): 37. https://doi.org/10.3389/fnhum.2018.00037.

Broad, William, and Nicholas Wade. *Betrayers of the Truth*. New York: Simon & Schuster, 1983.

Brookes, Paul S. "Going Beyond Faked Data . . . Faking the Raw Data!" November 10, 2020. https://psblab.org/?p=651.

———. "Internet Publicity of Data Problems in the Bioscience Literature Correlates with Enhanced Corrective Action." *PeerJ* 2 (April 2014): e313. https://doi.org/10.7717/peerj.313.

Brown, Alan S., and Dana R. Murphy. "Cryptomnesia: Delineating Inadvertent Plagiarism." *Journal of Experimental Psychology: Learning, Memory, and Cognition* 15, no. 3 (1989): 432–42. https://doi.org/10.1037/0278-7393.15.3.432.

Bülow, William, and Gert Helgesson. "Criminalization of Scientific Misconduct." *Medicine, Health Care and Philosophy* 22, no. 2 (June 2019): 245–52. https://doi.org/10.1007/s11019-018-9865-7.

Byrne, Jennifer A., Yasunori Park, Reese A. K. Richardson, Pranujan Pathmendra, Mengyi Sun, and Thomas Stoeger. "Protection of the Human Gene Research Literature from Contract Cheating Organizations Known as Research Paper Mills." *Nucleic Acids Research* 50, no. 21 (November 2022): 12058–70. https://doi.org/10.1093/nar/gkac1139.

Cabanac, Guilleamue, and Cyril Labbé. "Prevalence of Nonsensical Algorithmically Generated Papers in the Scientific Literature." *Journal of the Association for Information Science and Technology* 72, no. 12 (May 2021) 1461–76. https://doi.org/10.1002/asi.24495.

Callaway, Ewen. "Beat It, Impact Factor! Publishing Elite Turns Against Controversial Metric." *Nature* 535, no. 7611 (July 2016): 210–11. https://doi.org/10.1038/nature.2016.20224.

———. "Danish Court Quashes Ruling Against Physiologist." *Nature*, 2015. https://doi.org/10.1038/nature.2015.16960.

———. "Danish Neuroscientist Challenges Fraud Findings." *Nature*, 2012. https://doi.org/10.1038/nature.2012.11146.

———. "In Vino Non Veritas? Red Wine Researcher Implicated in Misconduct Case." *Nature*, January 12, 2012. https://blogs.nature.com/news/2012/01/in-vino-non-veritas-red-wine-researcher-implicated-in-misconduct-case.html.

Caron, Minal M., Sarah B. Dohan, Mark Barnes, and Barbara E. Bierer. "Defining 'Recklessness' in Research Misconduct Proceedings." *Journal of Accounting Research* 11 (September 2023): 1–23. https://doi.org/10.1080/08989621.2023.2256650.

Carreyrou, John. *Bad Blood: Secrets and Lies in a Silicon Valley Startup*. New York: Knopf, 2018.

Catanzaro, Michele. "Possible Misconduct in Papers from Italian Health Minister." *Science* 381, no. 6664 (September 2023): 1267–68. https://doi.org/10.1126/science.adk9441.

Chawla, Dalmeet Singh. "8 Percent of Researchers in Dutch Survey Have Falsified or Fabricated Data." *Nature*, July 2021. https://doi.org/10.1038/d41586-021-02035-2.

———. "How a Site Peddles Author Slots in Reputable Publishers' Journals." *Science* 376, no. 6590 (April 2022): 231–32. https://doi.org/10.1126/science.abq4276.

———. "Hundreds of 'Predatory' Journals Indexed on Leading Scholarly Database." *Nature*, February 2021. https://doi.org/10.1038/d41586-021-00239-0.

———. "Two-Thirds of Researchers Report 'Pressure to Cite' in Nature Poll." *Nature*, October 2019. https://doi.org/10.1038/d41586-019-02922-9.

———. "What's Wrong with the Journal Impact Factor in 5 Graphs. Scholars Love to Hate the Journal Impact Factor, but How Flawed Is It?" *Nature*, April 3 2018.

Cirino, Giuseppe, Csaba Szabo, and Andreas Papapetropoulos. "Physiological Roles of Hydrogen Sulfide in Mammalian Cells, Tissues, and Organs." *Physiological Reviews* 103, no. 1 (January 2023): 31–276. https://doi.org/10.1152/physrev.00028.2021.

Collins, Francis S., and Lawrence A. Tabak. "Policy: NIH Plans to Enhance Reproducibility." *Nature* 505, no. 7485 (January 2014): 612–13. https://doi.org/10.1038/505612a.

Cruz, Fernanda Ferreira, Patricia Rieken Macedo Rocco, and Paolo Pelosi. "Anti-Inflammatory Properties of Anesthetic Agents." *Critical Care* 21, no. 1 (March 2017): 67. https://doi.org/10.1186/s13054-017-1645-x.

Culliton, Barbara J. "Coping with Fraud: The Darsee Case." *Science* 220, no. 4592 (April 1983): 31–35. https://doi.org/10.1126/science.6828878.

Curcic, Dimitrije. "Academic Publishers Statistics." Words Rated, June 21, 2023. https://wordsrated.com/academic-publishers-statistics/.

Curtin, Nicola J., and Csaba Szabo. "Poly (ADP-Ribose) Polymerase Inhibition: Past, Present and Future." *Nature Reviews Drug Discovery* 19, no. 10 (October 2020): 711–36. https://doi.org/10.1038/s41573-020-0076-6.

Curtis, Michael J., Steve Alexander, Giuseppe Cirino, James R. Docherty, Christopher H. George, Mark A. Giembycz, Daniel Hoyer et al. "Experimental Design and Analysis and Their Reporting II: Updated and Simplified Guidance for Authors and Peer Reviewers." *British Journal of Pharmacology* 175, no. 7 (April 2018): 987–93. https://doi.org/10.1111/bph.14153.

Cyranoski, David. "Accusations Pile Up Amid Japan's Stem-Cell Controversy." *Nature*, 2014. https://doi.org/10.1038/nature.2014.15163.

———. "China Introduces 'Social' Punishments for Scientific Misconduct." *Nature* 564, no. 7736 (December 2018): 312. https://doi.org/10.1038/d41586-018-07740-z.

———. "How a Scientific Fraud Reinvented Himself." *New York Times*, June 21, 2023. https://www.nytimes.com/2023/06/21/opinion/cloning-science-fraud.html.

Dahlberg, Brett. "Cornell Finds Laboratory Head Falsified Data." WRVO, November 6, 2018. www.wrvo.org.

Dansinger, Michael. "Dear Plagiarist: A Letter to a Peer Reviewer Who Stole and Published Our Manuscript as His Own." *Annals of Internal Medicine* 166, no. 2 (January 2017): 143. https://doi.org/10.7326/M16-2551.

Davenas, Elisabeth, F. Beauvais, J. Amara, M. Oberbaum, B. Robinzon, A. Miadonna, A. Tedeschi et al. "Human Basophil Degranulation Triggered by Very Dilute Antiserum Against IgE." *Nature* 333, no. 6176 (June 1988): 816–18. https://doi.org/10.1038/333816a0.

Dobbs, David. "Marc Hauser, Monkey Business, and the Sine Waves of Science." *Wired*, August 11, 2010. https://www.wired.com/2010/08/marc-hauser-monkey-business-and-the-sine-waves-of-science/.

Dobosz, Paula, and Tomasz Dzieciątkowski. "The Intriguing History of Cancer Immunotherapy." *Frontiers in Immunology* 10 (December 2019): 2965. https://doi.org/10.3389/fimmu.2019.02965.

Dolgin, Elie. "PubMed Commons Closes Its Doors to Comments. The US National Institutes of Health Shutters Its Journal-Commenting Platform." *Nature*, February 2018. https://doi.org/10.1038/d41586-018-01591-4.

"Duke Officials Silenced Med Student Who Reported Trouble in Anil Potti's Lab." *Cancer Letter*, January 9, 2015. https://cancerletter.com/the-cancer-letter/20150109_1.

Eastwood, Susan, Pamela Derish, Evangeline Leash, and Stephen Ordway. "Ethical Issues in Biomedical Research: Perceptions and Practices of Postdoctoral Research Fellows Responding to a Survey." *Science and Engineering Ethics* 2, no. 1 (January 1996): 89–114. https://doi.org/10.1007/BF02639320.

Eaves-Pyles, Tonyia, Kanneganti Murthy, Lucas Liaudet, László Virág, Gary Ross, Francisco G. Soriano, Csaba Szabo, and A. L. Salzman. "Flagellin, a Novel Mediator of Salmonella-Induced Epithelial Activation and Systemic Inflammation: I Kappa B Alpha Degradation, Induction of Nitric Oxide Synthase, Induction of Proinflammatory Mediators, and Cardiovascular Dysfunction." *Journal of Immunology* 166, no. 2 (January 2001): 1248–60. https://doi.org/10.4049/jimmunol.166.2.1248.

Elliott, Carl. *White Coat, Black Hat: Adventures on the Dark Side of Medicine*. Boston: Beacon Press, 2011.

Else, Holly. "Dozens of Papers Co-Authored by Nobel Laureate Raise Concerns." *Nature* 611, no. 7934 (November 2022): 19–20. https://doi.org/10.1038/d41586-022-03032-9.

———. "Scammers Impersonate Guest Editors to Get Sham Papers Published." *Nature* 599, no. 7885 (November 2021): 361. https://doi.org/10.1038/d41586-021-03035-y.

———. "What Makes an Undercover Science Sleuth Tick?" *Nature* 608, no. 7923 (August 2022): 463. https://doi.org/10.1038/d41586-022-02099-8.

"Enhancing Reproducibility Through Rigor and Transparency." Policy & Compliance, NIH Grants & Funding. last modified January 2023. https://grants.nih.gov/policy/reproducibility/index.htm.

Erosheva, Elena A., Patricia Martinková, and Carole J. Lee. "When Zero May Not Be Zero: A Cautionary Note on the Use of Inter-Rater Reliability in Evaluating Grant Peer Review." *Journal of the Royal Statistical Society—Series A* 184, no. 3 (July 2021): 904–19. https://doi.org/10.1111/rssa.12681.

Errington, Timothy M., Elizabeth Iorns, William Gunn, Fraser Elisabeth Tan, Joelle Lomax, and Brian A. Nosek. "An Open Investigation of the Reproducibility of Cancer Biology Research." *eLife* 3 (December 2014): e21627. https://doi.org/10.7554/eLife.04333.

Errington, Timothy M., Maya Mathur, Courtney K. Soderberg, Alexandria Denis, Nicole Perfito, Elizabeth Iorns, and Brian A. Nosek. "Investigating the Replicability of Preclinical Cancer Biology." *eLife* 10 (December 2021): e71601. https://doi.org/10.7554/eLife.71601.

Fanelli, Daniele. "How Many Scientists Fabricate and Falsify Research? A Systematic Review and Meta-Analysis of Survey Data." *PLOS One* 4, no. 5 (May 2009): e5738. https://doi.org/10.1371/journal.pone.0005738.

Fang, Ferric C., and Arturo Casadevall. "Research Funding: The Case for a Modified Lottery." *mBio* 7, no. 2 (April 2016): e00422-16. https://doi.org/10.1128/mBio.00422-16.

———. "Why We Cheat." *Scientific American Mind* 24, no. 2 (May 2013): 30–37, 2013.

Fersht, Alan. "The Most Influential Journals: Impact Factor and Eigenfactor." *Proceedings of the National Academy of Sciences of the United States of America* 106, no. 17 (April 2009): 6883–84. https://doi.org/10.1073/pnas.0903307106.

Fire, Michael, and Carlos Guestrin. "Over-Optimization of Academic Publishing Metrics: Observing Goodhart's Law in Action." *GigaScience* 8, no. 6 (June 2019): 1–20. https://doi.org/10.1093/gigascience/giz053.

Flier, Jeffrey S. "Misconduct in Bioscience Research: a 40-Year Perspective." *Perspectives in Biology and Medicine* 64, no. 4 (Autumn 2021): 437–56. https://doi.org/10.1353/pbm.2021.0035.

Fox, Fiona. *Beyond the Hype: The Inside Story of Science's Biggest Media Controversies*. London: Elliott & Thompson, 2022.

Freedman, Leonard P., Iain M. Cockburn, and Timothy S. Simcoe. "The Economics of Reproducibility in Preclinical Research." *PLOS Biology* 13, no. 6 (June 2015): e1002165. https://doi.org/10.1371/journal.pbio.1002165.

Freedman, Leonard P., Mark C. Gibson, Stephen P. Ethier, Howard R. Soule, Richard M. Neve, and Yvonne A. Reid. "Reproducibility: Changing the Policies and Culture of Cell Line Authentication." *Nature Methods* 12, no. 6 (June 2015): 493–97. https://doi.org/10.1038/nmeth.3403.

Freedman, Leonard P., Gautham Venugopalan, and Rosann Wisman. "Reproducibility2020: Progress and Priorities," *F1000Research* 6 (May 2017): 604. https://doi.org/10.12688/f1000research.11334.1.

Furchgott, Robert F. "Endothelium-Derived Relaxing Factor: Discovery, Early Studies, and Identification as Nitric Oxide." In *Vasodilation, Vascular Smooth Muscle, Peptides and Endothelium*. R. M. Robertson, 401–14. New York: Raven Press, 1998.

Furchgott, Robert F., and John V. Zawadzki. "The Obligatory Role of Endothelial Cells in the Relaxation of Arterial Smooth Muscle by Acetylcholine." *Nature* 288, no. 5789 (November 1980): 373–76. https://doi.org/10.1038/288373a0.

Galton, David J. "Did Mendel Falsify His Data?" *QJM: An International Journal of Medicine* 105, no. 2 (February 2012): 215–16. https://doi.org/10.1093/qjmed/hcr195.

Gerö, Domokos, Petra Szoleczky, Katalin Módis, John P. Pribis, Yousef Al-Abed, Huan Yang, Sangeeta Chevan, Timothy R. Billiar, Kevin J. Tracey, and Csaba Szabo. "Identification of Pharmacological Modulators of HMGB1-Induced Inflammatory Response by Cell-Based Screening." *PLOS One* 8, no. 6 (June 2013): e65994. https://doi.org/10.1371/journal.pone.0065994.

Ghasemi, Asghar, Parvin Mirmiran, Khosrow Kashfi, and Zahra Bahadoran. "Scientific Publishing in Biomedicine: A Brief History of Scientific Journals." *International Journal of Endocrinology and Metabolism* 21, no. 1 (December 2022): e131812. https://doi.org/10.5812/ijem-131812.

Giles, Jim. "Taking on the Cheats." *Nature* 435 (2005): 258–59. https://doi.org/10.1038/435258a.

Goldacre, Ben. *Bad Pharma: How Drug Companies Mislead Doctors and Harm Patients: How Medicine Is Broken, and How We Can Fix It*. New York: Harper Collins, 2013.

Goodyear, Dana. "The Stress Test." *New Yorker*, February 29, 2016. https://www.newyorker.com/magazine/2016/02/29/the-stem-cell-scandal.

Greene, Jay, and John Schoof. "Indirect Costs: How Taxpayers Subsidize University Nonsense." *Heritage Foundation, Center for Education Policy*, no. 3681, January 18, 2022.

Grens, Kerry. "Oncologist Found Guilty of Misconduct. A Government Investigation Concludes That Anil Potti Faked Data on Multiple Grants and Papers." *The Scientist*, November 9, 2015. https://www.the-scientist.com/oncologist-found-guilty-of-misconduct-34523.

Grimes, David Robert, and James Heathers. "The New Normal? Redaction Bias in Biomedical Science." *Royal Society Open Science* 8, no. 12 (December 2021): 211308. https://doi.org/10.1098/rsos.211308.

Gross, Charles. "Scientific Misconduct." *Annual Review of Psychology* 67, no. 1 (2016): 693–711. https://doi.org/10.1146/annurev-psych-122414-033437.

Grove, Jack. "Nobelist: Scientific Success 'No Barrier' to Work-Life Balance." Times Higher Education, June 29, 2023. https://www.timeshighereducation.com/news/nobelist-scientific-success-no-barrier-work-life-balance.

Gunsalus, C. K., and Aaron D. Robinson. "Nine Pitfalls of Research Misconduct." *Nature* 557, no. 7705 (May 2018): 297–99. https://doi.org/10.1038/d41586-018-05145-6.

Gunsalus, C. K., Marcia K. McNutt, Brian C. Martinson, Larry R. Faulkner, and Robert M. Nerem. "Overdue: A US Advisory Board for Research Integrity." *Nature* 566, no. 7743 (February 2019): 173–75. https://doi.org/10.1038/d41586-019-00519-w.

Hamilton, Joan O'C. "Why We Cheat." *Stanford Magazine*, September 2015.

Hanson, Mark A., Pablo Gómez Barreiro, Paolo Crosetto, and Dan Brockington. "The Strain on Scientific Publishing." arXiv, September 2023. https://arxiv.org/ftp/arxiv/papers/2309/2309.15884.pdf.

Harris, Richard. *Rigor Mortis: How Sloppy Science Creates Worthless Cures, Crushes Hope, and Wastes Billions*. New York: Basic Books, 2017.

Hartgerink, Chris H. J., Jan G. Voelkel, Jelte Wicherts, and Marcel A. L. M. van Assen. "Detection of Data Fabrication Using Statistical Tools." PsyArXiv Preprints, August 19, 2019. https://osf.io/preprints/psyarxiv/jkws4/. https://doi.org/10.31234/osf.io/jkws4.

Hauser, Marc D., and Justin N. Wood. "Replication of 'Rhesus Monkeys Correctly Read the Goal-Relevant Gestures of a Human Agent.'" *Proceedings of the Royal Society B: Biological Sciences* 278, no. 1702 (January 2011): 158–59. https://doi.org/10.1098/rspb.2010.1441.

Head, Megan L., Luke Holman, Rob Lanfear, Andrew T. Kahn, and Michael D. Jennions. "The Extent and Consequences of p-Hacking in Science." *PLOS Biology* 13, no. 3 (March 2015): e1002106. https://doi.org/10.1371/journal.pbio.1002106.

Hensley, Scott. "UConn Claims Resveratrol Researcher Falsified Work." *NPR News*, January 12, 2012. https://www.npr.org/sections/health-shots/2012/01/12/145117068/uconn-claims-resveratrol-researcher-falsified-work.

Herb, Ulrich. "Das Wissenschaftliche Publikations—und Reputationssystem ist Gehackt." Telepolis, April 15, 2020. https://www.telepolis.de/features/Das-wissenschaftliche-Publikations-und-Reputationssystem-ist-gehackt-4701388.html.

Hirst, Stuart J., N. A. Hayes, J. Burridge, F. L. Pearce, and J. C. Foreman. "Human Basophil Degranulation Is Not Triggered by Very Dilute Antiserum Against Human IgE." *Nature* 366, no. 6455 (December 1993): 525–27. https://doi.org/10.1038/366525a0.

Hixson, Joseph. *The Patchwork Mouse*. New York: Anchor & Doubleday, 1976.

Horbach, Serge P. J. M., and Willem Halffman. "The Extent and Causes of Academic Text Recycling or 'Self-Plagiarism'." *Research Policy* 48, no. 2 (March 2019): 492–502. https://doi.org/10.1016/j.respol.2017.09.004.

Hulu. "The Dropout." Movie Mini Series made in 2022. https://www.imdb.com/title/tt10166622/.
"Hwang Indicted for Fraud, Embezzlement. Korean Prosecutors Also Charge Five Other Members of Stem Cell Team." *Science*, May 12, 2006. https://www.science.org/content/article/hwang-indicted-fraud-embezzlement.
Hylander, Bonnie L., Elizabeth A. Repasky, and Sandra Sexton. "Using Mice to Model Human Disease: Understanding the Roles of Baseline Housing-Induced and Experimentally Imposed Stresses in Animal Welfare and Experimental Reproducibility." *Animals (Basel)* 12, no. 3 (February 2022): 371. https://doi.org/10.3390/ani12030371.
Hyun-kyung, Kang. "Whistleblower Recalls Price Paid for Revealing Hwang Woo-suk's Scientific Misconduct." *Korea Times*, July 27, 2023. https://www.koreatimes.co.kr/www/nation/2024/04/113_355797.html.
Ioannidis, John P. A. "Why Most Published Research Findings Are False." *PLOS Medicine* 2, no. 8 (August 2005): e124. https://doi.org/10.1371/journal.pmed.0020124.
Isaacs, Tony. "A Tragic and Stunning Case of Scientific Fraud in Studies on Red Wine and Resveratrol." *Natural News*, March 22, 2012. https://www.naturalnews.com/035315_red_wine_resveratrol_scientific_fraud.html.
Jaslow, Ryan. "Red Wine Researcher Dr. Dipak K. Das Published Fake Data." *CBS News* (January 12, 2012). https://www.cbsnews.com/news/red-wine-researcher-dr-dipak-k-das-published-fake-data-uconn/.
Johnson, Carolyn Y. "Former Harvard Professor Marc Hauser Fabricated, Manipulated Data, US Says." Boston.com, September 5, 2012. https://www.boston.com/uncategorized/noprimarytagmatch/2012/09/05/former-harvard-professor-marc-hauser-fabricated-manipulated-data-us-says/.
———. "Harvard Investigation Finds Fraudulent Data in Papers by Heart Researcher." *Washington Post*, October 15, 2018. https://www.washingtonpost.com/science/2018/10/15/harvard-investigation-finds-fraudulent-data-papers-by-heart-researcher/.
———. "Harvard Report Shines Light on Ex-Researcher's Misconduct." *Boston Globe*, May 29, 2014. https://www.bostonglobe.com/metro/2014/05/29/internal-harvard-report-shines-light-misconduct-star-psychology-researcher-marc-hauser/maSUowPqL4clXrOgj44aKP/story.html.
Kahn, Shulamit, and Donna K. Ginther. "The Impact of Postdoctoral Training on Early Careers in Biomedicine." *Nature Biotechnology* 35, no. 1 (January 2017): 90–94. https://doi.org/10.1038/nbt.3766.
Kaiser, Jocelyn. "Potti Found Guilty of Research Misconduct. Federal Report Closes Long Cancer Research Saga." *Science*, November 9, 2015. https://www.science.org/content/article/potti-found-guilty-research-misconduct.
Kalichman, Michael W., and Paul J. Friedman. "A Pilot Study of Biomedical Trainees' Perceptions Concerning Research Ethics." *Academic Medicine* 67, no. 11 (November 1992): 769–75. https://doi.org/10.1097/00001888-199211000-00015.
Kane, Adam, and Bawan Amin. "Amending the Literature Through Version Control." *Biology Letters* 19, no. 1 (January 2023): 20220463. https://doi.org/10.1098/rsbl.2022.0463.
Kiank, Cornelia, Pia Koerner, Wolfram Kessler, Tobias Traeger, Stefan Maier, Claus-Dieter Heidecke, and Christine Schuett. "Seasonal Variations in Inflammatory Responses to Sepsis and Stress in Mice." *Critical Care Medicine* 35, no. 10 (October 2007): 2352–58. https://doi.org/10.1097/01.ccm.0000282078.80187.7f.

Knox, R. "The Harvard Fraud Case: Where Does the Problem Lie?" *JAMA* 249, no. 14 (April 1983): 1797–807.
Kolata, Gina. "He Promised to Restore Damaged Hearts. Harvard Says His Lab Fabricated Research." *New York Times*, October 29, 2018. https://www.nytimes.com/2018/10/29/health/dr-piero-anversa-harvard-retraction.html.
———. "The Price for 'Predatory' Publishing? $50 Million." *New York Times*, April 3, 2019.
Kozlov, Max. "'Disruptive' Science Has Declined—Even as Papers Proliferate." *Nature* 613, no. 7943 (January 2023): 225. https://doi.org/10.1038/d41586-022-04577-5.
———. "NIH Issues a Seismic Mandate: Share Data Publicly." *Nature* 602, no. 7898 (February 2022): 558–59. https://doi.org/10.1038/d41586-022-00402-1.
———. "NIH Plans Grant Review Overhaul to Reduce Bias." *Nature* 612, no.7941 (December 2022): 602–3. https://doi.org/10.1038/d41586-022-04385-x.
Kumar, Malhar N. "Dealing with Misconduct in Biomedical Research: A Review of the Problems and the Proposed Methods for Improvement." *Journal of Accounting Research* 16, no. 6 (November 2009): 307–30. https://doi.org/10.1080/08989620903328576.
Kuzma, Susan M. "Criminal Liability for Misconduct in Scientific Research." *University of Michigan Journal of Law Reform* 25, no 2 (1992): 357–421.
"Laboratory Water. Its Importance and Application." Division of Technical Resources, NIH. March 2013. https://orf.od.nih.gov/TechnicalResources/Documents/DTR%20White%20Papers/Laboratory%20Water-Its%20Importance%20and%20Application-March-2013_508.pdf.
Langin, Katie. "U.S. Labs Face Severe Postdoc Shortage." *Science* 376, no. 6600 (2022): 1369–70.
Larson, Richard C., Navid Ghaffarzadegan, and Yi Xue. "Too Many PhD Graduates or Too Few Academic Job Openings: The Basic Reproductive Number Ro in Academia." *Systems Research and Behavioral Science* 31, no. 6 (November-December 2014): 745–50. https://doi.org/10.1002/sres.2210.
Lauer, Mike. "Outcomes of Amended (A1) Applications." Extramural Nexus (blog), March 23, 2017. https://nexus.od.nih.gov/all/2017/03/23/outcomes-of-amended-a1-applications/.
Lee, C. "Slump in NIH Funding Is Taking Toll on Research." *Washington Post*, May 28, 2007.
Leistedt, Samuel J., and Paul Linkowski. "Fraud, Individuals, and Networks: A Biopsychosocial Model of Scientific Frauds." *Science & Justice* 56, no. 2 (March 2016): 109–12. https://doi.org/10.1016/j.scijus.2016.01.002.
Leys, Tony. "Ex-Scientist Sentenced to Prison for Academic Fraud." *USA Today*, July 1, 2015. https://eu.usatoday.com/story/news/nation/2015/07/01/ex-scientist-sentenced-prison-academic-fraud/29596271.
Linn, Steve C., Paul J. Morelli, Iris Edry, Susan E. Cottongim, Csaba Szabo, and Andrew L. Salzman. "Transcriptional Regulation of Human Inducible Nitric Oxide Synthase Gene in an Intestinal Epithelial Cell Line." *American Journal of Physiology* 272, no. 6, part 1 (June 1997): G1499-508. https://doi.org/10.1152/ajpgi.1997.272.6.G1499.
Liu, Yang, Yuanyuan Liu, Chao Sun, Lu Gan, Luwei Zhang, Aihong Mao, Yuting Du, Rong Zhou, and Hong Zhang. "Carbon Ion Radiation Inhibits Glioma and Endothelial Cell Migration Induced by Secreted VEGF." *PLOS One* 9, no. 6 (June 2014): e98448. https://doi.org/10.1371/journal.pone.0098448.
Lowe, Derek. "The Sirtris Compounds: Worthless? Really?" Science: In the Pipeline (blog), January 12, 2010.

Lyden, Patrick D., Márcio A. Diniz, Francesca Bosetti, Jessica Lamb, Karisma A. Nagarkatti, André Rogatko, Sungjin Kim et al. "A Multi-Laboratory Preclinical Trial in Rodents to Assess Treatment Candidates for Acute Ischemic Stroke." *Science Translational Medicine* 15, no. 714 (September 20, 2023): eadg8656. https://doi.org/10.1126/scitranslmed.adg8656.

Mackenzie, Ruairi J. "Inside a 'Fake' Conference: A Journey Into Predatory Science." Technology Networks, July 11, 2019. https://www.technologynetworks.com/tn/articles/inside-a-fake-conference-a-journey-into-predatory-science-321619.

Mallapaty, Smriti. "China Bans Cash Rewards for Publishing Papers." *Nature* 579, no. 7797 (March 2020): 18. https://doi.org/10.1038/d41586-020-00574-8.

Manca, Andrea, Lucia Cugusi, Andrea Cortegiani, Giulia Ingoglia, David Moher, and Franca Deriu. "Predatory Journals Enter Biomedical Databases Through Public Funding." *British Medical Journal* 371 (December 2020): m4265. https://doi.org/10.1136/bmj.m4265.

Mansoury, Morva, Maya Hamed, Rashid Karmustaji, Fatima A. L. Hannan, and Stephen T. Safrany. "The Edge Effect: A Global Problem. The Trouble with Culturing Cells in 96-Well Plates." *Biochemistry and Biophysics Reports* 26 (March 2021): 100987. https://doi.org/10.1016/j.bbrep.2021.100987.

Marchant, Jo. "Powerful Antibiotics Discovered Using AI. Machine Learning Spots Molecules That Work Even Against 'Untreatable' Strains of Bacteria." *Nature* (February 20, 2020). https://doi.org/10.1038/d41586-020-00018-3.

Marcus, Adam. "Braggadacio, Information Control, and Fear: Life Inside a Brigham Stem Cell Lab Under Investigation." Retraction Watch, May 30, 2014. https://retractionwatch.com/2014/05/30/braggadacio-information-control-and-fear-life-inside-a-brigham-stem-cell-lab-under-investigation/.

———. "Years Later, Researcher at Center of Highly Publicized Case Has Another Paper Retracted." Retraction Watch, September 24, 2020. https://retractionwatch.com/2020/09/24/years-later-researcher-at-center-of-highly-publicized-case-has-another-paper-retracted/.

Marcus, Adam, and Ivan Oransky. "Retraction Watch Hijacked Journal Checker." Retraction Watch (blog). https://retractionwatch.com/the-retraction-watch-hijacked-journal-checker/.

Markowitz, David M., and Jeffrey T. Hancock. "Linguistic Obfuscation in Fraudulent Science." *Journal of Language and Social Psychology* 35, no. 4 (2016): 435–45. https://doi.org/10.1177/0261927X15614605.

Martin, Edward. "Deceit at Duke: How Fraud at a University Research Lab Prompted a $112M Fine." *Business North Carolina*, August 2019. https://businessnc.com/deceit-at-duke-how-fraud-at-a-university-research-lab-prompted-a-112m-fine/.

Masicampo, E. J., and Daniel R. Lalande. "A Peculiar Prevalence of p Values Just Below .05." *Quarterly Journal of Experimental Psychology (Hove)* 65, no. 11 (2012): 2271–79. https://doi.org/10.1080/17470218.2012.711335.

Mastroianni, A. "Experimental History." ongoing newsletter since 2022. https://www.experimental-history.com/.

Mazières, David, and Eddie Kohler. "Get Me Off Your Fucking Mailing List." Stanford Secure Computer Systems Group, February 4, 2005. https://www.scs.stanford.edu/~dm/home/papers/remove.pdf.

McCarthy, Michael. "Former Duke University Oncologist Is Guilty of Research Misconduct, US Officials Find." *British Medical Journal* 351 (November 2015): h6058. https://doi.org/10.1136/bmj.h6058.

McCook, Alison. "Copenhagen Revokes Degree of Controversial Neuroscientist Milena Penkowa." Retraction Watch, September 12, 2017. https://retractionwatch.com/2017/09/12/copenhagen-revokes-degree-controversial-neuroscientist-milena-penkowa/.

———. "Whistleblower Sues Duke, Claims Doctored Data Helped Win $200 Million in Grants. Alleging That Faked Pulmonary Research Led to Grants from NIH and Other Federal Agencies, Scientist Could Win Huge Payout from False-Claims Suit and Set a Precedent." *Science*, September 16, 2016. https://www.science.org/content/article/whistleblower-sues-duke-claims-doctored-data-helped-win-200-million-grants.

Medawar, Peter. "Is the Scientific Paper a Fraud?" In *Experiment: A Series of Scientific Case Histories First Broadcast in the BBC Third Programme*, ed. D. Edge, 7–13. London: BBC, 1964.

Miyakawa, Tsuyoshi. "No Raw Data, No Science: Another Possible Source of the Reproducibility Crisis." *Molecular Brain* 13, no. 1 (February 2020): 24. https://doi.org/10.1186/s13041-020-0552-2.

Mobley, Aaron, Suzanne K. Linder, Russell Braeuer, Lee M. Ellis, and Leonard Zwelling. "A Survey on Data Reproducibility in Cancer Research Provides Insights Into Our Limited Ability to Translate Findings from the Laboratory to the Clinic." *PLOS One* 8, no. 5 (May 2013): e63221. https://journals.plos.org/plosone/article?id=10.1371/journal.pone.0063221.

Mom, Charlie, and Peter van den Besselaar. "Do Interests Affect Grant Application Success? The Role of Organizational Proximity." *arXiv* 2206 (2022): 03255. https://doi.org/10.48550/arXiv.2206.03255.

Moore, Andrew. "Bad Science in the Headlines. Who Takes Responsibility When Science Is Distorted in the Mass Media?" *EMBO Reports* 7, no. 12 (December 2006): 1193–96. https://doi.org/10.1038/sj.embor.7400862.

Morgenstern, Justin. "Bias in Medical Research." First 10 EM, July 2, 2021. https://first10em.com/bias/.

Mosimann, James E., Claire V. Wiseman, and Ruth E. Edelman. "Data Fabrication: Can People Generate Random Digits?" *Accountability in Research* 4, no. 1 (1995): 31–55. https://doi.org/10.1080/08989629508573866.

Moss, Ralph W. *Free Radical: Albert Szent-Györgyi and the Battle Over Vitamin C*. New York: Paragon, 1988.

Motulsky, Harvey J., and Martin C. Michel. "Commentary on the BJP's New Statistical Reporting Guidelines." *British Journal of Pharmacology* 175, no. 18 (September 2018): 3636–37. https://doi.org/10.1111/bph.14441.

Mullard, Asher. "Half of Cancer Studies Fail High-Profile Replication Test." *Nature* 600, no. 7889 (December 2021): 368–69. https://doi.org/10.1038/d41586-021-03691-0.

Munafò, Marcus R., Brian A. Nosek, Dorothy V. M. Bishop, Katherine S. Button, Christopher D. Chambers, Nathalie Percie du Sert, Uri Simonsohn, Eric-Jan Wagenmakers, Jennifer J. Ware, and John P. A. Ioannidis. "A Manifesto for Reproducible Science." *Nature Human Behaviour* 10, no. 1 (January 2017): 21. https://doi.org/10.1038/s41562-016-0021.

National Academies of Sciences, Engineering, and Medicine. *Fostering Integrity in Research*. Washington, DC: National Academies Press, 2017. https://doi.org/10.17226/21896.

National Institutes of Health. "Welcome to the NIH Data Book." https://report.nih.gov/nihdatabook/.

Nicholson, Joshua M., and John P. A. Ioannidis. "Research Grants: Conform and Be Funded." *Nature* 492, no. 7427 (December 2012): 34–36. https://doi.org/10.1038/492034a.

Nosek, Brian A., and Timothy M. Errington, "The Best Time to Argue About What a Replication Means? Before You Do It." *Nature* 583, no. 7817 (July 2020): 518–20. https://doi.org/10.1038/d41586-020-02142-6.

Nuzzo, Regina. "How Scientists Fool Themselves—and How They Can Stop." *Nature* 526, no. 7572 (October 2015): 182–85. https://doi.org/10.1038/526182a.

Office of Research Integrity at the National Institutes of Health. "Case Summaries." https://ori.hhs.gov/content/case_summary.

———. "Research Misconduct." https://ori.hhs.gov.

Oransky, Ivan. "Retractions Are Increasing, but Not Enough." *Nature* 608, no. 7921 (August 2022): 9. https://doi.org/10.1038/d41586-022-02071-6.

Osherovich, Lev. "Hedging Against Academic Risk." *Science-Business eXchange* 4 (April 2011): 416. https://doi.org/10.1038/scibx.2011.416.

Oza, Anil. "Reproducibility Trial: 246 Biologists Get Different Results from Same Data Sets. Wide Distribution of Findings Shows How Analytical Choices Drive Conclusions." *Nature*, October 2023. https://doi.org/10.1038/d41586-023-03177-1.

Pain, Elisabeth. "Paul Brookes: Surviving as an Outed Whistleblower." *Science*, March 10, 2014. https://www.science.org/content/article/paul-brookes-surviving-outed-whistleblower. https://doi.org/10.1126/science.caredit.a1400061.

Palmer, Richard M. J., A. G. Ferrige, and Salvador Moncada. "Nitric Oxide Release Accounts for the Biological Activity of Endothelium-Derived Relaxing Factor." *Nature* 327, no. 6122 (June 1987): 524–26. https://www.nature.com/articles/327524a0.

Panagaki, Theodora, Laszlo Pecze, Elisa B. Randi, Anni I. Nieminen, and Csaba Szabo. "Role of the Cystathionine β-Synthase / H_2S Pathway in the Development of Cellular Metabolic Dysfunction and Pseudohypoxia in Down Syndrome." *Redox Biology* 55 (July 2022): 102416. https://doi.org/10.1016/j.redox.2022.102416.

Panagaki, Theodora, Elisa B. Randi, Fiona Augsburger, and Csaba Szabo. "Overproduction of H_2S, Generated by CBS, Inhibits Mitochondrial Complex IV and Suppresses Oxidative Phosphorylation in Down Syndrome." *Proceedings of the National Academy of Sciences of the United States of America* 116, no. 38 (September 2019): 18769–71. https://doi.org/10.1073/pnas.1911895116.

Papapetropoulos, Andreas, and Csaba Szabo. "Inventing New Therapies Without Reinventing the Wheel: The Power of Drug Repurposing." *British Journal of Pharmacology* 175, no. 2 (January 2018): 165–67. https://doi.org/10.1111/bph.14081.

"Paper Mills and Research Misconduct: Facing the Challenges of Scientific Publishing." Hearing of the U.S. House of Representatives, Committee on Science, Space and Technology, Subcommittee on Investigations and Oversight. July 20, 2022. https://www.congress.gov/117/meeting/house/115022/documents/HHRG-117-SY21-20220720-SD002.pdf.

Park, Michael, Erin Leahey, and Russell J. Funk. "Papers and Patents Are Becoming Less Disruptive Over Time." *Nature* 613, no. 7942 (January 2023): 138–44. https://doi.org/10.1038/s41586-022-05543-x.

Parker, Lisa, Stephanie Boughton, Rosa Lawrence, and Lisa Bero. "Experts Identified Warning Signs of Fraudulent Research: A Qualitative Study to Inform a Screening Tool." *Journal of Clinical Epidemiology* 151 (November 2022): 1–17. https://doi.org/10.1016/j.jclinepi.2022.07.006.

Paulos, John Allen. *Innumeracy*. New York: Penguin, 2000.

Pickett, Justin T., and Sean Patrick Roche. "Questionable, Objectionable or Criminal? Public Opinion on Data Fraud and Selective Reporting in Science." *Science and Engineering Ethics* 24, no. 1 (February 2018): 151–71. https://doi.org/10.1007/s11948-017-9886-2.

Pier, Elizabeth L., Markus Brauer, Amarette Filut, Anna Kaatz, Joshua Raclaw, Mitchell J. Nathan, Cecilia E. Ford, and Molly Carnes. "Low Agreement Among Reviewers Evaluating the Same NIH Grant Applications." *Proceedings of the National Academy of Sciences of the United States of America* 115, no. 12 (March 2018): 2952–57. https://doi.org/10.1073/pnas.1714379115.

Piller, Charles. "Blots on a Field? A Neuroscience Image Sleuth Finds Signs of Fabrication in Scores of Alzheimer's Articles, Threatening a Reigning Theory of the Disease." *Science* 377, no. 6604 (July 2022): 358–63. https://doi.org/10.1126/science.add9993.

———. "Picture Imperfect." *Science* 385, no. 6716 (September 2024): 1406–1412. https://doi.org/10.1126/science.adt3535.

Piper, Kelsey. "Science Funding Is a Mess. Could Grant Lotteries Make It Better? We Might Be Better Off Funding Scientific Research by Choosing Projects at Random. Here's Why." *Vox*, January 18, 2019.

Pitt, Richard N, Yasemin Taskin Alp, and Imani A. Shell. "The Mental Health Consequences of Work-Life and Life-Work Conflicts for STEM Postdoctoral Trainees." *Frontiers in Psychology* 12 (November 2021): 750490. https://doi.org/10.3389/fpsyg.2021.750490.

Prebble, John, and Bruce Weber. *Wandering in the Gardens of the Mind: Peter Mitchell and the Making of Glynn*. Oxford: Oxford University Press, 2003.

Price, Michael. "A Shortcut to Better Grantsmanship." *Science*, June 4, 2013. https://www.science.org/content/article/shortcut-better-grantsmanship.

Prinz, Florian, Thomas Schlange, and Khusru Asadullah. "Believe It or Not: How Much Can We Rely on Published Data on Potential Drug Targets?" *Nature Reviews Drug Discovery* 10, no. 9 (August 2011): 712. https://doi.org/10.1038/nrd3439-c1.

Pulloor, Niyas Kudukkil, Sajith Nair, Kathleen McCaffrey, Aleksandar D. Kostic, Pradeep Bist, Jeremy D. Weaver, Andrew M. Riley et al. "Human Genome-Wide RNAi Screen Identifies an Essential Role for Inositol Pyrophosphates in Type-I Interferon Response." *PLOS Pathog* 10, no 2 (February 2014): e1003981. https://doi.org/10.1371/journal.ppat.1003981.

Pupovac, Vanja, and Daniele Fanelli. "Scientists Admitting to Plagiarism: A Meta-Analysis of Surveys." *Science and Engineering Ethics* 21, no. 5 (October 2015): 1331–52. https://doi.org/10.1007/s11948-014-9600-6.

Quan, Wei, Bikun Chen, and Fei Shu. "Publish or Impoverish: An Investigation of the Monetary Reward System of Science in China (1999–2016)." *Aslib Journal of Information Management* 69, no. 5 (2017): 486–502. https://doi.org/10.1108/AJIM-01-2017-0014.

Rasko, John, and Carl Power. "What Pushes Scientists to Lie? The Disturbing but Familiar Story of Haruko Obokata." *The Guardian*, February 18, 2015. https://www.theguardian.com/science/2015/feb/18/haruko-obokata-stap-cells-controversy-scientists-lie.

Reardon, Sara. "A Mouse's House May Ruin Experiments." *Nature* 530, no. 7590 (February 2016): 264. https://doi.org/10.1038/nature.2016.19335.

———. "US Vaccine Researcher Sentenced to Prison for Fraud." *Nature* 523 (2015): 138–39. https://doi.org/10.1038/nature.2015.17660.

Resnick, Brian. "Trump Wants to Cut Billions from the NIH. This Is What We'll Miss Out On if He Does." *Vox*, March 11, 2019.

Riddiford, Nick. "A Survey of Working Conditions Within Biomedical Research in the United Kingdom." *F1000Research* 6 (March 2017): 229. https://doi.org/10.12688/f1000research.11029.2.

Ritchie, Stuart. *Science Fictions*. New York: Metropolitan, 2021.

Roberts, A. "Nobel Laureate, Hopkins Researcher Retracts Additional Articles, Bringing Total to Six in Two Years." *Baltimore Sun*, August 1, 2023.

Roy, Siddhartha, and Marc A. Edwards. "NSF Fellows' Perceptions About Incentives, Research Misconduct, and Scientific Integrity in STEM Academia." *Scientific Reports* 13, no. 1 (April 2023): 5701. https://doi.org/10.1038/s41598-023-32445-3.

Sabel, Bernhard A., Emely Knaack, Gerd Gigerenzer, and Mirela Bilc. "Fake Publications in Biomedical Science: Red-Flagging Method Indicates Mass Production." MedRxiv, October 18, 2023. https://doi.org/10.1101/2023.05.06.23289563.

Sample, Ian. "Stem Cell Scientist Haruko Obokata Found Guilty of Misconduct." *The Guardian*, April 1, 2014. https://www.theguardian.com/science/2014/apr/01/stem-cell-scientist-haruko-obokata-guilty-misconduct-committee.

Saraei, Raedeh, Heshu Sulaiman Rahman, Masoud Soleimani, Mohammad Asghari-Jafarabadi, Adel Naimi, Ali Hassanzadeh, and Saeed Solali. "Kaempferol Sensitizes Tumor Necrosis Factor-Related Apoptosis-Inducing Ligand-Resistance Chronic Myelogenous Leukemia Cells to Apoptosis." *Molecular Biology Reports* 49, no. 1 (January 2022): 19–29. https://doi.org/10.1007/s11033-021-06778-z.

Schekman, Randy. "How Journals Like Nature, Cell and Science Are Damaging Science." *The Guardian*, December 9, 2013.

Schillebeeckx, Maximiliaan, Brett Maricque, and Cory Lewis. "The Missing Piece to Changing the University Culture." *Nature Biotechnology* 31, no. 10 (October 2013): 938–41. https://doi.org/10.1038/nbt.2706.

Science-Publisher. "Co-Authorship in Scopus, Web of Science Research Papers." Science-Publisher, September 07, 2021. https://web.archive.org/web/20210907141327/https://science-publisher.org/coauthorship/articles/.

Seifert, Roland. "How Naunyn-Schmiedeberg's Archives of Pharmacology Deals with Fraudulent Papers from Paper Mills." *Naunyn Schmiedebergs Archives of Pharmacology* 394, no. 3 (March 2021): 431–36. https://doi.org/10.1007/s00210-021-02056-8.

Selye, Hans. *The Stress of Life*. New York: McGraw-Hill, 1956.

Seok, Junhee, H. Shaw Warren, Alex G. Cuenca, Michael N. Mindrinos, Henry V. Baker, Weihong Xu, Daniel R. Richards et al. "Genomic Responses in Mouse Models Poorly Mimic Human Inflammatory Diseases." *Proceedings of the National Academy of Sciences of the United States of America* 110, no. 9 (February 2013): 3507–12. https://doi.org/10.1073/pnas.1222878110.

Servick, Kelly. "$10 Million Settlement Over Alleged Misconduct in Boston Heart Stem Cell Lab. Brigham and Women's Hospital to Pay Government After Probe of Manipulated and Fabricated Data." *Science*, April 2017. https://www.science.org/content/article/10-million-settlement-over-alleged-misconduct-boston-heart-stem-cell-lab.

Shih, Tommy. "Recalibrated Responses Needed to a Global Research Landscape in Flux." *Journal of Accounting Research* 2 (August 2022): 1–7. https://doi.org/10.1080/08989621.2022.2103410.

Sholto, David. "A Quantitative Study of Inappropriate Image Duplication in the Journal *Toxicology Reports*." bioRxiv, September 10, 2023. https://www.biorxiv.org/content/10.1101/2023.09.03.556099v2. https://doi.org/10.1101/2023.09.03.556099.

Siler, Kyle. "Demarcating Spectrums of Predatory Publishing: Economic and Institutional Sources of Academic Legitimacy." *Journal of the Association for Information Science and Technology* 71 (February 2020): 1386–401. https://doi.org/10.1002/asi.24339.

Silver, Andrew. "Controversial Website That Lists 'Predatory' Publishers Shut Down." *Nature*, January 2017. https://doi.org/10.1038/nature.2017.21328.

Simonsohn, Uri. "Just Post It: The Lesson from Two Cases of Fabricated Data Detected by Statistics Alone." *Psychological Science* 24, no. 10 (October 2013): 1875–88. https://doi.org/10.1177/0956797613480366.

Sinha, Gunjan. "Critics Challenge Data Showing Key Lipids Can Curb Inflammation." *Science* 376, no. 6593 (May 2022): 565–66. https://doi.org/10.1126/science.abq8439.

Smith, Richard. "Time to Assume That Health Research Is Fraudulent Until Proven Otherwise?" BMJ (blog), July 5, 2021. https://blogs.bmj.com/bmj/2021/07/05/time-to-assume-that-health-research-is-fraudulent-until-proved-otherwise/.

Sorge, Robert E., Loren J. Martin, Kelsey A. Isbester, Susana G. Sotocinal, Sarah Rosen, Alexander H. Tuttle, Jeffrey S. Wieskopf et al. "Olfactory Exposure to Males, Including Men, Causes Stress and Related Analgesia in Rodents." *Nature Methods* 11, no. 6 (June 2014): 629–32. https://doi.org/10.1038/nmeth.2935.

Soriano, Francisco Garcia, László Virág, Prakash Jagtap, Éva Szabó, Jon G. Mabley, Lucas Liaudet, Anita Marton et al. "Diabetic Endothelial Dysfunction: The Role of Poly (ADP-Ribose) Polymerase Activation." *Nature Medicine* 7, no. 1 (January 2001): 108–13. https://doi.org/10.1038/83241.

Spitzer, Kirk. "Science Scandal Triggers Suicide, Soul-Searching in Japan." *Time Magazine*, August 8, 2014. https://time.com/3091584/japan-yoshiki-sasai-stem-cells-suicide-haruko-obokata/.

Stefan, Angelica M., and Felix D. Schönbrodt. "Big Little Lies: A Compendium and Simulation of p-Hacking Strategies." *Royal Society Open Science* 10, no. 2 (February 2023): 220346. https://doi.org/10.1098/rsos.220346.

Steinfath, Matthias, Silvia Vogl, Norman Violet, Franziska Schwarz, Hans Mielke, Thomas Selhorst, Matthias Greiner, and Gilbert Schönfelder. "Simple Changes of Individual Studies Can Improve the Reproducibility of the Biomedical Scientific Process as a Whole." *PLOS One* 13, no. 9 (September 2018): e0202762. https://doi.org/10.1371/journal.pone.0202762.

Stemwedel, Janet D. "Life After Misconduct: Promoting Rehabilitation While Minimizing Damage." *Journal of Microbiology & Biology Education* 15, no. 2 (December 2014): 177–80. https://doi.org/10.1128/jmbe.v15i2.827.

Stern, Andrew M., Arturo Casadevall, R. Grant Steen, and Ferric C. Fang. "Financial Costs and Personal Consequences of Research Misconduct Resulting in Retracted Publications." *Elife* 3 (August 2014): e02956. https://doi.org/10.7554/eLife.02956.

Stone, Ken. "Scripps Research Institute to Pay 10M for Improper Use of NIH Grants." *Times of San Diego*, September 11, 2020. https://timesofsandiego.com/tech/2020/09/11/scripps-research-institute-to-pay-10m-over-improper-use-of-nih-grants/.

Sun, Kexuan, Zhiqiang Qiu, Abel Salinas, Yuzhong Huang, Dong-Ho Lee, Daniel Benjamin, Fred Morstatter, Xiang Ren, Kristina Lerman, and Jay Pujara. "Assessing Scientific Research Papers with Knowledge Graphs." *SIGIR '22: Proceedings of the 45th International ACM SIGIR Conference on Research and Development in Information Retrieval* (July 2022): 2467–72. https://doi.org/10.1145/3477495.3531879.

Susco, Sara G., Sulagna Ghosh, Patrizia Mazzucato, Gabriella Angelini, Amanda Beccard, Victor Barrera, Martin H. Berryer et al. "Molecular Convergence Between Down Syndrome and Fragile X Syndrome Identified Using Human Pluripotent Stem Cell Models." *Cell Reports* 40, no. 10 (September 2022): 111312. https://doi.org/10.1016/j.celrep.2022.111312.

Szabo, Csaba. "Hungary: Academy Needs More Than Internal Reform." *Nature* 442, no. 7101 (July 2006): 353. https://doi.org/10.1038/442353b.

———. "Hydrogen Sulphide and Its Therapeutic Potential." *Nature Reviews Drug Discovery* 6, no. 11 (November 2007): 917–35. https://doi.org/10.1038/nrd2425.

Szabo, Csaba, Ciro Coletta, Celia Chao, Katalin Módis, Bartosz Szczesny, Andreas Papapetropoulos, and Mark R Hellmich. "Tumor-Derived Hydrogen Sulfide, Produced by Cystathionine-β-Synthase, Stimulates Bioenergetics, Cell Proliferation, and Angiogenesis in Colon Cancer." *Proceedings of the National Academy of Sciences of the United States of America* 110, no. 30 (July 2013): 12474–79. https://doi.org/10.1073/pnas.1306241110.

Szabo, Csaba, Eörs Dóra, Mária Faragó, Ildikó Horváth, and Arisztid G. Kovách. "Noradrenaline Induces Rhythmic Contractions of Feline Middle Cerebral Artery at Low Extracellular Magnesium Concentration." *Blood Vessels* 27, no. 6 (1990): 373–77.

Szabo, C., G. S. Scott, L. Virág, G. Egnaczyk, A. L. Salzman, T. P. Shanley, and G. Haskó. "Suppression of Macrophage Inflammatory Protein (MIP)-1Alpha Production and Collagen-Induced Arthritis by Adenosine Receptor Agonists." *British Journal of Pharmacology* 125, no. 2 (September 1998): 379–87. https://doi.org/10.1038/sj.bjp.0702040.

Szabo, Csaba, Basilia Zingarelli, Michael O'Connor, and Andrew L. Salzman. "DNA Strand Breakage, Activation of Poly (ADP-Ribose) Synthetase, and Cellular Energy Depletion Are Involved in the Cytotoxicity of Macrophages and Smooth Muscle Cells Exposed to Peroxynitrite." *Proceedings of the National Academy of Sciences of the United States of America* 93, no. 5 (March 1996): 1753–58. https://doi.org/10.1073/pnas.93.5.1753.

Szilagyi, D. Emerick. "The Elusive Target: Truth in Scientific Reporting. Comments on Error, Self-Delusion, Deceit, and Fraud." *Journal of Vascular Surgery* 1, no. 2 (March 1984): 243–53. https://doi.org/10.1016/0741-5214(84)90055-7.

Takao, Keizo, and Tsuyoshi Miyakawa. "Correction for: Genomic Responses in Mouse Models Greatly Mimic Human Inflammatory Diseases." *Proceedings of the National Academy of Sciences of the United States of America* 112, no. 10 (February 2015): E1163-7. https://doi.org/10.1073/pnas.1502188112.

———. "Genomic Responses in Mouse Models Greatly Mimic Human Inflammatory Diseases." *Proceedings of the National Academy of Sciences of the United States of America* 112, no. 4 (August 2014): 1167–72. https://doi.org/10.1073/pnas.1401965111.

Tang, Gengyan, and Jingyu Peng. "Are the Lists of Questionable Journals Reasonable: A Case Study of Early Warning Journal Lists." *Journal of Accounting Research* 22 (September 2023): 1–24. https://doi.org/10.1080/08989621.2023.2261846.

Tauginienė, Loreta, and Inga Gaižauskaitė. "Jumping with a Parachute—Is Promoting Research Integrity Meaningful?" *Journal of Accounting Research* 30, no. 8 (December 2023): 548–73. https://doi.org/10.1080/08989621.2022.2044318.

Taylor, Marisa, and Brad Heath. "Years After Brigham-Harvard Scandal, U.S. Pours Millions Into Tainted Stem-Cell Field." *Reuters Investigation*. June 21, 2022. https://www.reuters.com/investigates/special-report/health-hearts-stem-cells/.

"The Proximal Origin of a Cover-Up: Did the "Bethesda Boys" Downplay a Lab Leak?" U.S. House Committee on Oversight and Accountability. Interim Staff Report. July 12, 2023. https://oversight.house.gov/report/interim-staff-report-the-proximal-origin-of-a-cover-up-did-the-bethesda-boys-downplay-a-lab-leak/.

The Royal Society, "The Scientific Century: Securing Our Future Prosperity." The Royal Society Policy Document 02/10, March 2010, DES1768.

"Timeline of a Controversy. A Chronology of Woo Suk Hwang's Stem-Cell Research." *Nature*, December 19, 2005. https://doi.org/10.1038/news051219-3.

University of Stanford. "Instances of Scientific Misconduct." https://bps.stanford.edu/?page_id=7155.

U.S. Attorney's Office, Southern District of Iowa. "Former Iowa State Researcher Sentenced for Making False Statements." (Press release) July 1, 2015. https://www.justice.gov/usao-sdia/pr/former-iowa-state-researcher-sentenced-making-false-statements.

U.S. House Committee on Oversight and Accountability, Interim Staff Report. "The Proximal Origin of a Cover-Up: Did the 'Bethesda Boys' Downplay a Lab Leak?" July 12, 2023. https://oversight.house.gov/report/interim-staff-report-the-proximal-origin-of-a-cover-up-did-the-bethesda-boys-downplay-a-lab-leak/.

Vane, John R. "Adventures and Excursions in Bioassay: The Steppingstones to Prostacyclin." Nobel lecture, December 8, 1982. https://www.nobelprize.org/uploads/2018/06/vane-lecture.pdf.

Vanhoutte, Paul M. "Could the Absence or Malfunction of Vascular Endothelium Precipitate the Occurrence of Vasospasm?" *Journal of Molecular and Cellular Cardiology* 18, no. 7 (July 1986): 679–89. https://doi.org/10.1016/s0022-2828(86)80940-3.

Volpato, Viola, and Caleb Webber. "Addressing Variability in iPSC-Derived Models of Human Disease: Guidelines to Promote Reproducibility." *Disease Models & Mechanisms* 13, no. 1 (January 2020): dmm042317. https://doi.org/10.1242/dmm.042317.

Wadman, Meredith. "Despite Sexual Harassment Shadow, Biologist David Sabatini Lands Job at Top Czech Institute." *Science*, November 2023. https://www.science.org/content/article/despite-sexual-harassment-shadow-biologist-david-sabatini-lands-job-top-czech-institute.

———. "NIH Mulls Rules for Validating Key Results." *Nature* 500, no. 7460 (August 2013): 14–16. https://doi.org/10.1038/500014a.

Wang, Haichao, Ona Bloom, Minghuang Zhang, Jaideep M. Vishnubhakat, Michael Ombrellino, Jiantu Che, Svetlana Ivanova et al. "HMGB-1 as a Late Mediator of Endotoxin Lethality in Mice." *Science* 285, no. 5425 (July 1999): 248–51. https://doi.org/10.1126/science.285.5425.248.

Wansink, Brian. "The Grad Student Who Never Said No." web.archive (blog), November 21, 2016. https://web.archive.org/web/20170312041524/http:/www.brianwansink.com/phd-advice/the-grad-student-who-never-said-no.

Weintraub, Pamela. "The Doctor Who Drank Infectious Broth, Gave Himself an Ulcer, and Solved a Medical Mystery." *Discover*, April 8, 2010.

Williams, Shawna. "Cell Biologist David Sabatini Fired for Sexual Harassment." *The Scientist*, August 22, 2021. https://www.the-scientist.com/news-opinion/cell-biologist-david-sabatini-fired-for-sexual-harassment-69117.

Winston, Robert. *Human Instinct: How Our Primeval Impulse Shapes Our Modern Lives.* London: Bantam, 2003.

Wise, Nick. nhwise.bsky.social X account. Started April 2012. https://twitter.com/nickwizzo.
Woolston, Chris. "PhDs: The Tortuous Truth." *Nature* 575, no. 7782 (November 2019): 403–6. https://doi.org/10.1038/d41586-019-03459-7.
Xie, Yu, Kai Wang, and Yan Kong. "Prevalence of Research Misconduct and Questionable Research Practices: A Systematic Review and Meta-Analysis." *Science and Engineering Ethics* 27, no. 4 (June 2021): 41. https://doi.org/10.1007/s11948-021-00314-9.
Xin, Ling. "Young China Science Journal 'The Innovation' Joins International Rankings." *South China Morning Post*, July 3, 2023.
Yang, Zequan, Basilia Zingarelli, and Csaba Szabo. "Crucial Role of Endogenous Interleukin-10 Production in Myocardial Ischemia/Reperfusion Injury." *Circulation* 101, no. 9 (March 2000): 1019–26. https://doi.org/10.1161/01.cir.101.9.1019.
Yeo-Teh, Nicole Shu Ling, and Bor Luen Tang. "An Active Aigiarism Declaration for Manuscript Submission." *Journal of Accounting Research* 8 (March 2023): 1–2. https://doi.org/10.1080/08989621.2023.2185776.
——. "Research Data Mismanagement—from Questionable Research Practice to Research Misconduct." *Journal of Accounting Research* 14 (January 2023): 1–8. https://doi.org/10.1080/08989621.2022.2157268.
You, Dahui. "Work-Life Balance: Can You Actually Make That Happen?" *Frontiers in Pediatrics* 6, no. 3 (January 2016): 117. https://doi.org/10.3389/fped.2015.00117.
Zeinalzadeh, Elham, Alexey Valerievich Yumashev, Heshu Sulaiman Rahman, Faroogh Marofi, Navid Shomali, Hossein Samadi Kafil, Solali S. et al. "The Role of Janus Kinase/STAT3 Pathway in Hematologic Malignancies with an Emphasis on Epigenetics." *Frontiers in Genetics* 12 (December 2021): 703883. Retracted September 4, 2023. https://doi.org/10.3389/fgene.2021.703883.
Zhu, Cassie Shu, Xiaoling Qiang, Weiqiang Chen, Jianhua Li, Xiqian Lan, Huan Yang, Jonathan Gong et al. "Identification of Procathepsin L-Neutralizing Monoclonal Antibodies to Treat Potentially Lethal Sepsis." *Science Advances* 9, no. 5 (February 2023): eadf4313. https://doi.org/10.1126/sciadv.adf4313.
Zimring, James C. *Partial Truths*. New York: Columbia University Press, 2022.
Zucker, Irving, and Annaliese K. Beery. "Males Still Dominate Animal Studies." *Nature* 465, no. 7299 (June 2010): 690. https://doi.org/10.1038/465690a.

INDEX

Page numbers in *italics* indicate figures or tables.

Abalkina, Anna, 171–72
Abe, Koji, 184
abstracts, AI creating, 233–34
academic career progression, publication and, *13*
academic misconduct: of Anversa, 118–19; by Darsee, 117–18; by Das, 120–21; by Han, *121*; by Hauser, 119–20; by Hwang, 119; by Obokata, *122*; of Penkowa, 118; by Potti, *120*; by Potts-Kant, *121*; by Summerlin, 117
academic position, advancement of, *14*
academic pressures, 10
academic recruitment, *16*
accidental error, 267n30
Accountability in Research (publication), 204
Aczél, Balázs, 154
Aczél, György, 2
Adobe Photoshop, 135, 137–38
Advice to a Young Scientist (Medawar), 80
Ahluwalia, Amrita, 107
AI. *See* artificial intelligence
albino BALB/c mice, 95
Allen, Paul G., 30
Allen Institute for Brain Science, 30
Alliance Academy of Science (organization), 172
Alter, Harvey, 10
alternative career paths, in biomedical science, 28

Altman, Douglas G., 209
Alzheimer's disease, therapy for, 228
American Association for the Advancement of Science, 158
American Heart Association, 213
American Immigration Council, 22
American Type Culture Collection (ATCC), 67
Amgen (biotechnology firm), x, 86
animal experiments: errors during, 102–3; Institutional Animal Care and Use Committees regulating, 98–99; n-numbers in, 107. *See also* mice
animal handling, variability introduced during, 97
animal models, 71–72, 253n12
animal studies, drug administration during, 98
Aniston, Jennifer, 168
anonymous grant reviewers, 274n9
antibodies, 93–94
Antioxidants and Redox Signaling (journal), *120*
Anversa, Piero, 26, 118–19, 125, 127–30, 263n3, 265n10
appeal process, 47–48
applicability, of experimental findings, 83–84
applications, grant. *See* grant applications
Armstead, William, 59
Arnold, Frances, 273n64

302 INDEX

Arnold Foundation, 87
article corrections, *186*
artificial intelligence (AI), 207–8; abstracts created using, 233–34; fraud and, 167; scientific papers generated by, 270n30
astroturfing, 192
ATCC. *See* American Type Culture Collection
Atlas Venture, 86–87
authors, order of, 150
authorship, 151, 278n52; honorary, *152*; paper mills selling, 171; retractions and, *172*
awards, fraud and, 272n61

Baltimore Sun (newspaper), 131
Balwani, Sunny, 31
baseline differences, 110
basic research, biomedical science founded on, ix–x
Bayer (pharmaceutical giant), x
Beall, Jeffrey, 164–65
beallslist.net, 164–65
beautification of data, 58
Beck, Mihály, 4
believers, 128–29
bench scientists, 17
Benveniste, Jacques, 104, 128
Berberi, Ilias, 215
Berger, Nathan, 52
Bertagnolli, Monica, 222
Beutler, Bruce, 10, 70, 273n64
Bezos, Jeff, 248n9
Bhutta, Zulfiqar, 203
biases: in biomedical research, 254n30; citation, 176–81; dissemination, 76; redaction, 108; reputational, 45, 249n24
Bik, Elizabeth, 173, 199, 200
billionaires, research institutes founded by, 30
Bimler, David, 199
biological experiments, variability in, 104–5
biological variables, 108
biology, vascular, 5–6
biomedical literature, x–xi, 178

biomedical research, 38; biases in, 254n30; Galveston Shriners Institute supporting, 248n19; loss of trust in, 86; reproducibility in, 205, *206*; technological advancements depended on by, 75
biomedical science, 7–8, 10; advancement of, 246n30; alternative career paths in, 28; basic research as foundation of, ix–x
biomedical scientific literature, as irreproducible, 84
Bishop, Dorothy, 199
blinded experiments, 107
blind studies, double, 79
blood vessels, 65–66
Blood Vessels (journal), 6
Bohannon, John, 167
Boldt, Joachim, 195
Bolli, Roberto, 127
Booth, Bruce, 86–87
Boston Biomedical Research Institute, 38
Boston Children's Hospital, 125
Box, George E. P., 252n1
brain drain, 22–28
Braunwald, Eugene, 124
Brazilian Reproducibility Initiative, 87
Breaking Through (Karikó), 245n25
Brenner, Sydney, 159
Brigham and Women's Hospital, 117–18
British Journal of Pharmacology, 107, 176
British Medical Journal, 145, 203
Brookes, Paul, 155, 194, 195, 229, 267n26
Buck, Linda, 273n64
bulk retractions, paper mills exposed during, 173
Burris, Thomas, 41–42

C57BL/6 mice, 95, 96
Caenorhabditis elegans (worm), 71
calendar days, experimental days differentiated from, 255n31
Cambridge Crystallographic Data Centre, 173
cancer researchers, pressure experienced by, 27
Candidate of Science degree, 244n13
carbon copy papers, *177*

INDEX 303

Cardiovascular Health Center, at University of Connecticut, 120
career paths, alternative, 28
Carver, Raymond, 268n39
Case Western Reserve University, 125
catecholamine levels, 110
CBS. *See* cystathionine beta-synthase
Cell (journal), 159, 218
cell culture model system, 66–71, 67, 90, 99
cell cultures, edge effect in, 101–2
cell-derived cell projects, 99–100
cell lineage problem, 94
cell lines: immortalized, 67–68; *Mycoplasma* contaminating, 101; research laboratories using, 94; stem, 123
Cell Press, 159
Center for Scientific Integrity, 195
Centers of Excellence, NIH funding, 212–13
Cerami, Anthony, 70
Cerebrovascular Research Center, at University of Pennsylvania, 32
Chair of department, 15–16
Chan-Zuckerberg Biohub, 30
ChatGPT, 167, 233–34
cheating, competition for limited resources creating, 145–46
chemical synthesis, 258n71
Chinese institutions, publication rewarded by, 18
chronic unpredictable mild stress (CUMS), 97
Churchill, Winston, xi
citation, 180, 271n42; impact factor influenced by, 156; paper mills offering, 271n45; referees pushing for, 272n54; reviewers and, 154; selective, 179
citation bias, cryptomnesia and plagiarism and, 176–81
clinical development programs, retractions and, 256n52
Clinical Experimental Research Department and Second Institute of Physiology, at Semmelweis University, 4
clinical laboratories, in county hospitals, 246n30

clinical observation, 73
clinical research, reproducibility lacked by, 228
clinical trials, 201–2
Clinton administration, 37
Clyde, Smut, 173–74, 199
Cognition (Hauser), 119
Collins, Francis, 205
Columbia University Press, 280n2
Committee on Publication Ethics (COPE), 214
communism, in Hungary, 1–2
Competitive Renewals, 51, 55
compound libraries, screening campaign using, 257n70
conceptual replication, 85
conferences, scamferences and, 181–85
confirmatory studies, 53, 88
"Conform and Be Funded" (Nicholson and Ioannidis), 61
contaminants, water, 89–90
controls, positive and negative, 76
COPE. *See* Committee on Publication Ethics
Copenhagen University, 118
Cornell University, 113
corticosteroid levels, 110
Corvina Press, 280n2
county hospitals, clinical laboratories in, 246n30
COX-2 inhibitors. *See* cyclooxygenase-2 inhibitors
Crane, Julian, 203
criminal publishing gangs, 168
cross-reactivity, 93
crude simplification, 82
cryptomnesia, citation bias and plagiarism and, 176–81
C3H/HeJ mice, LPS resisted by, 91
CUMS. *See* chronic unpredictable mild stress
Current Biology (journal), 272n45
Current Contents (journal), 278n53
cyclooxygenase-2 inhibitors (COX-2 inhibitors), 222, 279n58
cystathionine beta-synthase (CBS), 100
cytometry, 140–41
Czeizel, Endre, 2

304 INDEX

Dana Farber Cancer Research Institute, 199
Dansinger, Michael, 268n6
Darsee, John "JR Hughes-Darsee," 117–18, 124, 125, 129, 262n3
Das, Dipak, 120–21, 128–29, 264n3
data, 129, 255n36; beautification of, 58; grant applications doctoring, 58–59; interpretation of, 81
data collection, enhancement to processes of, 212
data detectives, 198–200, 218
data integrity, enhancement to processes of, 212
data manipulation, 114
data policies, open, 215
data sharing, 216
David, Sholto, 142
deflection, 144–45
Derailed (Stapel), 145
desk rejection, 152
diagnosis, doctors presenting, 281n7
diffusible biological regulatory molecules, 244n12
direct replication, 84
Discover (magazine), 254n24
dissemination bias, 76
DNA, biological functions of, 254n24
DNA and Cell Biology (journal), 173
doctors, diagnosis presented by, 281n7
Dolly (sheep), 254n21
Dóra, Eörs, 5, 6
double blind studies, 79
dredging, 114
drug administration, during animal studies, 98
drug repurposing, 51–52
Duke University, 62, 120, 121, 125

early-career scientists, 3, 9
Eastern Bloc, 1–2
Eaves-Pyles, Tonyia, 90–91
edge effect, in cell cultures, 101–2
editorial boards, of journals, 162–63
Editorial Expression of Concern, 186

EDRF. *See* Endothelium-Derived Relaxing Factor
Eigenfactor, 269n12
Einstein, Albert, 49–50
Eisen, Michael, 217, 277n49
eLife (journal), 215, 217
Elsevier (publishing company), 170, 197
Emory University, 125
Endothelium-Derived Relaxing Factor (EDRF), 5–7, 181–82
endotoxin (LPS), 89, 90–91, 257n66
environment, equipment and, 249n24
Eötvös University, 154
equipment, 247n4, 249n24
Escherichia coli (bacteria), 69
estrogens, 89
EU. *See* European Union
European Commission, 63, 178
European Union (EU), 34, 63, 252n66
Europe PubMed Central (Europe PMC), ix
Evilicious (Hauser), 119
"exceptional scientist," nationalization of, 22–28
experimental days, calendar days differentiated from, 255n31
experimental findings: applicability of, 83–84; mice impacting, 94–96; variability in, 72
experimental models, 252n1
experimental results, hypothesis supported by, 34
experiments, blinded, 107
external pressures, 147
extramural grants: costs covered by, 14; NIH awarding, 38; PI bringing in, 18–19

Facilities & Administrative costs (F&A), 37–38
faculty-level appointments, visa applications for, 23–24
faculty members, grant money and, 245n27
faculty positions, junior, 13–14, 245n24
False Claims Act, 125
false positives, 104

falsifying results, *123*
Fauci, Anthony, 10
faux cytometry, 141
FBS. *See* fetal bovine serum
federal employees, NIH researchers as, 29
Federal Trade Commission, 167
Federation of American Societies for Experimental Biology, 55
fetal bovine serum (FBS), 69, 257n65
Fibiger, James, 250n38
fishing, 51, 62, 261n111
flagella, 91
flaw cytometry, 141
Flier, Jeffrey, 262n3
flow cytometry, 140–41
ForBetterScience.com, 197–99
Foster, William Michael, *121*
fraud, *111*, 149, 166, 278n51; AI and, 167; awards and, 272n61; financial motivation for, 202; guest editors and, 170; image, 133–34, 136–37, *142*; intentional, 143–45; journals fighting manipulation and, 155; *Nature* revealing, 170; scientific, 26, 146, 194; triangle of, 147
fraud cases, high profile, 115
Free Academic Paper Generators, 167
Freedman, Leonard, 205
Free University of Berlin, 171
French National Center for Scientific Research, 195
Frontiers Genetics (journal), 172
Fujii, Yoshitaka, 195
Fulda, Simone, 199, 274n9
Furchgott, Robert, 5, 7, 66, 182, 272n58

Galveston Shriners Institute, 248n19
game theory, 41
Gates, Bill, 248n9
Gay, Claudine, 179
Genentech (biotech firm), 130
genetic drift, in mice, 254n20
"Get Me Off Your Fucking Mailing List" (Kohler and Mazieres), 166–67
ghostwriting, 191–92

Giffey, Franziska, 178
Global Biological Standards Institute, 205
GLP. *See* good laboratory practice
Good, Robert, 124
Goodhart's Law, 157
good laboratory practice (GLP), 102
"Grad Student Who Never Said No, The" (Wansink), 113
gradual improvements, traditional suggestions for, 204–8
Graham, David, 279n58
grant applications, 34–44, *42*, 55, 247nn3–4, 249n23, 249n26; amount of time spent writing and editing, 248n17; data doctored in, 58–59; rejection rates for, 40, 60; R01, 37, 39, 43
grant cycle, in theory and practice, *56*
granting mechanisms, investigators evaluating, 249n21
grant money, faculty members and, 245n27
grant reviewers, 45, 49, 52–53, *231–32*, 250n28; anonymous, 276n30; Dansinger plagiarized by, 268n6; ultraconservative attitude of, 61–62
grant reviewing, randomness in, 60–61
grants, 33, 43, 248n20; calculating chance of receiving, 248n11; in EU, 63, 252n66; extramural, 14; from NIH, 34–37, 41–42, 251n43, 252n63. *See also* extramural grants
Graphpad (software company), 107
Grassley, Charles, *121*, 201
green cards, 24–25, 247n42
guest editors, fraud and, 170
Gunsalus, C. K., 145, 207
Guttenberg, Karl-Theodor zu, 178
Győr (Hungary), 1–2

Han, Dong-Pyou, *121*, 125, 201, 202, 264n3, 265n6
Handler, Philip, 266n20
HARKing. *See* Hypothesizing After the Results are Known
Harvard University, *117–18*, *119*, 129–30, 199
Hauser, Marc, *119–20*, 263n3

306　INDEX

Helicobacter pylori (bacteria), 73–74
HGMB1 mechanism, 70
Hindawi (publishing company), 173, 271n41
Holmes, Elizabeth, 31
homeopathy, 104, 260n95
H-1-lottery, 24
H-1 visas, 23–25, 246n40
honest errors, 100–104
Horizon Europe, 63
Horizon Program, by European Union, 34
Howard Hughes Medical Institute, 30
H2. *See* hydrogen gas
H2S. *See* hydrogen sulfide
"Human Basophil Degranulation Triggered by Very Dilute Antiserum Against IgE" (Benveniste), 104
human factors, 103
human inflammatory diseases, mouse models not mimicking, 106
human trails, animal models contrasted with, 253n12
Hungarian Academy of Sciences, 7, 33, 247n3
Hungarian-born immigrants, Manhattan Project worked on by, 246n39
Hungary: communism in, 1–2; medical genetics in, 243n3; Nobel Prize winners from, 4, 243n4; scientists revered by, 3–4
Hwang, Woo-Suk, *119*, 129, 263n3
hybrid publication model, 149
hydrogen gas (H2), inhalation of, 109
hydrogen sulfide (H2S), 91, 98, 100
hype, 254n29, 256n46
hypothesis: experimental results supporting, 34; null, 255n38; working, 73, 77
Hypothesizing After the Results are Known (HARKing), 112

Ignarro, Louis, 66
IgNobel Prize, 113
IL-10-/- mice. *see* interleukin 10 knockout mice
image fraud, 133–34, 136–37, *142*
image manipulations, 97, 197, *218*
immortalized cell line, 67–68
immunohistochemical staining process, 135

immunohistochemistry, 93, 258n72
impact factor, 249n25, 269n10, 269n12, 278n55; artificial inflation of, 166; citation influencing, 156; of journals, 189; of *Science Advances*, 150
incentives, in life science research, *19*
independent peer review, *153*
independent researchers, 209
Indian Journal of Pharmacy and Drug Studies (journal), 176
industry scientists, publication and, 31
industry track, postdocs choosing, 21
inflammatory diseases, human, 106
Information Sciences Institute, of University of Southern Carolina, 208
Innovation, The (journal), 156
INO-1001 (PARP inhibitor), 92
Inotek (company), 92
Institute of Scientific Information, 159
Institutional Animal Care and Use Committees, animal experiments regulated by, 98–99
institutional inertia, 116–17
institutions, equipment funded by, 247n4
integrated medical education, 64
integrity, research, 193–94
intentional fraud, in scientific literature, 143–45
interleukin 10 knockout mice (IL-10-/- mice), 96–97, 258n78
International Journal of Advanced Computer Technology (journal), 166–67
interpretation of data, 81
intestinal flora, 96
intraperitoneal injections, 103
investigators, granting mechanisms evaluated by, 249n21
in vitro, in vivo vs., 254n19
Ioannidis, John, ix, 61, 85, 226
Iowa State University, *121*
iron-catalyzed Fenton reaction, 89
irreproducibility, in life science research, 227

Janelia Farm (research center), 30
JCI Insight (journal), 159

John Maddox Prize, 200
Johns Hopkins University, 62, 131, 132
J-1 visas, 23–25, 246n40
Journal of Clinical Epidemiology, 155
Journal of Clinical Investigation, 159
Journal of Hepatology, 197
Journal of the American Medical Association, 279n58
journals, 148, 157, 160, 193; editorial boards of, 162–63; impact factors of, 189; mandatory reporting agreed to by, 219; manipulation and fraud fought by, 155; open access, 161–62, 165; predatory and fake, 164–68; publishers and, 156; specialization of, 163; submission and acceptance to, 188–89. *See also* impact factor; *specific journals*
Joyce, James, 268n39
Judo, Pete, 200
junior faculty positions, 13–14, 245n24

Kádár, János, 1
Kahneman, Daniel, 255n41
Karikó, Katalin, 245n25, 280n1
Kloner, Robert, 118
Koch-Mehrin, Silvana, 178
Kohler, Eddie, 166–67
Kornberg, Roger, 62
Kovách, A. G. B., 5
Kuzma, Susan M., 203

lab consumables, 15
laboratory-based research, 65
laboratory hierarchy, 16
Lalande, Daniel R., 112
Large Hadron Collider, 255n44
Leyen, Ursula von der, 178
life science publications, 160
life science research: as career advancement tool, 225; incentives in, 19; irreproducibility in, 227
Lindau Nobel Laureate Meeting, 22
linguistic obfuscation, 181
lipopolysaccharide (LPS), 69
literature, unreliability of, 83, 225, 226

litigation, murkiness and, 129–30
living animals, variability introduced by, 94
loading controls, 139, 142
London Tercentenary Foundation, 33
Louisiana State University, 20
LPS. *See* endotoxin; lipopolysaccharide
Lyden, Patrick, 85

Mackenzie, Ruari, 184
Maddox, John, 104
mandatory misconduct detection and reporting system, 219
mandatory reporting, journals agreeing to, 219
Manhattan Project, 22, 246n39
manipulation: data, 114; easily detectable methods of, 132; image, 97, 197, 218; journals fighting fraud and, 155; *Nature* analyzing for, 218; of Western blots, 140, 141
manuscript, to publication, 148
manuscript evaluation system, 272n52
Marcus, Adam, 195
Marie Curie grant mechanism, 34
Marshall, Barry, 10, 73–74, 254n24
Masicampo, E. J., 112
Massachusetts Institute of Technology, 196
Mastroianni, Adam, 200
mathematical calculation errors, 102
Mazieres, David, 166–67
McNutt, Marcia, 207
MD Anderson Cancer Center, 27
MDs, PhDs contrasted with, 10
Medawar, Peter, 80, 255n34
media, 253nn6–7
medical genetics, in Hungary, 243n3
Mendel, Gregor, 57–58
Message of the Genes, The (Sullivan), 3
meta-analysis, 190–91
methotrexate, 51
mice, 103, 106, 108–10, 258n80, 259n84, 261n108; albino BALB/c, 95; C57BL/6, 95, 96; C3H/HeJ, 91; experimental findings impacted by, 94–96; genetic drift in, 254n20; identification of, 261n109; interleukin 10 knockout, 96–97

misconduct, 143, 272n52, 279n58; clinical trials, patient care, or patient lives impacted by, 201–2; detection and reporting system for, 219; research, 146. *See also* academic misconduct; fraud; paper mills; retractions; scientific misconduct
Mitchell, Peter, 57, 251n49
Miyakawa, Tsuyoshi, 277n47
mM (unit of measure), 260n95
model systems, 66, 71
model validation, 76
Molecular Brain (journal), 277n47
molecular libraries, 92
Moncada, Salvador, 6–7, 182, 244n11, 272n58
Moniz, Egas, 250n38
Montagnier, Luc, 104
mouse models, 106, 261n108
Musk, Elon, 30
Mycoplasma (bacterium), cell lines contaminated by, 101

Nadal-Ginard, Bernardo, 125
NaHS. *See* sodium hydrogen salt
National Cancer Institute (NCI), 39
National Institute of General Medical Sciences (NIGMS), 55
National Institutes of Health (NIH), 29–30, 45–48, 46, 53–54, 62–63, 205–7; budget of, 37, 39; Centers of Excellence funded by, 212–13; extramural grants awarded by, 38; grants from, 34–37, 41–42, 251n43, 252n63; institutions fined by, 125–26; NSF contrasted with, 247n5; ORI at, 58–59, 121, 193, 265n3; OSI at, xi, 28; reproducibility guidelines published by, 87; screening-based efforts invested in by, 250n32; water contaminants published on by, 89–90. *See also* R01; Study Section
nationalization, of the "exceptional scientist," 22–28
National Journal of Pharmaceutical Sciences (journal), 176
National Public Radio, 174
Nature (journal), 104, 118, 122, 125, 144, 180; fraud revealed in, 170; impact factor of, 156; manipulation analyzed for by, 218; rejection rate of, 150; Springer Nature publishing, 157
Nature Communications (journal), 158
Nature Medicine (journal), 191
Naunyn-Schmiedeberg's Archives of Pharmacology (journal), 170, 271n38
NCATS (library), 250nn31–32
NCI. *See* National Cancer Institute
negative controls, 76
Neumann, John von, 41
Neuralink (project), 30
New England Journal of Medicine (journal), 279n58, 280n2
New York Times (newspaper), 131
Nicholson, Joshua, 61
NIGMS. *See* National Institute of General Medical Sciences
NIH. *See* National Institutes of Health
NIH administrators, funding determined by, 53–54
"NIH Plans to Enhance Reproducibility" (Collins), 205
NIH RePORTER site, 53
NIH researchers, as federal employees, 29
Nikolaev, Anatoly, 130–31
96-well plates, 101–2, 107, 259n94
nitric oxide (NO), 5–7, 94, 181–82
n-numbers, in animal experiments, 107
NO. *See* nitric oxide
Nobel Prize winners, 4, 243n4
nontenure track, tenure track distinguished from, 14
"no-reject publishing" model, 217
Nosek, Brian, 55, 87–88
NSF. *See* U.S. National Science Foundation
null hypothesis, 255n38

Obokata, Haruko, 122, 127–28, 264n3
Office of Research Integrity (ORI), at NIH, 58–59, 121, 193, 265n3
Office of Scientific Integrity (OSI), at NIH, xi, 28
OMICS group, 167, 184, 220, 253n10
"omics" study, 112

123mi.ru (paper mill), 171
open access, 161–62, *165*, *216*
open access journals, 161–62, *165*
open data policies, 215
opioid crisis, 228
Oransky, Ivan, 26, 195, 200
Organic Chemistry and Biochemistry Prague, 274n5
ORI. *See* Office of Research Integrity
OSI. *See* Office of Scientific Integrity
outlier exclusion, 108–9, 115, 261n110
oversight, by PI, 193
OxyContin, 280n4

paper mills, 168–70, *169*, 175; authorship sold by, 171; bulk retractions exposing, 173; citation offered by, 271n45; Sabel identifying, 174; tortured language revealing, 181
papers, carbon copy, *177*
Parker Institute for Cancer Immunotherapy, 30
PARP. *See* poly (ADP-ribose) polymerase
PARP (enzyme), 92
Parton, Dolly, 254n21
pathologically narcissistic behavior, 146–47
peer review, *153*, *164*, *215*, 250n28, 271nn35–36
Penkowa, Milena, *118*, *127*, 262n3
personal penalties, 126–27
Pfizer (pharmaceutical company), 279n58
p-hacking, 110–12
pharmaceutical companies, COX-2 inhibitors developed by, 279n58
Pharmaceutics and Pharmacology Research (journal), 166
PhD degrees, 244nn16–17
PhDs, MDs contrasted with, 10
PhD students, 11; teaching and laboratory duties forced onto, 10–11; on training visas, 23; work/life balance managed by, 21
PhD training, 210–11
Photoshop, Adobe, 135, 137–38
PI. *See* principal investigator
Pier, Elizabeth, 60
pilot experiment, 76
pipetting, 103–4

plagiarism, cryptomnesia and citation bias and, 176–81
plate readers, 101–2
PLOS One (mega journal), 159
PNAS (journal), 60, 106, 159, 217
PNAS Nexus (journal), 159
Poehlman, Eric, 202
politicians, plagiarism by, 178–79
poly (ADP-ribose) polymerase (PARP), 51
Popper, Karl, 254n22
positive controls, 76
postdoc shortages, 12, 245n20
postdoctoral scientists (postdocs): industry track chosen following, 21; publication of, 12–13; on training visas, 23
Potti, Anil, 116, *120*, 125, 129, 263n3
Potts-Kant, Erin, 58, 62–63, *121*, 125, 264n3
power analysis, 110–11
power calculation, 77
precarious employment, 26
preclinical research, money lost to, 85–86
pressure, cancer researchers experiencing, 27
pressure, supervisory, 27
pressure to publish, *18*
principal investigator (PI), 15–17; extramural grants brought in by, 18–19; oversight by, 193; "principle" investigators and, 245n28, 248n6
private initiatives, science cleaned up through, 200
professors, tenured full, 15–16
Program Project (joint grant), 43
Progress Reports, 50–51
Puara, Jay, 208
the public, research paid for by, 221
publication, *157*, *169*, *204*, *214*; academic career progression and, *13*; Chinese institutions rewarding, 18; of confirmatory studies, 53; declining impact of scientific, 159; industry scientists and, 31; manuscript to, 148; of postdoctoral scientists, 12–13; referees game influencing, 151–52; submission/evaluation process before, 149. *See also* authorship; journals; scientific literature; scientific papers

publication process, replication emphasized by, 219
publications, life science, 160
publish, pressure to, 18
publishers, 156, 164–68
publishing industry, 214, 220
PubMed (database), ix, 166, 187, 217, 218–19, 220
PubMed Commons (database), 200, 205
PubPeer (journal club), 135–36, 190, 195–97, 199, 265n12, 266n22
Purdue Pharma, 280n4
purity, of reagents, 91–92
Putin, Vladimir, 178
pyrogens, 257n61

QC. *See* quality control
QRP. *See* questionable research practice
quality control (QC), 102
questionable research practice (QRP), 204

Randi, James, 104
randomization, 78–79
randomness, in grant reviewing, 60–61
Raven Press, 272n58
RDMM. *See* research data mismanagement
Reagan, Ronald, 279n57
reagents: purity of, 91–92; storage conditions impacting, 102; variability in, 88
redaction bias, 108
red-flag method, 174, *175*
referees, 154, 170, 174, 218–19, 270n34, 271n35; citation pushed for by, 272n54; scientific misconduct and, 155; triage process recommended by, 36, 57. *See also* Study Section
referees game, publication influenced by, 151–52
reforms, 208, 223
replication, 87, 256n53, 257n57; conceptual, 85; direct, 84; publication process emphasizing, 219
replication crisis, 86, 250n37
replication efforts, 52–53

replication studies, 84, 87–88
reporting system, mandatory misconduct detection and, 219
reproducibility, 84, 227, 257n57; in biomedical research, 205, *206*; clinical research lacking, 228; scientific literature lacking, 225–26, 229
Reproducibility 2020 Initiative, 205
reproducibility crisis, x, 82–88, 104–5, 143, 193
reproducibility guidelines, NIH publishing, 87
Reproducibility Project: Cancer Biology, x, 87
reputational bias, 45, 249n24
Requests for Application (RFAs), 36
research, 30, 204, 208; basic, ix–x; clinical, 228; laboratory-based, 65; money lost funding, 256n50; preclinical, 85–86; the public paying for, 221; stem cell, 265n10; writing articles and, *151*. *See also* biomedical research; citation; experimental findings; life science research
Research (journal), 159
research data mismanagement (RDMM), 204
researchers: independent, 209; NIH, 29; pressure experienced by cancer, 27. *See also specific researchers*
research institutes, billionaires founding, 30
research integrity, 193–94, 207
Research Integrity Organization, 207
research laboratories, 94, 266n23
research materials, variability in, 88
research misconduct, pitfalls of, *146*
Research & Reviews (journal), 176
responses, to "flagging" of papers, 238–40
responsibility, admission of personal, 127–28
resveratrol (natural compound), 86
retractions, *187*, 228, 266n18; authorship and, *172*; bulk, 173; clinical development programs and, 256n52; consequences of, 185–88; from Hindawi, 271n41; of Nobel Prize winners, 4; in *Science*, 129, 131
Retraction Watch, 26, 166, 195, 268n6
review articles, 221

Reviewed Preprints, 217
reviewers, citation and, *154*
review panels, *50*
RFAs. *See* Requests for Application
R56. *See* Short-Term Project Awards
Riken Center for Developmental Biology, *122, 127*
Risk-Eraser (organization), *120*
Roche, Dominique G., 215
Royal Society in England, 33
Royal Society of Chemistry, 13, 173
R21 (grant), 43
R29 (grant), 37
Rubik, Ernő, 243n2
Rubik's cube, 243n2
runaway trains, scientific systems and, 130
R01 grant applications, 37, 39, 43
R01 "RO1" (grant), *54*, 248n6; direct budget provided by, 37; impact scoring system for, *46, 52*; rejection rate of applications for, 40–41

Sabatini, David "mTORman," 196, 274n5
Sabel, Bernhard, paper mills identified by, 174
Salzman, Andy, 90
Sasai, Yoshiki, *122*
SBIR. *See* Small Business Innovation Research
scamferences, 181–85, 234, *235–38*
Schavan, Annette, 178
Schekman, Randy, 22, 150, 215, 217
Schillaci, Orazio, 132
Schmitt, Pál, 178
Schneider, Leonid, 198–99, 274n9
Schönbrodt, Felix D., 113
science, 200, 262n1. *See also* biomedical science; life science research
Science (journal), 12, 167, 278n49; American Association for the Advancement of Science publishing, 158; impact factor of, 156; rejection rate of, 150; retractions in, 129, 131
science, technology, engineering, and mathematics (STEM), working in, 8, 9
Science Advances (journal), 150
Science and Engineering Ethics (journal), 144
Science–Fake Science (Beck), 4
science-fraud.org, 194
science-publisher.org, 172
Science Quality Officers, 211
Science Translational Medicine (journal), 84–85
scientific career tracks, 209–10
scientific fraud, 26, 146, 194
scientific integrity, *210, 211*, 229–30
Scientific Integrity (degree), 211
scientific literature: intentional fraud in, 143–45; reproducibility lacked by, 225–26, 229; retroactive "cleaning up" of, *218*; skepticism and, 243n1. *See also* journals
scientific meetings, 181–83, 244n11
scientific misconduct, 113–15, *123, 126, 128, 198*, 265n3; criminalization of, 201–4; databases or initiated websites dealing with, 194; psychological aspects of, 145–47; referees and, 155
scientific news organizations, 221
scientific papers: AI generating, 270n30; classical format of, 149; order of authors of, 150; reading, 189–92
scientific publications, incremental advances proposed to improved, 214
scientific publishing industry, 161
Scientific Reports (journal), 158
scientific systems, runaway trains and, 130
scientists, *191*; bench, 17; early-career, 3, 9; Hungary revering, 3–4; industry, 31. *See also* PhD students; postdoctoral scientists; principal investigator (PI)
SCIPO trial, 127
Scopus (database), 166
SCORE. *See* systematizing confidence in open research and evidence
screening campaign, compound libraries used by, 257n70
Scripps Institute, 41–42
Secrets of Inheritance (television), 2
selective citation, 179
Selye, Hans, 97, 259n81

Semenza, Gregg, 131, 266n18, 273n64
Semmelweis University, 4
Seoul National University, 119
septic shock, 66–67, 69–70
shenaniganometry, 240–42
Shock Society, 55
Short-Term Project Awards (R56), 54
Shriners Foundation, 43
Sigma, 92
significance level, 79–80
significant fields of inquiry, 116
simplification, 82–83
SIRT1 (enzyme), 86
Sirtris Pharmaceuticals, 86
60 Minutes (TV program), 116
skepticism, scientific literature and, 243n1
Slim by Design (Wansink), 113
Sloan-Kettering, 117
Small Business Innovation Research (SBIR), 43
Small Business Technology Transfer (STTR), 43
Smith, Richard, 145
SNSF. *See* Swiss National Science Foundation
Snyder, Solomon, 10, 52
sociopathic/psychopathic behavior, 146–47
sodium hydrogen salt (NaHS), 92
solitary genius scientist (myth), 276n37
Somers, Scott, 55
Sooam Biotech Research Foundation, 119
Soros, George, 248n9
Soviet Bloc, 1
SPAN Network, 84–85
Special Study Section, 249n22
Spirli, Carlo, 58
splicing, 138
Springer Nature (publishing company), 157–58, 170
staining process, immunohistochemical, 135
stakeholders, 209, 211–12, 214, 221
standard laboratory techniques, antibodies relied on by, 93
standard model, 255n44
Stanford Daily (newspaper), 130

"Stanford President to Step Down Despite Probe Exonerating Him of Research Misconduct" (article), 131
"Stanford President Will Resign After Report Found Flaws in His Research" (article), 131
Stanford University, 117, 130–31
Stanley, Wendell Meredith, 250n38
STAP. *See* stimulus triggered acquisition of pluripotency
Stapel, Diederick, 145
State Department (U.S.), 23
statistical convention, 109
statistical significance, 109, 255n35
Stefan, Angelica M., 113
Steinfath, Matthias, 210
Stell, Brandon, 195–96
STEM. *See* science, technology, engineering, and mathematics
stem cell lines, 123
stem cell research, 265n10
steroid treatments, 110
stimulus triggered acquisition of pluripotency (STAP), 122, 127, 128
storage conditions, reagents impacted by, 102
STTR. *See* Small Business Technology Transfer
Students' Scientific Circle (organization), 4
Study Section, 44–45, 47, 53–54, 58, 251n44, 251n49; funding and, 57; meeting of, 48; novelty and innovation emphasized by, 52; Special, 249n22; true discovery frowned upon by, 51
subcutaneous injections, 103
subject variability, 72
substandard experimentation practices, variability caused by, 100–104
success rate, of R01 grant applications, 39
success rate, of replication studies, 84
Sudeikis, Jason, 168
Südhof, Thomas, 273n64
Sullivan, Navin, 3
Summerlin, William, 4, 116, 117, 127, 262n3
supervisors, of research laboratories, 266n23

supervisory pressure, 27
Swiss National Science Foundation (SNSF), 34, 214, 247n4
systematizing confidence in open research and evidence (SCORE), 208
Szent-Györgyi, Albert, 4, 8, 252n4
Szostak, Jack, 273n64

Tabak, Lawrence, 205, 222
taxes, 38, 221, 222, 225, 246n29
technological advancements, biomedical research depending on, 75
tenured full professors, 15–16
tenure track, nontenure track distinguished from, 14
Tessier-Lavigne, Marc, 124, 130–31, 196, 266n21, 280n59
thalidomide, 51
That Day (Obokata), 127
theory-induced blindness, Kahneman on, 255n41
Theranos, 31, 253n10
Thomas, Joe, *121*
3-aminobenzamide (3-AB), 92
3T (doctrine), 2
tissue sections, as unique, 133–34
TLR4 (membrane proteins), 91
TNFα. *See* tumor-necrosis factor alfa
Topol, Eric, 279n58
tortured language, paper mills revealed in, 181
Toxicology Reports (journal), 142
Tracey, Kevin, 10, 70
training visas, 23
Transplantation Proceedings (Summerlin), *117*
triage process, 36, 49, 57, 251n44
triangle of fraud, 147
deTriusce, Frances, 194
Trump, Donald, 38
tumor growth, in mouse models, 261n108
tumor-necrosis factor alfa (TNFα), 70
tumors, 108–9
23rd International Conference on Neurology & Neurophysiology and 24th International Conference on Neurosurgery & Neuroscience (meeting), 184

unethical behavior, 125
United States (U.S.), 22, 23, 35, 195. *See also* National Institutes of Health
University of California, 62, 144
University of Cincinnati, 246n40
University of Colorado, 164–65
University of Connecticut, *120–21*, 129
University of Debrecen, *120*
University of Fribourg, 7
University of Iowa, 125
University of London, 7
University of Lund, 32
University of Pennsylvania, 5, 32
University of Pittsburgh, 62
University of Rochester, 155, 194
University of Southern Carolina, 208
University of Texas, 110
unreliability, of literature, 83, 225, 226
U.S. *See* United States
U.S. House Committee on Science, Space, and Technology, 195
U.S. National Science Foundation (NSF), 8, 35, 247n5

VA. *See* Veteran's Administration
Vacanti, Charles, 127, 128
Vane, John, 7, 32, 66, 268n1
variability, 89–93, 98–99; animal handling introducing, 97; in biological experiments, 104–5; in experimental findings, 72; living animals introducing, 94; in reagents, 88; in research materials, 88; substandard experimentation practices causing, 100–104
vascular biology, 5–6
vascular ring system, 88–89
vascular strips, 66
Vasodilatation, Vascular Smooth Muscle, Peptides, and Endothelium (Raven Press), 272n58

Veteran's Administration (VA), 43
Vioxx (pharmaceutical company), 279n58
virology lab, *192*
visa applications, for faculty-level appointments, 23–24
visas: H-1, 23–25, 246n40; J-1, 23–25, 246n40; training, 23; working, 22–28
Voluntary Settlement Agreement, 59–60

Wakayama, Teruhiro, 127–28
Wallace, Alfred Russel, 57
Wang, Haichao, 70
Wansink, Brian, 112–13, 261n117
Waseda University, 122
wasted money, 125–26
water contaminants, 89–90
"Water Memory" debacle, 103–4
We're the Millers (comedy), 168
Western blots, 136–38, *137*, *142*, 258n72, 274n6; cross-reactivity causing misinterpretation of, 93; loading controls for, *139*; manipulation of, *140*, *141*

Whitehead Institute, at Massachusetts Institute of Technology, 196, 274n5
Whiteman, Matt, 92
Why Do Trees Have Leaves? (book), 2
"Why Most Published Research Findings Are False" (Ioannidis), ix
Wiley (publishing company), 271n41
William Harvey Research Institute, 7, 32, 246n40
Wilmut, Ian, 254n21
Wise, Nick, 199
Wise Owl (book series), 2
Wonders of Blood, The (book series), 2
working hypothesis, 73, 77
working visas, 22–28
work/life balance, 20–22
World War II, 1
writing articles, research and, *151*

You, Dahui, 20

zombie papers, 188
Zuckerberg, Mark, 248n9

μM (unit of measure), 260n95

GPSR Authorized Representative: Easy Access System Europe, Mustamäe tee 50, 10621 Tallinn, Estonia, gpsr.requests@easproject.com